A Complete 'O' Level Chemistry

G. N. Gilmore

PhD CChem M.R.I.C.

Lecturer in Chemistry,
Bury College of Further Education

Stanley Thornes (Publishers) Ltd

© 1978 G. N. Gilmore

All rights reserved. No part of this publication may be reproduced, stored in a retrieval system, or transmitted in any form or by any means, electronic, mechanical, photocopying, recording or otherwise, without the prior written consent of the copyright holder.

First published 1978 by Stanley Thornes (Publishers) Ltd.,
EDUCA House, Liddington Estate, Leckhampton Road,
CHELTENHAM GL53 0DN
Reprinted 1979

ISBN 085950 089 6

Typesetting by Malvern Typesetting Services Ltd.
Printed in Great Britain by
The Pitman Press, Bath

A Complete 'O' Level Chemistry

Also by Dr. G. N. Gilmore
A MODERN APPROACH TO COMPREHENSIVE CHEMISTRY
for 'A' Level J.M.B. and University of London
syllabuses

Preface

Since the raising of the school leaving age to sixteen, students studying for G.C.E. 'O' Level at Colleges of Further Education have had to cover the entire course in a little over eight months. This has led to the need for a concise text keeping strictly to the syllabus requirements. This book is an attempt to meet that need and, in order to keep unrequired material to a minimum, it has been written primarily for the J.M.B. and A.E.B. syllabuses. Much of it, however, will be applicable to other Examining Boards.

The theoretical work is firmly based on the electronic theory and the electrochemical and activity series. Ionic equations are generally given in addition to molecular equations. The similarity in the properties of an anion with different cations has been emphasised. However, the compounds of the metals have been dealt with in a fairly traditional way in the belief that this enables their uses to be given in a clearer manner. Detailed practical exercises have been included so that the text presents a complete 'O' Level Course.

The nomenclature is that recommended by the Association for Science Education. SI units have generally been used, the exceptions being those recognised and employed by the Examining Boards.

The Author wishes to thank the Joint Matriculation Board and Associated Examining Board for permission to reproduce a considerable number of their examination questions and to update the nomenclature and terminology where necessary.

Contents

		Page
Chapter 1	Physical and Chemical Changes, Elements, Compounds and Mixtures	1
Chapter 2	Atoms, Molecules, Formulae and Equations, Kinetic Theory	14
Chapter 3	Atomic Structure, Valency, Crystal Structures, Periodic Table	27
Chapter 4	Air and Water	44
Chapter 5	Quantitative Relationships: Laws of Chemical Combination, Relative Atomic and Molecular Masses, Molar Volume	60
Chapter 6	Acids, Bases and Salts	69
Chapter 7	Electrolysis	86
Chapter 8	Oxidation and Reduction, Simple Cells, Electrochemical Series	98
Chapter 9	Hydrogen, Oxygen and Hydrogen Peroxide	107
Chapter 10	Thermochemistry, Rate of Change and Equilibria	121
Chapter 11	The Chemistry of some Metals	135
Chapter 12	The Chemistry of some Non-Metals	177
Chapter 13	Organic Chemistry	225
Chapter 14	Practical Exercises	243
	General Questions	268
	Relative Atomic Masses	279
	Periodic Table	280
	Answers to Numerical Questions	281
	Logarithms	284
	Index	288

1 Physical and Chemical Changes, Elements, Compounds, and Mixtures

INTRODUCTION

Chemistry involves the study of the nature and properties of the substances making up the universe. The various data is then classified and, where possible, generalisations are drawn so that laws can be obtained. Attempts are made to explain the laws and so this gives rise to theories. The theories not only attempt to explain the experimental facts but also suggest how other substances would behave in similar circumstances. Thus, chemistry is an ever expanding subject as new theories lead to new reactions and new facts.

STATES OF MATTER

Substances can generally exist in three states or phases according to the temperature and pressure, i.e. as gases, liquids, or solids. Most substances can be readily placed in one of these categories under normal conditions.

A gas has no definite volume or shape since it diffuses or spreads itself uniformly throughout the containing vessel. Its volume is greatly affected by even relatively small changes in temperature and pressure, the change in volume being remarkably similar from gas to gas. Liquids, like gases, have no fixed shape; they take up the shape of the vessel in which they are placed. However, they do have a fixed volume under a given set of conditions and this volume is considerably less affected, by changes in temperature and pressure, than that of gases. Solids have definite volumes and shapes and these are fairly resistant to change – they are said to be rigid. When solids are heated they generally expand by a very small amount which is characteristic of them.

TYPES OF CHANGE

Substances around us are continually undergoing change. For example, a change obviously occurs when an electric light is switched on or when a gas, liquid, or solid burns. Chemists classify the changes as being either physical or chemical.

Consider a block of ice. On heating this will turn initially into water and then to steam. Cooling the steam will give water which on further cooling gives ice again.

$$\text{Ice} \underset{\text{cool}}{\overset{\text{heat}}{\rightleftharpoons}} \text{Water} \underset{\text{cool}}{\overset{\text{heat}}{\rightleftharpoons}} \text{Steam}$$

These are regarded as physical changes for the following reasons.

(a) No new substances are produced – ice, water, and steam are the solid, liquid, and gaseous forms respectively of the same chemical substance.

(b) There is no change in mass, e.g. 10 g of ice give 10 g of water which in turn give 10 g of steam.

(c) The changes are readily reversed.

The magnetisation of a steel bar by stroking it with a magnet is a physical change because the mass, composition, and chemical properties are unaltered and the change is easily reversed by heating the bar. Similarly, dissolving sodium chloride (salt) in water is a physical change because the solid can be recovered by evaporating the water.

In physical changes, no new substance is formed but a change of state or physical form may occur. The changes are readily reversed.

If a test tube containing a little mercury(II) oxide is heated, the orange solid slowly disappears and globules of a silver coloured liquid appear on the cooler parts of the tube. This is classed as a chemical change because cooling does not reproduce the orange solid, there has been a loss in mass, and the properties of the orange solid and silver coloured liquid are quite different. The mercury(II) oxide has, in fact, decomposed:

$$\text{mercury(II) oxide} \rightarrow \text{mercury} + \text{oxygen}$$

A further example of a chemical change is seen when a piece of magnesium ribbon is heated in air. The silver coloured metal burns with an intense white flame leaving a white ash. There is an increase in mass during the process and the ash will not liberate hydrogen from dilute sulphuric acid as the magnesium will. The magnesium has, in fact, combined with oxygen from the air:

$$\text{magnesium} + \text{oxygen} \rightarrow \text{magnesium oxide}$$

The differences between physical and chemical changes may be summarised as follows.

Physical changes	*Chemical changes*
1. The composition of the substance(s) is unaltered.	—Substances with a different composition are produced.
2. The physical but not the chemical properties change.	—The physical and chemical properties are altered.
3. The change is easily reversible.	—The process cannot be reversed by physical means.
4. The mass of the substance(s) is unaltered.	—The total mass is unaltered but the masses of the individual substances alter.
5. Very little heat is absorbed or evolved.	—The heat changes may be large.

In any chemical change the masses of the individual substances change. For example, when mercury(II) oxide is heated:

$$\text{mercury(II) oxide} \rightarrow \text{mercury} + \text{oxygen}$$

it is found that 1 g of mercury(II) oxide produces 0.971 g mercury. However, if the evolved oxygen is weighed, it is found that 0.029 g is produced. The total mass of the products (mercury and oxygen) is therefore the same as the total mass of the reactants (mercury(II) oxide).

Similarly, when magnesium is burnt in air:

$$\text{magnesium} + \text{oxygen} \rightarrow \text{magnesium oxide}$$

it is found that 1 g of magnesium gives 1.666 g magnesium oxide. The gain in mass of 0.666 g represents the mass of oxygen from the air which combined with the magnesium. Thus, again the total mass of the reactants (magnesium and oxygen) is the same as the total mass of the products (magnesium oxide). This relationship is encountered in every reaction and it gives rise to the *law of conservation of mass* which states that matter cannot be created or destroyed in a chemical reaction.

ELEMENTS, COMPOUNDS, AND MIXTURES

In chemistry, the term *element* is used to describe a substance which cannot be split into anything simpler by normal chemical means. Obviously mercury(II) oxide cannot be an element because on heating it decomposes into mercury and oxygen (see page 2). However, the mercury and oxygen cannot be split into other substances and so they are elements.

There are ninety-two naturally occurring elements but about another twenty have been made artificially in atomic reactors. Examples of important elements are aluminium, calcium, carbon, chlorine, chromium, copper, gold, hydrogen, iodine, iron, lead, mercury, nitrogen, oxygen, silicon, sodium, sulphur, tin and zinc. Most of the elements are solids but mercury and bromine are liquids whilst a few are gases, e.g. nitrogen, oxygen, and chlorine.

The most abundant element is oxygen; it makes up about 50% by mass of the earth's crust, water, and air. Silicon, which is a major constituent of soil, sand, and rocks comes next at about 25%. This is followed by aluminium (7.3%), iron (4.2%), calcium (3.2%), sodium (2.4%), potassium (2.3%), magnesium (2.1%), and hydrogen (1%). The remaining eighty-three naturally occurring elements comprise about 1.8%

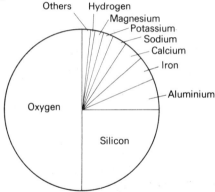

Figure 1.1. Abundance of the elements

It has been found very useful (page 15) to represent elements by symbols, the symbols being a letter or, in some cases, two letters. The letter is the first one of the name of the element but this is sometimes complicated by the fact that the Latin name is used. Since several elements have the same first letter, e.g. carbon, chlorine, and calcium, the first letter is followed by another one from the name. The symbol is always written with a capital letter but, if there are two letters, the second one is always small. Some common elements and their symbols are given below but a complete list is given on page 279.

Aluminium	Al	Mercury	Hg
Bromine	Br	Nitrogen	N
Calcium	Ca	Oxygen	O
Carbon	C	Phosphorus	P
Chlorine	Cl	Potassium	K
Copper	Cu	Silicon	Si
Hydrogen	H	Silver	Ag
Iodine	I	Sodium	Na
Iron	Fe	Sulphur	S
Lead	Pb	Tin	Sn
Magnesium	Mg	Zinc	Zn

Compounds are formed when two or more elements combine together in definite proportions by mass. For example, water is formed by oxygen and hydrogen combining in the ratio of 8 g of oxygen to 1 g of hydrogen. Other examples of compounds and the elements which they contain are given below.

	Compound	*Constituent elements*
1.	Magnesium oxide	— magnesium and oxygen.
2.	Sulphuric acid	— hydrogen, sulphur and oxygen.
3.	Calcium carbonate	— calcium, carbon and oxygen.
4.	Sugar	— carbon, hydrogen and oxygen.
5.	Salt	— sodium and chlorine.
6.	Nylon	— carbon, hydrogen, oxygen and nitrogen.

The properties of compounds are usually very different from those of the constituent elements. For example, salt (sodium chloride) is a compound of sodium and chlorine, and it exists as a white solid with a familiar taste. Sodium, however, is an extremely reactive silver coloured metal which reacts vigorously with water whilst chlorine is a yellow green gas with a choking smell.

Mixtures contain two or more substances which are not chemically combined. They differ from compounds in three main ways.

1. The ingredients of a mixture retain their own properties but, when in a compound, the properties are completely different. For example, air is a mixture containing approximately four-fifths nitrogen and one-fifth oxygen whilst dinitrogen oxide, a compound of these two elements, is an anaesthetic sometimes known as laughing gas.

2. Mixtures may have any composition but compounds always contain the same elements in the same proportions by mass. For instance, a mixture of copper powder and sulphur could contain anywhere between 0 and 100 per cent of copper with the sulphur making up the balance. In contrast, the compound, copper(II) sulphide, always contains copper and sulphur in the ratio of 63.5 g copper to 32 g sulphur.

3. The constituents of a mixture can be separated by physical methods but, generally, those of a compound can be separated only by chemical or electrical methods. Thus, mixtures of salt and sand can be separated by adding water to dissolve the salt and then removing the sand by filtration. The mixture of nitrogen and oxygen in air can be separated by cooling the air until it liquefies and then allowing the liquid air to warm up slowly – the nitrogen evaporates before the oxygen (page 113). On the other hand, a liquid compound such as water evaporates as one substance. In order to split water into its constituent elements, it is necessary to pass an electric current through it (page 49).

SEPARATION OF MIXTURES

a. Liquid-solid mixtures

(i) Liquids containing non-volatile impurities may be separated by a process known as *distillation*. The liquid is vaporised and subsequently condensed whilst the impurities remain behind in the distillation flask. The apparatus used is illustrated in Figure 1.2.

In order to prevent splashing over, the flask is only about half-filled. A few pieces of unglazed porcelain are added to provide nuclei for the vapour bubbles to form, the object being to promote steady boiling and minimise

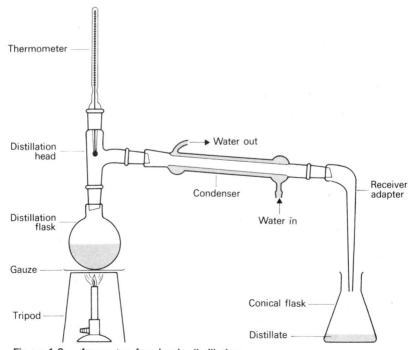

Figure 1.2. Apparatus for simple distillation

superheating or 'bumping'. The thermometer bulb is level with the side-arm so that the temperature of the vapour is measured as it leaves the flask; this gives the boiling point of the liquid at the prevailing atmospheric pressure. If the liquid boils below about 140 °C, a water condenser is used, but above this temperature an air condenser, i.e. a straight length of tube with no jacket, is employed to eliminate the risk of cracking. The flask is heated by a water-bath if low boiling point flammable liquids are being distilled or by an oil-bath or bunsen for high boiling liquids. Note that the flask is not heated to dryness since this could cause the flask to crack or it may result in decomposition of the solid residue.

If the solid component of the mixture is required, the distillation is continued until a concentrated solution is obtained. When the hot concentrated solution is allowed to cool, crystals of the solid are deposited (see also page 9).

If in aqueous solutions, i.e. solutions in water, only the solid is required, the solution may simply be evaporated to dryness in an evaporating basin on a boiling water-bath.

(ii) Solutions of solids can often be separated by chromatography, the separation being readily apparent if coloured substances are involved.

Screened methyl orange (page 79) can be separated by *column chromatography* using the apparatus illustrated in Figure 1.3.

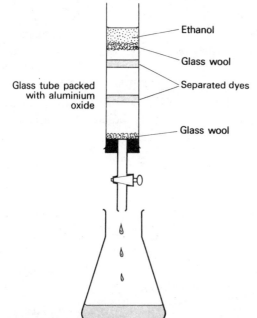

Figure 1.3. Column chromatography

The tube is three-quarters filled with aluminium oxide (alumina) suspended in ethanol. The tap is opened and, when the solvent level is just above the aluminium oxide, a few drops of screened methyl orange are added at the top of the column and allowed to pass down. Ethanol is added and runs through the column until the constituent indicators have separated, care being taken to see that the aluminium oxide is always covered by the ethanol. This method of separation is an example of *adsorption chromatography*. It is based on the fact that the various components of the mixture are adsorbed by the aluminium oxide (attracted to its surface) to different extents, the most weakly attracted substances passing down the column the fastest. If necessary, the elution can be continued until each component has passed through the column and been collected. This process may sometimes be speeded up by changing the solvent for the elution of the most strongly adsorbed components.

An alternative separation is provided by *paper chromatography*, the apparatus being shown in Figure 1.4. A small drop of black ink is placed about 1 cm from the edge of a rectangular sheet of filter paper. The ink is allowed to dry and the paper folded into the shape of a cylinder and secured with paper clips. The paper is then placed in a tall beaker containing a solvent or solvent mixture, the level of the solvent being below the sample of ink. The beaker is covered with a watch glass and left undisturbed. The solvent rises up the filter paper by capillary action and a number of coloured spots emerge in its wake. The separation is an example of *partition chromatography*, and it depends on the differential partition of the various dyes between the solvent (or solvent mixture) and the water loosely held by the cellulose in the paper, i.e. the components dissolve in the water and solvent to different extents, those most soluble in water moving the shortest distance up the paper.

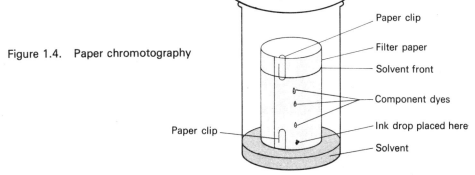

Figure 1.4. Paper chromotography

(iii) Liquids containing suspended solids can be separated by filtration (Figure 1.5). The solids are retained by the filter paper and they can be washed and dried. The liquid which passes through the filter paper is known as the filtrate.

Figure 1.5. Separation of solids and liquids by filtration

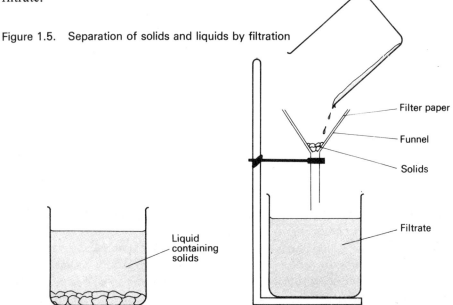

b. Liquid-liquid mixtures

(i) Mixtures of miscible liquids, for example methanol and water, are often separated by *fractional distillation*. The apparatus differs from that required for simple distillation (Figure 1.2) in that a fractionating column is inserted between the distillation flask and the take-off head (Figure 1.6). The fractionating column is generally a tube packed with broken glass or glass beads. When the mixture is heated, the vapour of both liquids passes up the column. However, the vapour of the higher boiling component will tend to condense in the fractionating column and run back down into the distillation flask. The liquid running back down the column scrubs or washes the vapour on the way up and in this way the ascending vapour becomes progressively richer in the lower boiling component. If the boiling points are not too close, good separations can be achieved.

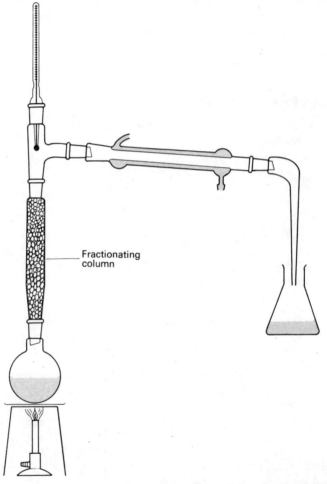

Figure 1.6. Fractional distillation

(ii) Immiscible liquids, such as tetrachloromethane and water, are separated by means of a separating funnel (Figure 1.7). The denser liquid is allowed to run out at the bottom. Since the funnel tapers just before the tap, the separation can be performed down to the last drop.

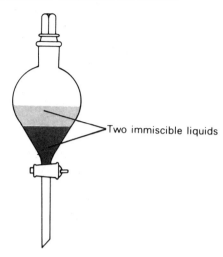

Figure 1.7. Separation of immiscible liquids

c. Solid–solid mixtures

(i) The solubility of most substances increases as the temperature of the solvent is raised and this fact is utilised in the separation of solid mixtures by crystallisation. The process of *crystallisation* involves finding a solvent which will readily dissolve the mixture when hot but only to a small extent when cold. The mixture is then dissolved in the minimum volume of the boiling solvent and the hot solution filtered, if necessary, to remove any insoluble impurities. The resulting solution is cooled. Ideally, the required product will crystallise out and the impurities will remain in solution. The crystals are filtered and dried and, if necessary, the process is repeated until a pure product is obtained.

The process of crystallisation may be clarified by a numerical example. The solubilities of potassium chloride and potassium chlorate are shown by the table:

Compound	Mass dissolved by 100 g water at	
	20 °C	100 °C
Potassium chloride	34.5 g	56.5 g
Potassium chlorate	7.5 g	54.0 g

Now consider a mixture consisting of 10 g of potassium chloride and 40 g of potassium chlorate. If this mixture is dissolved in 100 g of boiling water and the solution is then cooled to 20 °C, no potassium chloride will come out of solution because its solubility is not exceeded, but 32.5 g of potassium chlorate crystals will be deposited (the water can only dissolve 7.5 g of potassium chlorate at 20 °C and so $40 - 7.5 = 32.5$ g come out of solution).

Note that, in crystallisation, some of the substance is always lost. This is because the filtrate is saturated with the substance which has crystallised and the filtrate is discarded since it also contains the impurity. The reason for using the minimum volume of boiling solvent is therefore apparent.

(ii) All solids have a vapour pressure (some, such as moth balls, can be smelt) and the vapour pressure increases with temperature. In a few cases, the vapour pressure reaches atmospheric pressure before they melt and so the solid turns to vapour without an intermediate liquid stage: such substances are said to *sublime*. Examples of common substances of this type are carbon dioxide, iodine, ammonium chloride, and naphthalene. A small number of solid mixtures can therefore be separated by sublimation, the apparatus being illustrated in Figure 1.8. The mixture, e.g. salt and iodine, is placed in an evaporating basin and this is covered by a filter paper in which small perforations have been made. An inverted filter funnel rests on the filter paper and the mixture is heated very gently with a small flame. The iodine sublimes, passes through the perforations in the filter paper, and condenses on the cool parts of the funnel. Eventually, only salt remains in the basin.

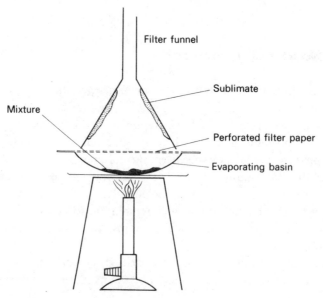

Figure 1.8. Separation of solids by sublimation

DRYING OF SOLIDS, LIQUIDS, AND GASES

Solids can be dried in an oven at 100 °C if they do not melt and are stable at this temperature. Many organic solids, however, are dried by squeezing them between layers of filter paper. The last traces of moisture are removed by leaving the solid in a desiccator (Figure 1.9 (a)) for a few days over a drying agent such as phosphorus(V) oxide (phosphorus pentoxide), silica gel, or concentrated sulphuric acid. This process can be speeded up by use of a vacuum desiccator (Figure 1.9 (b)): the pressure is reduced as low as possible so that evaporation occurs more rapidly.

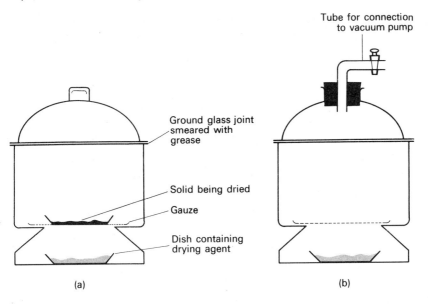

Figure 1.9. (a) Desiccator (b) Vacuum desiccator

Liquids are dried by standing them over a drying agent with which they do not react. Common drying agents for liquids are anhydrous magnesium sulphate, calcium sulphate, and calcium chloride. Sodium is used to dry hydrocarbons (page 226) and ethers. Often, organic liquids look cloudy when they are wet and so they are left standing over the drying agent, with occasional shaking, until they go clear.

Many acidic and neutral gases such as hydrogen chloride, sulphur dioxide, carbon dioxide, hydrogen, and chlorine are dried by bubbling them through concentrated sulphuric acid contained in a Dreschel bottle. However, some gases react with this acid and so other drying agents are necessary. For example, ammonia is dried by passing it up a tower containing calcium oxide (quick lime), whilst hydrogen is sometimes dried by passing it through U-tubes containing anhydrous calcium chloride. The various types of apparatus are illustrated in Figure 1.10.

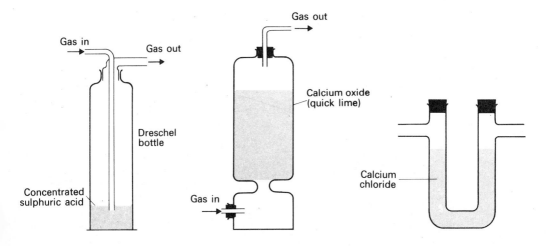

Figure 1.10. Apparatus for drying gases

MELTING POINTS AND BOILING POINTS AS CRITERIA OF PURITY

Pure solids generally melt over a range of 1 °C or less whereas impure solids melt gradually over a range of several degrees and always lower than the pure substance. Sharpness of melting point is therefore a useful criterion of purity.

A pure liquid has a sharp and constant boiling point at any particular pressure but the boiling point rises as the atmospheric pressure increases. Pure liquids also distil over completely leaving no residue, whereas impure liquids often show a steady rise in boiling point as a greater proportion of the lower boiling components distil off. However, a few mixtures have a sharp and constant boiling point at constant pressure. These constant boiling mixtures can be detected by repeating the distillation at a different pressure because the composition of the distillate is then different. A sharp boiling point is therefore not quite as good a criterion of purity as is a sharp melting point.

Some examples of constant boiling point mixtures are given below, the boiling points being those at one atmosphere.

Nitric acid/water mixture containing 68% nitric acid, b.p. 120.5 °C
Hydrochloric acid/water containing 20.2% hydrochloric acid, b.p. 108.6 °C
Ethanol/water mixture containing 95.6% ethanol, b.p. 78.1 °C

ELEMENTS AS METALS AND NON-METALS

The elements can be divided into metals and non-metals according to their physical and chemical properties. The main differences between metals and non-metals are summarised below but it must be stressed that they are general properties and exceptions occur. Thus some elements appear to be intermediate between metals and non-metals whilst other elements differ from the list in just one or two cases.

Physical properties

Metals	*Non-Metals*
1. Usually have high densities (sodium and potassium are exceptions).	— Low densities
2. Shiny and can be polished.	— Dull, cannot be polished.
3. Malleable and ductile, i.e. can be hammered into sheets and drawn into wire.	— Brittle.
4. High tensile strength.	— Low tensile strength.
5. High melting points (sodium and potassium are exceptions and mercury is a liquid).	— Low melting points: several are gases.
6. Good conductors of heat and electricity.	— Poor conductors of heat and electricity (but carbon in the form of graphite does conduct electricity).

Chemical properties

Metals	Non-Metals
1. Give basic oxides, i.e. react with acids.	Give acidic oxides.
2. Replace the hydrogen in acids giving salts.	Do not react with acids in this manner.
3. Form positive ions (pages 29–30).	Form negative ions (pages 29–30).
4. Form electrovalent chlorides stable to water (pages 39–40).	Form Covalent chlorides which react with water (pages 39–40).
5. Do not readily react with hydrogen.	Form stable compounds with hydrogen.

Questions

1. (a) State three ways in which a physical change differs from a chemical change.
   (b) Classify the following as physical or chemical changes, giving reasons in each case:
   (i) the rusting of iron;
   (ii) the setting of a jelly;
   (iii) the switching on of an electric light;
   (iv) the action of heat on iodine.

2. State whether the following are physical or chemical changes, giving reasons in each case:
   (i) the heating of a solution of salt;
   (ii) the lighting of a match;
   (iii) the cooling of a hot concentrated solution of potassium chloride so that crystals are deposited;
   (iv) the examination of ink by paper chromatography;
   (v) the baking of a cake.

3. (a) State the essential differences between compounds and mixtures.
   (b) If you were supplied with a green powder, explain how you would attempt to discover whether it was a compound or a mixture.

4. Describe briefly how you would separate a pure sample of the first named substance from each of the following mixtures:
   (i) ammonium chloride and sodium chloride;
   (ii) trichloromethane (boiling point 62 °C) and tetrachloromethane (boiling point 77 °C);
   (iii) sodium chloride contaminated with a little potassium chloride;
   (iv) petrol and water;
   (v) sugar and sand.

5. (a) When grass is ground up with propanone (a volatile organic solvent), a green coloured solution is obtained. Describe how you would show that the green colour is not due to a single substance.
   (b) Explain briefly how you would obtain a sample of pure water from sodium chloride solution. How would you show that your separation had been successful?

6. (a) Iron is a metal whilst sulphur is a non-metal. State five pieces of evidence in support of this statement.
   (b) When iron powder is heated with sulphur, a grey-black solid mass is obtained. Explain how you would attempt to show that this product is a compound and not a mixture.

2 Atoms, Molecules, Formulae, and Equations

MATTER CONSISTS OF SMALL PARTICLES

Potassium manganate(VII) (potassium permanganate) is a purple black solid which dissolves in water to give a purple solution. As the solution is progressively diluted, the colour gradually changes to pink. Now, if 0.1 g of potassium manganate(VII) (potassium permanganate) (an amount roughly equal to the size of a third of a pea) is weighed out and dissolved in water, it is found that the pink colour can still just be observed in 300 litres of solution. Since 1 cm^3 of solution contains about 20 drops, then 300 litres contain 6 million drops. If each drop contains only one particle of potassium manganate(VII) (potassium permanganate), then it is apparent that there will be 6 million particles in 0.1 g of the crystals. Obviously more than one particle is required to impart colour to the drop and so 0.1 g of potassium manganate(VII) (potassium permanganate) must contain a very large number of particles. Similar experiments can be performed with other coloured substances.

The experiment above indicates that substances are made up of an extremely large number of very small particles. The particles are, in fact, too small to be seen even with the most powerful optical microscope. Nevertheless, some indication of the size is given by the oil drop experiment described below.

A 1 cm^3 pipette is filled with camphorated oil which is then allowed to drip out, the number of drops being counted; the volume of one drop is therefore found. A large pneumatic trough is now filled with water and lightly dusted with talc. A drop of camphorated oil is placed on the surface of the water at the centre of the trough and the resultant oil film pushes back the talc. The area of the oil film is estimated and so the approximate depth can be calculated. The results of an experiment are given below.

$$
\begin{aligned}
&\text{Number of oil drops contained in 1 cm}^3 = 44 \\
&\therefore \text{Volume of 1 drop} = \tfrac{1}{44}\,\text{cm}^3 \\
&\text{Diameter of oil film} = 13\,\text{cm} \\
&\text{Area of oil film } (\pi r^2) = 3.142 \times 6.5^2\,\text{cm}^2 \\
&\text{Depth of oil film} = \frac{\text{volume}}{\text{area}} \\
&\qquad\qquad\qquad\quad = \frac{1}{44 \times 3.142 \times 6.5^2}\,\text{cm} \\
&\qquad\qquad\qquad\quad = 1.7 \times 10^{-4}\,\text{cm}
\end{aligned}
$$

Thus, the maximum length of an oil particle is 1.7×10^{-4} cm. The method is not very accurate since it is known that the film is actually several particles thick.

ATOMS AND MOLECULES

At the beginning of the nineteenth century, John Dalton made a number of suggestions which are known as the atomic theory. The main points are summarised below.

(a) Elements are made up of minute indivisible and indestructible particles called atoms.

(b) The atoms of an element are all identical but different from the atoms of all other elements.

(c) 'Compound-atoms' are formed by the combination of small whole numbers of atoms. 'Compound-atoms' are now called molecules. With a few modifications (page 27), the theory still holds today.

An atom is defined as the smallest part of an element that can take part in a chemical change. The size of an atom may be imagined from the fact that cigarette paper is about 100 000 atoms thick, or a full stop would be covered by about 2 000 000 atoms.

Although single atoms of an element take part in chemical changes, they do not necessarily exist in the free state. For example, oxygen atoms do not normally have a separate existence – they exist in pairs and these pairs are called molecules of oxygen. Similarly, the smallest particle of a compound must obviously contain at least two atoms because it contains at least two elements. The smallest particle of a compound is therefore called a molecule of the compound. *A molecule is defined as the smallest particle of an element or compound which can exist by itself.*

The sizes of atoms and molecules can now be determined very accurately by X-ray and electron diffraction, but the details of these processes are beyond the level of this work.

FORMULAE

Since atoms are represented by symbols, it is possible to represent molecules by formulae. A molecule of oxygen contains two atoms and so its formula is written as O_2. Similarly, a molecule of hydrogen is written as H_2.

A molecule of water contains two atoms of hydrogen and one atom of oxygen and so its formula is H_2O. Since a molecule of copper sulphate contains one atom of copper, one of sulphur, and four of oxygen its formula is $CuSO_4$. Note that, if a molecule contains only one atom of an element, no number is put in the formula after the symbol of that element. Further, the general rule when writing formulae is to put the most metallic element first.

In the examples above, the number of atoms of each element in the molecule have been stated. It should be stressed that this information is obtained as a result of analysis; an illustration of this is given on pages 225 to 227.

When more than one molecule of a substance is being considered, the number involved is placed in front of the formula, e.g. $2CuSO_4$ means two molecules of copper sulphate, and so we are concerned with two atoms of copper, two of sulphur, and eight of oxygen.

VALENCY

It is found that when atoms combine to form molecules, they have a definite combining power or valency. Hydrogen is the simplest element and it is said to have a valency of one. *The valency of an element is then defined as the number of atoms of hydrogen which one atom of the element will combine with or displace.*

It has already been stated that, in water, two atoms of hydrogen combine with one of oxygen; hence, according to the definition, the valency of oxygen must be two. Now oxygen combines with most of the other elements and so, if these compounds are analysed and their formulae found, the valency of the elements may be deduced. Some elements and their valencies are given below.

Valency 1 (monovalent) : H   Na   K   Cl   Br   I   Ag
Valency 2 (divalent)   : Ca   Mg   Zn   Cu   Pb   O   S
Valency 3 (trivalent) : Al   N
Valency 4 (tetravalent): C   Si

Some elements have valencies of 5, 6, 7, 8, or 0 but these are not important at this stage. Also, some elements can have more than one valency, for example, iron can have valencies of 2 or 3 and this leads to such compounds as iron(II) chloride, $FeCl_2$ and iron(III) chloride, $FeCl_3$. Note that roman numerals in brackets, after the name of an element, indicate its valency. This procedure is adopted when the element has more than one common valency.

RADICALS

Some groups of atoms often occur together and remain together during chemical reactions. These groups of atoms are known as radicals and, like elements, they have a definite valency although they cannot exist in isolation. Some of the common radicals are listed below.

Name	Formula	Valency	Examples
Carbonate	CO_3	2	$CaCO_3$, Na_2CO_3
Nitrate	NO_3	1	$NaNO_3$, HNO_3
Sulphate	SO_4	2	H_2SO_4, K_2SO_4
Sulphite	SO_3	2	H_2SO_3, $MgSO_3$
Hydroxide	OH	1	KOH
Ammonium	NH_4	1	NH_4OH, NH_4Cl
Phosphate	PO_4	3	H_3PO_4, Na_3PO_4

With the exception of ammonium, these radicals combine with hydrogen or metals. The ammonium radical can combine with the other radicals.

It should be noted that, when two atoms or radicals combine, all the valencies must be satisfied and so the relative numbers of each must be adjusted until this is the case. For example, sodium and carbonate have valencies of one and two respectively and so two sodiums are required to satisfy one carbonate, i.e. the formula of sodium carbonate is Na_2CO_3.

Until proficiency is gained in working out formulae, it may be advantageous to indicate the valencies as links from the atoms or radicals involved and then to ensure that all the links are joined. For example, the valency of aluminium is 3, and of sulphate 2, and so the formula of aluminium sulphate is derived as below.

$$Al\!\!\equiv\; +\; {=}SO_4 \rightarrow Al\!\!\equiv\!\!SO_4$$

The aluminium has a spare valency and so another sulphate must be added.

$$Al\!\!\equiv\!\!SO_4\; +\; {=}SO_4 \rightarrow Al\!\!\equiv\!\!\genfrac{}{}{0pt}{}{SO_4}{SO_4}$$

Now a sulphate has a spare valency and so another aluminium is required.

$$Al\!\!\equiv\!\!\genfrac{}{}{0pt}{}{SO_4}{SO_4}\; +\; Al\!\!\equiv \rightarrow \genfrac{}{}{0pt}{}{Al}{Al}\!\!\equiv\!\!\genfrac{}{}{0pt}{}{SO_4}{SO_4}$$

Addition of a further sulphate now utilises all the valencies.

$$\genfrac{}{}{0pt}{}{Al}{Al}\!\!\equiv\!\!\genfrac{}{}{0pt}{}{SO_4}{SO_4}\; +\; {=}SO_4 \rightarrow \genfrac{}{}{0pt}{}{Al}{Al}\!\!\equiv\!\!\genfrac{}{}{0pt}{}{SO_4\; SO_4}{SO_4}$$

When more than one radical is present in a compound, the radical is placed in brackets with the number outside as a subscript. Since aluminium sulphate contains 2 aluminiums and 3 sulphates, its formula is written as $Al_2(SO_4)_3$.

Further examples of formulae are:

Pb valency 2, NO_3 valency 1   ∴ lead nitrate is $Pb(NO_3)_2$
Al valency 3, OH valency 1   ∴ aluminium hydroxide is $Al(OH)_3$
NH_4 valency 1, SO_4 valency 2   ∴ ammonium sulphate is $(NH_4)_2SO_4$

EQUATIONS

Substances can be represented by formulae and so chemical reactions can be represented by equations. Thus, the reaction

copper(II) oxide + sulphuric acid → copper(II) sulphate + water

is given by the equation

$$CuO + H_2SO_4 \rightarrow CuSO_4 + H_2O$$

It is now becoming common practice to include the physical states of the various substances in the equation and so the above reaction may be given as

$$CuO(s) + H_2SO_4(aq) \rightarrow CuSO_4(aq) + H_2O(l)$$

where (s) = solid, (l) = liquid, and (aq) = aqueous (i.e. a solution in water). Gases are indicated by (g). Note that the equation gives no information about the conditions required, e.g. the temperature or the concentration of the acid.

When potassium chlorate is heated, it decomposes into potassium chloride and oxygen and so the equation for the reaction may be expected to be

$$KClO_3 \rightarrow KCl + O_2$$

However, this equation is not balanced, i.e. on the left hand side there are three oxygen atoms whereas there are only two on the right hand side. Now, according to the law of conservation of mass (page 3), matter cannot be created or destroyed in a reaction and so the total number of atoms of each element in the reactants must be the same as in the products. It is necessary, therefore, to adjust the relative amounts of the substances so that the equation does balance. Hence the above equation should read

$$2KClO_3 \rightarrow 2KCl + 3O_2$$

It must be stressed that it is the number of molecules of the substances which are adjusted; the formulae cannot be changed.

Further examples of the balancing of equations are given below.

Unbalanced	$Ca + H_2O \rightarrow Ca(OH)_2 + H_2$
Balanced	$Ca + 2H_2O \rightarrow Ca(OH)_2 + H_2$
Unbalanced	$K_2CO_3 + HNO_3 \rightarrow KNO_3 + H_2O + CO_2$
Balanced	$K_2CO_3 + 2HNO_3 \rightarrow 2KNO_3 + H_2O + CO_2$
Unbalanced	$Cu + H_2SO_4 \rightarrow CuSO_4 + SO_2 + H_2O$
Balanced	$Cu + 2H_2SO_4 \rightarrow CuSO_4 + SO_2 + 2H_2O$

THE KINETIC THEORY

Substances may exist as solids, liquids, or gases, depending upon the conditions. In the solid state, a substance has a definite volume and shape and it is said to be rigid. However, in the liquid state, the shape is determined by the containing vessel but the volume is fixed if the temperature and pressure are constant. Finally, in the gaseous state there is no definite shape or volume since the gas diffuses (spreads out) so that it uniformly occupies the whole volume available to it.

It is found that solids and liquids generally expand by their own characteristic amount when they are heated. In contrast, most gases change their volumes by similar amounts if they undergo the same temperature or pressure change. This similarity in the behaviour of gases led to the kinetic theory.

THE KINETIC THEORY OF GASES

The kinetic theory was put forward to explain experimental observations on gases; it makes the following assumptions about them.

(a) Gases consist of a very large number of extremely small particles (molecules) which are in a state of continuous rapid motion.

(b) The molecules move in straight lines until they collide with one another or with the walls of the container.

(c) The collisions are perfectly elastic, i.e. there is no change in the total energy as a result of a collision.

(d) The distances between the molecules are large compared to their size. This supposition is indicated by the great compressibility of gases.

(e) The molecules are so far apart that the attractive forces between them are negligible.

The assumptions above relate to *ideal gases* but in practice *real gases* never fully achieve this 'ideal' behaviour. Thus, conditions (d) and (e) are not often completely attained.

DIFFUSION OF GASES

The movement of gaseous molecules can be demonstrated using bromine. Thus, a gas jar of air is inverted over a gas jar of bromine vapour (Figure 2.1). Despite the fact that bromine vapour is much denser than air, the red-brown vapour is seen to slowly occupy the whole volume available to it.

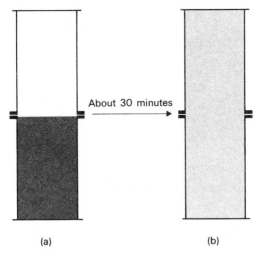

Figure 2.1. The diffusion of bromine vapour

The velocities of molecules are continually changing as a result of their collisions. However, calculations show that, at 20 °C, the average velocity of oxygen molecules is about 1000 m.p.h. By contrast, bromine molecules, which are five times as heavy as oxygen molecules, have an average velocity of a little under 500 m.p.h. at 20 °C. Different gases therefore diffuse at different rates and this can be demonstrated using the apparatus shown in Figure 2.2. If a gas jar of hydrogen is placed over the porous pot (Figure 2.2 (a)), a stream of bubbles issues from the tube in the water. This is because hydrogen, which is less dense than air, diffuses *into* the porous pot faster than the air can diffuse *out* thus causing an increase in pressure in the pot. However, if the porous pot is placed in a gas jar of carbon dioxide (Figure 2.2(b)), the water level in the tube rises. This is due to the air diffusing *out* of the pot faster than the more dense carbon dioxide can diffuse *in*. The water level in the tube therefore rises as the pressure in the pot falls.

The behaviour of diffusing gases is summed up by Graham's law of diffusion which states that *the rate of diffusion of a gas is inversely proportional to the square root of its density*. Thus, as seen above, the rates of diffusion of gases increase as their densities decrease.

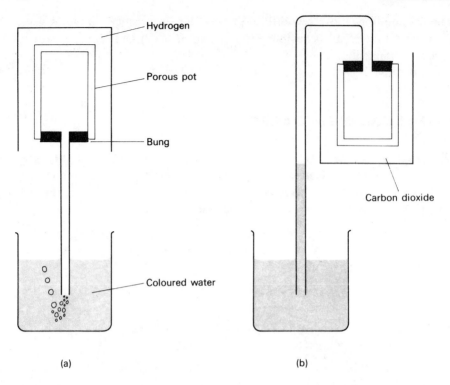

Figure 2.2. Diffusion of (a) light and (b) heavy gases

BOYLE'S LAW

If a fixed mass of a gas is placed in a closed vessel, then the gas will exert a pressure on the vessel due to the molecules bombarding it. If the volume of the vessel is now increased, the molecules will travel further between collisions with the walls, i.e. there will be fewer collisions in a given time and so the pressure will fall. This explains Boyle's law which states that *the volume of a fixed mass of gas is inversely proportional to the pressure provided that the temperature remains constant.* The 'inversely proportional to' is the mathematical way of saying that, as one variable is increased, the other decreases in the same proportion, e.g. if the volume of a given mass of gas is doubled, then the pressure is halved. The law may be expressed mathematically as

$$P \propto \frac{1}{V} \quad \text{at constant temperature}$$

or $\quad PV = \text{constant} \quad$ where $\quad P = $ pressure
and $\quad V = $ volume

CHARLES' LAW

If a fixed mass of gas in a closed vessel is heated, the molecules gain more energy and so their velocity increases. There will therefore be more collisions with the walls of the vessel in a given time and the pressure will increase. If the pressure is to be kept the same, the volume will obviously have to be increased. This relationship is embodied in Charles' law which states that *the volume of a fixed mass of gas is directly proportional to the absolute temperature provided that the pressure remains constant.* The 'directly proportional to' is the mathematical way of saying that the volume increases in the same proportion as the absolute temperature increases. The law may be expressed mathematically as

$$V \propto T \quad \text{at constant pressure}$$

or

$$\frac{V}{T} = \text{constant} \quad \text{where} \quad V = \text{volume}$$
$$\text{and} \quad T = \text{absolute temperature}$$

The absolute temperature scale arises as follows. Charles found that, when the pressure is kept constant, the volume of a gas decreases by $\frac{1}{273}$ of its volume at 0 °C for every °C the temperature is lowered. Theoretically, therefore, if 1 cm^3 of a gas at 0 °C is cooled, then at −273 °C the gas will have no volume. This is obviously impossible. What in fact happens is that, as the temperature is lowered, the molecules have less kinetic energy (energy of motion) and so they move more and more slowly. At −273 °C the molecules would come to rest and so the volume would be the volume of the static molecules. Thus, −273 °C is the lowest possible temperature that could be achieved and it is called 'absolute zero'. This has given rise to the absolute temperature scale in which 0 °C is taken as 273 degrees absolute, i.e. 273 K (Kelvin). Any temperature on the Celsius scale is converted to the absolute scale by adding 273 to it, e.g. 50 °C is 273 + 50 = 323 K, −20 °C is 273 + (−20) = 253 K.

If the temperature of a gas is lowered, the molecules move more slowly and eventually, if it is cooled enough, it will liquefy because the molecules become so slow that intermolecular attractions become the predominating factor. Liquefaction can also be brought about by increasing the pressure, i.e. forcing the molecules closer together.

COMBINATION OF BOYLE'S AND CHARLES' LAWS

Consider a fixed mass of gas. According to Boyle's law,

$$V \propto \frac{1}{P} \quad \text{at constant temperature}$$

whilst Charles' law states that

$$V \propto T \quad \text{at constant pressure}$$

Combination of these two equations gives

$$V \propto \frac{T}{P} \quad \text{or} \quad V = k\frac{T}{P}$$

which on rearrangement gives

$$\frac{PV}{T} = k \quad \text{where} \quad k \text{ is a constant.}$$

Now, suppose the fixed mass of gas has pressure P_1 and volume V_1 at absolute temperature T_1, then

$$\frac{P_1 V_1}{T_1} = k$$

If the same mass of the same gas has pressure P_2 and volume V_2 at absolute temperature T_2, then

$$\frac{P_2 V_2}{T_2} = k$$

It therefore follows that

$$\frac{P_1 V_1}{T_1} = \frac{P_2 V_2}{T_2}$$

This equation is very important since it allows the state of a gas to be calculated under different conditions. This may be illustrated by the examples below.

Example A gas occupies 200 cm^3 at 25 °C and 745 mm pressure; what will be its volume at −15 °C and 738 mm pressure?

$$\frac{P_1 V_1}{T_1} = \frac{P_2 V_2}{T_2} \quad \text{where} \quad \begin{aligned} P_1 &= 745 \text{ mm} \\ V_1 &= 200 \text{ cm}^3 \\ T_1 &= 25 + 273 = 298 \text{ K} \end{aligned}$$

and
$$\begin{aligned} P_2 &= 738 \text{ mm} \\ V_2 &= ? \text{ cm}^3 \\ T_2 &= -15 + 273 = 258 \text{ K} \end{aligned}$$

i.e.
$$\frac{745 \times 200}{298} = \frac{738 \times V_2}{258}$$

Therefore $\quad V_2 = \dfrac{745 \times 200 \times 258}{298 \times 738} = 174.8 \text{ cm}^3$

Example A gas occupies 90 cm^3 at 11 °C and 550 mm pressure. At what temperature will it have a volume of 80 cm^3 if the pressure is adjusted to 750 mm?

$$\frac{P_1 V_1}{T_1} = \frac{P_2 V_2}{T_2} \quad \text{where} \quad \begin{aligned} P_1 &= 550 \text{ mm} \\ V_1 &= 90 \text{ cm}^3 \\ T_1 &= 11 + 273 = 284 \text{ K} \\ P_2 &= 750 \text{ mm} \\ V_2 &= 80 \text{ cm}^3 \\ T_2 &= ? \text{ K} \end{aligned}$$

i.e.
$$\frac{550 \times 90}{284} = \frac{750 \times 80}{T_2}$$

Therefore $\quad T_2 = \dfrac{284 \times 750 \times 80}{550 \times 90} = 344.2 \text{ K}$

Hence the temperature in degrees Celsius = 344.2 − 273 = 71.2 °C.

STANDARD TEMPERATURE AND PRESSURE (S.T.P.)

The volumes of gases vary with temperature and pressure and so, if volumes have to be compared, it would obviously be helpful to have them quoted at some standard temperature and pressure. The standard pressure is taken as 760 mm mercury, i.e. one atmosphere or 101 325 pascal (Pa), and the standard temperature as 0 °C or 273 K.

A volume of gas at any temperature and pressure can be converted to s.t.p. using the equation

$$\frac{P_1 V_1}{T_1} = \frac{P_2 V_2}{T_2}$$

Example At −10 °C and 745 mm pressure, a gas occupies 1500 cm³; what will be its volume at s.t.p.?

$$\frac{P_1 V_1}{T_1} = \frac{P_2 V_2}{T_2}$$

where $P_1 = 745$ mm
$V_1 = 1500$ cm³
$T_1 = -10 + 273 = 263$ K
$P_2 = 760$ mm
$V_2 = ?$ cm³
$T_2 = 0 + 273 = 273$ K

i.e. $\dfrac{745 \times 1500}{263} = \dfrac{760 \times V_2}{273}$

Therefore $V_2 = \dfrac{745 \times 1500 \times 273}{263 \times 760} = 1526$ cm³

KINETIC THEORY OF LIQUIDS

The main difference between liquids and gases is that the molecules in a liquid are much closer together. As a result, the intermolecular forces are much greater and so the molecules are far less free to move about independently of each other. The molecules in the body of the liquid will be subject to equal forces all round. However, the molecules near the surface will be attracted by greater forces from below than above (there are fewer molecules above) and this will tend to pull them inwards. This accounts for the surface tension of a liquid.

The velocities of the molecules in a liquid will be continually changing as a result of their collisions. Nevertheless, many molecules will have about the same average velocity whilst some will have greater velocities and others slower velocities. If the very fast molecules are near the surface of the liquid and moving in an upward direction, they may escape from the liquid mass and exist as vapour molecules, i.e. vaporisation or evaporation occurs. This results in liquids having a vapour pressure. (The term vapour is the name given to the gaseous form of a substance which is normally a solid or a liquid at room temperature.)

Energy is related to temperature and so the temperature of a liquid will fall when evaporation takes place since it is the high energy molecules which escape. *The amount of heat which must be supplied to prevent the temperature falling during the evaporation of one mole (see page 63) of a liquid is known as the enthalpy, or molar latent heat, of vaporisation.* Note that the vapour produced is not hotter than the liquid because the escaping molecules use their extra energy in overcoming the attractive forces at the surface of the liquid.

DIFFUSION IN LIQUIDS

Diffusion of solids in liquids is apparent if a crystal of ammonium dichromate(VI) is added to a beaker of water; the orange colour gradually spreads throughout the water. Diffusion in liquids is, however, slower than diffusion of gases. The diffusion may be speeded up by using hot water, the reason being that the average velocity of the molecules increases as the temperature is raised.

The motion of solvent molecules is also illustrated by the Brownian movement. Thus, the botanist, Brown, using an ultramicroscope, observed that small particles of pollen suspended in water followed a continuous, random, zig-zag path. The random motion must be the result of uneven bombardment of the pollen by the water molecules. This phenomenon, in fact, occurs with all colloidal solutions, i.e. liquids which contain very small particles in suspension (particles with diameters between about 10^{-7} to 10^{-5} cm).

BOILING POINTS

The boiling point of a liquid is the temperature at which its vapour pressure equals the external pressure. In order to standardise the conditions, a pressure of 760 mm of mercury (1 atmosphere) is generally chosen and so the boiling points of liquids can be readily compared. The boiling point of a liquid is therefore taken to be the temperature at which its vapour pressure is equal to 760 mm of mercury unless some other pressure is stated.

Boiling under reduced pressure is readily demonstrated as follows. A Pyrex flask is half filled with water and the latter is boiled for a few minutes to expel dissolved air. The heat is then turned off and a bung fitted in the flask (Figure 2.3). The flask is cooled under water, some of the vapour condenses,

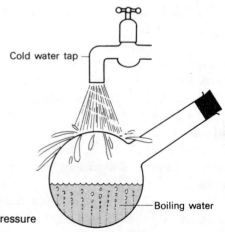

Figure 2.3. Water boiling under reduced pressure

and so the pressure is reduced. The vapour pressure of the water will then be equal to the pressure in the flask and so the water will reboil. This process may be continued and the water will boil even when its temperature is in the low thirties.

Before a liquid boils, bubbles are seen to form at the bottom of the container where it is hot. However, these bubbles disappear before they reach the surface: they condense due to the colder liquid above and the pressure. But, when a liquid is boiling, the bubbles reach the surface of the liquid because their pressure equals atmospheric pressure. If porous pot, glass beads, etc. are not added to provide nuclei for the bubbles of vapour to form, superheating or 'bumping' will probably occur.

It has already been stated that the kinetic energy of molecules is controlled by the temperature. Now kinetic energy is related to mass and velocity by the equation

$$K.E. = \tfrac{1}{2} mv^2 \quad \text{where} \quad \begin{aligned} K.E. &= \text{kinetic energy} \\ m &= \text{mass of a molecule} \\ v &= \text{velocity of a molecule} \end{aligned}$$

It therefore follows that, at any given temperature, heavy molecules will have lower average velocities than lighter ones. Thus, at a given temperature, liquids consisting of heavy molecules will have fewer of them with sufficient energy to escape as vapour than those consisting of lighter molecules. Hence, it is a general rule that the heavier the molecules of a liquid the lower the vapour pressure and the higher the boiling point. Exceptions to the rule are not uncommon owing to the tendency of some molecules to move about in clusters (see page 52).

KINETIC THEORY OF SOLIDS

If a liquid is cooled, the velocity of the molecules becomes slower and slower. Eventually, the intermolecular attractions become the predominating factor, the liquid solidifies, and the molecules become fixed in a definite pattern in the crystal lattice. The molecules can then only vibrate about these fixed positions, they do not have the relatively free movement of molecules in a liquid. Further reduction in the temperature results in the vibrations becoming smaller. On the other hand, if the solid is heated, the vibrations of the molecules become greater until eventually they overcome the attractive forces of the crystal lattice and the solid melts. Since all the molecules in a solid have practically the same energy, the melting point is usually sharp and definite. *Melting point is defined as that temperature at which the solid and liquid forms of a substance can co-exist indefinitely under a pressure of one atmosphere.*

The amount of heat required to convert 1 mole (page 63) of a solid to a liquid without change in temperature, is known as the *enthalpy, or molar latent heat, of fusion*. The degree of intermolecular attraction in substances will be indicated by the size of their enthalpies (molar latent heats) of vaporisation and fusion since energy is required to separate the molecules, e.g.

the enthalpies of fusion of hydrogen sulphide (small intermolecular attractions) and water (appreciable intermolecular attractions) are 2.43 and 5.93 kJ mol⁻¹ respectively.

Questions

1. Write balanced equations for the following reactions:
   (a) Zinc oxide + nitric acid (HNO_3) → zinc nitrate + water.
   (b) Sodium hydroxide + sulphuric acid (H_2SO_4) → sodium sulphate + water.
   (c) Lead(IV) oxide + hydrochloric acid (HCl) → lead(II) chloride + water + chlorine.
   (d) Potassium + water → potassium hydroxide + hydrogen.
   (e) Ammonium chloride + calcium hydroxide → calcium chloride + water + ammonia (NH_3).

2. Write balanced equations for the following reactions:
   (a) Iron + chlorine → iron(III) chloride.
   (b) Aluminium oxide + sulphuric acid (H_2SO_4) → aluminium sulphate + water.
   (c) Copper(II) nitrate → copper(II) oxide + nitrogen dioxide (NO_2) + oxygen.
   (d) Iron(III) chloride + calcium hydroxide → iron(III) hydroxide + calcium chloride.
   (e) Magnesium nitride (Mg_3N_2) + water → magnesium hydroxide + ammonia (NH_3).

3. (a) Explain why a gas in an enclosed vessel exerts a pressure on the vessel and how the pressure alters when the gas is cooled.
   (b) Describe briefly how you would show that hydrogen diffuses faster than chlorine.
   (c) Explain how and why the temperature of a liquid changes as evaporation takes place.

4. A certain mass of gas occupies a volume of 530 cm³ at 770 mm pressure and 15 °C. Calculate its volume at 745 mm pressure and 20 °C.

5. A certain mass of carbon dioxide occupies a volume of 190 cm³ at 755 mm pressure and 18 °C. Calculate its volume at s.t.p.

6. A gas at 19 °C and 750 mm is sealed in a glass tube with a capacity of 200 cm³. What pressure will the gas exert if the temperature is raised to 150 °C?

7. (a) Describe a demonstration which shows that *a gas of low density diffuses more rapidly than a gas of high density*. In your description give:
   (i) a labelled diagram of the apparatus you would use,
   (ii) an account of the method employed, and
   (iii) details of the observations you would make which would lead you to the conclusion which has been printed in italics.
   (b) Use the kinetic theory to explain:
   (i) the difference between a solid and a gas,
   (ii) the evaporation of water from a puddle, and
   (iii) the change in pressure of a fixed volume of gas as it is cooled.

[J.M.B.]

3 Atomic Structure

SUB-ATOMIC PARTICLES

When Dalton put forward his atomic theory, he thought that atoms were hard, extremely small indivisible balls, but this is now known to be incorrect. Atoms are, in fact, made up of several smaller particles, the three most important being protons, electrons, and neutrons. The properties of these sub-atomic particles are summarised below.

Particle	Relative mass	Relative charge
Proton	1 unit	Positive
Electron	$\frac{1}{1836}$ unit	Negative
Neutron	1 unit	Electrically neutral

Atoms are electrically neutral and so they must contain the same number of protons as electrons. *The number of protons or electrons in an atom is known as the atomic number of the element.* Hydrogen has the simplest atom – it consists simply of one proton and one electron. This is followed by

Helium	2 protons		2 electrons	
Lithium	3	"	3	"
Beryllium	4	"	4	"
Boron	5	"	5	"
Carbon	6	"	6	"

Uranium contains 92 protons and electrons.

The atoms of the elements other than hydrogen also contain some neutrons, but there is no definite relationship between the number of protons or electrons and the number of neutrons in an atom. However, this is not important to the chemist since it is the number of protons and electrons which decide the identity and properties of the element whereas the neutrons just affect the mass.

Hydrogen atoms have a mass of one unit, i.e. 1 proton + 1 electron of negligible mass. Helium atoms consist of 2 protons, 2 neutrons, and 2 electrons and so they have a mass of 4 units. Dalton thought that all the atoms of an element were identical but this has been found to be untrue. For example, there are two types of chlorine atom – one consists of 17 protons, 17 electrons, and 18 neutrons and so has a mass of 35 units, and the other consists of 17 protons, 17 electrons, and 20 neutrons, and so their mass is 37 units. *Atoms of an element which have different masses are called isotopes of the element.* It is now known that most elements exist as mixtures of isotopes. The isotopes of chlorine are represented as $^{35}_{17}Cl$ and $^{37}_{17}Cl$, the upper number being the mass number, i.e. the mass to the nearest whole number, and the lower number is the atomic number.

The volume occupied by an atom is mostly space, i.e. the volume of the protons, electrons, and neutrons is small compared to the volume of the atom as a whole. Atoms consist of a nucleus, which contains the neutrons and protons, surrounded a relatively large distance away by the electrons. The radius of the nucleus is in fact about 1/10 000 of the radius of an atom.

A hydrogen atom may be represented as and a helium atom as

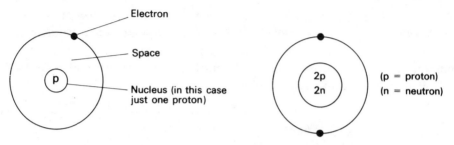

ELECTRONIC STRUCTURE OF ATOMS

The electrons move round the nucleus in definite shells or energy levels. The closer the shell is to the nucleus the lower its energy. Each shell or energy level can only contain a limited number of electrons: the first shell can hold up to 2 electrons and the second shell up to 8 electrons. The third and subsequent shells initially behave as if they are full when they contain 8 electrons but this number can be exceeded (this is discussed in more advanced texts).

The electron distributions in some of the simple atoms are given below.

Atom	Atomic number	Electrons in			
		1st shell	2nd shell	3rd shell	4th shell
H	1	1			
Li	3	2	1		
C	6	2	4		
Ne	10	2	8		
Na	11	2	8	1	
Cl	17	2	8	7	
Ar	18	2	8	8	
K	19	2	8	8	1

Electronic structures are important since they determine the valency of atoms. Thus, the valency of elements with atomic numbers of 1 to 20 can be determined by using the relationship: *the valency is the same as the number of electrons in the outer shell if the number does not exceed 4; and 8 minus the number of electrons in the outer shell if the number is greater than 4.* The relationship is illustrated below.

Element	Li	Be	B	C	N	O	F	Ne
Electronic structure	2.1	2.2	2.3	2.4	2.5	2.6	2.7	2.8
Valency	1	2	3	4	3	2	1	0

Valency = number of electrons in outer shell   Valency = 8 − number of electrons in outer shell

If the atomic number of the element is greater than 20, predicting the valency is more complicated and beyond the level of this work.

STABILITY OF THE NOBLE GAS ELECTRONIC STRUCTURE

Helium, neon, argon, krypton, xenon, and radon are a group of elements which used to be known as the inert gases because they appeared to be extremely unreactive and form no compounds. In recent years, however, it has been found that some of these elements do form a number of compounds and so they are now generally referred to as the noble gases.

A look at the electronic structures of the noble gases shows that, with the exception of helium, they have 8 electrons in their outer shell and so it is apparent that this is a very stable arrangement. Helium has only two electrons but these fill the first shell and, in view of the complete lack of helium compounds, this is also a very stable arrangement. It should come as no great surprise, therefore, that all elements try to achieve the noble gas electronic structure when they combine. Different ways of achieving the noble gas electronic structure lead to different types of bond.

ELECTROVALENCY

Sodium has atomic number 11 and so its electronic structure is 2.8.1 whilst chlorine has an atomic number of 17 and electronic structure 2.8.7. Thus, sodium has one more electron than the nearest noble gas whilst chlorine has one less. When these two elements combine together to give sodium chloride, the sodium therefore gives its outer electron to the chlorine. Now, if a sodium atom loses an electron, the remaining entity has 11 protons (each positively charged) and only 10 electrons (each negatively charged) and consequently is positively charged. Since the chlorine atom gains an electron the resultant particle will be negatively charged. The particles resulting from atoms gaining or losing electrons are known as *ions*, the number of charges on the ion being the same as the valency. The changes occurring in the formation of sodium chloride can be summarised as below.

$$\text{Na} \longrightarrow \text{Na}^+ + e^- \quad (e^- = \text{an electron})$$

sodium atom   sodium ion
2.8.1   2.8
             (electronic structure
             of the noble gas
             neon)

$$\text{Cl} + e^- \longrightarrow \text{Cl}^-$$

chlorine atom   electron from   2.8.8
2.8.7   sodium atom   chlorine ion
                                   2.8.8
                                   (electronic structure of
                                   the noble gas argon)

The equation can be written as

$$Na + \tfrac{1}{2}Cl_2 \rightarrow Na^+Cl^-$$

The Na^+ and Cl^- ions are held together purely by electrostatic attraction – there is no bond between them.

Some elements need to gain or lose two electrons to achieve the noble gas electronic structure, but this occurs less readily than when only one electron has to be lost or gained. Thus, removal of an electron from an atom gives a positively charged ion and so the protons will attract the remaining electrons more strongly. Removal of more electrons will therefore become progressively more difficult. Similarly, when an atom accepts an electron a negatively charged ion is produced. Further addition of electrons will become progressively more difficult since the mutual repulsion of the negatively charged ion and the electron will have to be overcome.

In spite of what has been said above, a considerable number of elements form di-positive or di-negative ions. For example, when magnesium reacts with oxygen to give magnesium oxide, the electronic changes taking place are

$$Mg \rightarrow Mg^{2+} + 2e^-$$
$$2.8.2 \quad\quad 2.8$$
$$O + 2e^- \rightarrow O^{2-}$$
$$2.6 \quad\quad 2.8$$

i.e. $\quad\quad Mg + \tfrac{1}{2}O_2 \rightarrow Mg^{2+}O^{2-}$

In this case, both the ions attain the electronic structure of the noble gas neon, but it must be stressed that they are completely different chemically.

A considerable amount of energy is required for an element to lose or gain three electrons and so it is not very common for an element to attain the noble gas structure in this manner. However, aluminium does lose three electrons to give Al^{3+} in the formation of its oxide

$$Al \rightarrow Al^{3+} + 3e^-$$
$$2.8.3 \quad\quad 2.8$$
$$O + 2e^- \rightarrow O^{2-}$$
$$2.6 \quad\quad 2.8$$

i.e. $\quad\quad 2Al + 1\tfrac{1}{2}O_2 \rightarrow (Al^{3+})_2(O^{2-})_3$

Note that when ionic or electrovalent compounds are formed, it is the metals which form the positive ions whilst the non-metals form the negative ions. Also, elements such as carbon (2.4) and silicon (2.8.4) cannot attain the noble gas electronic structure either by gaining or losing electrons since too much energy would be required.

COVALENCY

Hydrogen atoms have just one electron in their outer shell. They can achieve the noble gas electronic structure of helium by pairs of them sharing their electrons in the formation of hydrogen molecules.

2 hydrogen atoms   hydrogen molecule

In the hydrogen molecules, a covalent bond joins the two atoms, both the atoms having a share of two electrons. The two electrons are held in position by the attraction of the two nuclei. Covalent bonds are often represented by a line joining the two atoms involved, e.g. H—H.

Chlorine atoms (electronic structure 2.8.7) can similarly achieve the noble gas electronic structure by each sharing an electron to give chlorine molecules, Cl—Cl.

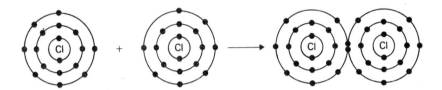

Both chlorine atoms now have the electronic structure of argon.

Oxygen atoms have six electrons in their outer shell and so they need to share two electrons to attain the electronic structure of the nearest noble gas, neon.

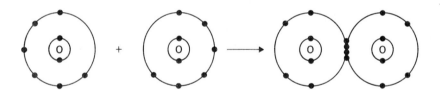

Two covalent bonds therefore join the two atoms and so the molecule may be represented as O=O. Note that oxygen, in common with a number of other elements, can form ions and can take part in covalent bonds.

Covalent bonds are not restricted to atoms of the same element; many compounds are covalent. For example, methane is formed by a carbon atom (carbon 2.4) forming covalent bonds with four hydrogen atoms (hydrogen 1).

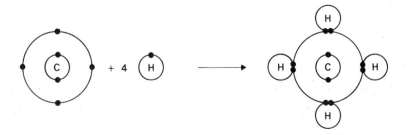

The molecule may be represented as

$$\begin{array}{c} H \\ | \\ H-C-H \\ | \\ H \end{array}$$

The carbon and hydrogen atoms now have the electronic structure of neon and helium respectively.

CO-ORDINATE OR DATIVE COVALENCY

Co-ordinate or dative covalent bonds are similar to normal covalent bonds in that a pair of electrons are shared between two atoms: the difference is that both the electrons come from the same atom.

An ammonia molecule consists of a nitrogen atom combined with three hydrogen atoms

The nitrogen in the ammonia is said to have a *lone pair of electrons*, i.e. a pair of electrons in the outer shell which is not involved in bond formation. If ammonia is added to hydrochloric acid (made up of H^+ and Cl^- ions), the lone pair of electrons on the nitrogen attracts the proton (H^+) and a co-ordinate or dative covalent bond is formed. The resultant ammonium ion attracts the Cl^- to give ammonium chloride.

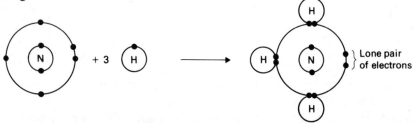

Co-ordinate bonds are often indicated by arrows as above, the head of the arrow pointing to the atom which is receiving a share of the two electrons. Co-ordinate bonds are identical to normal covalent bonds once they have been formed – it is the method of formation which differs.

A further common example of co-ordinate bond formation is when an acid is added to water. Protons from the acid are attracted by lone pairs of electrons on the oxygens of the water molecules and this results in the formation of oxonium ions.

$$H-\ddot{O}-H \;+\; H^+ \longrightarrow \begin{array}{c} H-\ddot{O}^+-H \\ | \\ H \end{array}$$

Although the oxygen has two lone pairs of electrons, only one pair is utilised in forming co-ordinate bonds.

CRYSTAL STRUCTURES

1. Ionic crystals

In ionic compounds, e.g. sodium chloride, the ions do not exist in pairs but in large numbers arranged in regular patterns in a crystal lattice. The sodium chloride lattice is illustrated in Figure 3.1 (a). Each sodium ion is surrounded equidistantly by six chlorine ions whilst each chlorine ion is surrounded equidistantly by six sodium ions (Figure 3.1 (b)). This arrangement is the same in all directions throughout the crystal. It is therefore incorrect to regard two particular Na^+ and Cl^- ions as being combined together; each Na^+ is equally associated with six Cl^- ions and *vice versa*. Other ionic compounds may have different types of lattice but in all cases an ion is surrounded by a definite number of oppositely charged ions and this holds them firmly in position.

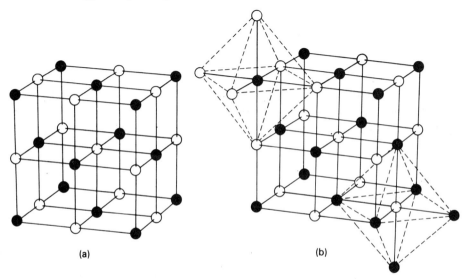

Figure 3.1. (a) The sodium chloride lattice.
(b) Extra ions added to show that each Na^+ ion is surrounded equidistantly by 6 Cl^- ions and each Cl^- is surrounded equidistantly by 6 Na^+ ions

2. Molecular crystals

When covalent compounds solidify, the molecules take up definite positions in a crystal lattice but, in this case, they are held in place only by relatively weak forces known as van der Waals forces. A plan of the iodine crystal is illustrated in Figure 3.2.

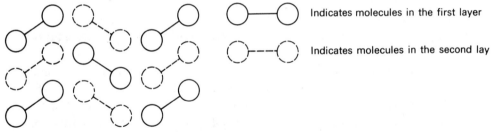

Figure 3.2. Plan view of the iodine lattice

The differences in properties between ionic and covalent compounds are summarised below.

Ionic compounds	Covalent compounds
1. Generally have high melting or boiling points – a lot of energy is needed to destroy the crystal lattice.	— Usually volatile, i.e. have low melting or boiling points because their crystal lattices are weak.
2. Conduct an electric current when molten or dissolved in water (see pages 89–91).	— Non-conductors of electricity.
3. Insoluble in organic solvents such as ethers, alcohols, and benzene.	Soluble in organic solvents.
4. Hard crystals.	— Soft crystals.

3. Macromolecular crystals

In this type of lattice, a very large number of atoms are joined together by covalent bonds. Diamond is a typical example, each carbon atom being joined by covalent bonds to four other carbon atoms situated symmetrically round it (Figure 3.3). Diamonds are therefore known as macromolecules since they contain millions of carbon atoms joined together in this manner. Substances like this are usually very hard and are non-volatile because a large number of covalent bonds must be broken to destroy the crystal lattice. They are insoluble in water and non-conductors of electricity.

Figure 3.3. Diamond lattice

4. Metallic lattices

The nature of metallic lattices is not fully understood. It is thought that the atoms are packed tightly together, as illustrated in Figure 3.4, but that the valency electrons can move freely throughout the lattice. This accounts for why metals are good conductors of electricity. Consider a metal wire joining the terminals of a battery, Figure 3.5. At the negative and positive terminals of a battery there are an excess and deficiency of electrons respectively. When the wire joins the terminals, some of the excess electrons on the negative terminal push the loose valency electrons along the wire to the positive terminal.

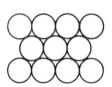 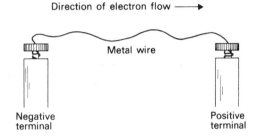

Figure 3.4. Close packing of atoms   Figure 3.5. Metal wires as conductors

Metallic lattices are generally strong because of the attraction of the nuclei for the delocalised valency electrons and so high melting and boiling points result. Metals are malleable because, even when the lattice is distorted by planes of atoms sliding over one another (Figure 3.6), the valency electrons can readily move between the planes to maintain the bonding.

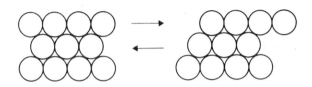

Figure 3.6. Planes of atoms in a metallic lattice sliding over one another

SHAPES OF MOLECULES

The shapes of molecules are dependent on the electron pairs in the valency shells of the various atoms involved. Since like charges repel one another, the electron pairs will try to get as far apart as possible. Consider, for example, the methane molecule, CH_4. Repulsion of the four pairs of electrons making up the four covalent bonds will result in the molecule having the tetrahedral configuration, i.e. the bonds will point to the four corners of a regular tetrahedron with the carbon at the centre (Figure 3.7). The bonds do not point towards the corners of a square since the H—C—H bond angles would then be only 90° instead of the 109.5° in the tetrahedron structure.

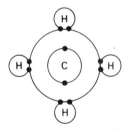 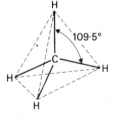

Figure 3.7. The methane molecule

The nitrogen atom in an ammonia molecule also has four pairs of electrons in its outer shell – 3 pairs involved in covalent bonds and one lone pair. The molecule is therefore similar to methane but the tetrahedron is slightly distorted. The distortion occurs because the lone pair is closer to the nucleus of the nitrogen than the bonding pairs and so gives a stronger repulsion. The atoms are in a trigonal pyramid shape (Figure 3.8) and the H—N—H bond angles are 107°.

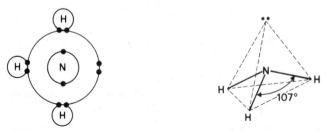

Figure 3.8. The ammonia molecule

The oxygen atom in a water molecule has two lone pairs of electrons as well as two pairs of electrons involved in covalent bonds. In this case, therefore, the electron pairs point towards the corners of a slightly more distorted tetrahedron. The shape of the molecule is angular (Figure 3.9) and the H—O—H bond angle is 104.5°.

Figure 3.9. The water molecule

The carbon dioxide molecule (Figure 3.10) has a carbon atom joined by double bonds to two oxygen atoms. Mutual repulsion of the double bonds results in the molecule being linear.

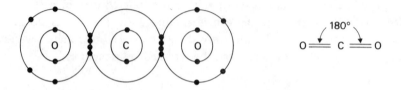

Figure 3.10. The carbon dioxide molecule

In an ethene molecule, $CH_2=CH_2$, the carbon atoms are joined by a double bond and they both form single covalent bonds with two hydrogen atoms. Repulsion of the pairs of electrons in the bonds results in a planar molecule (Figure 3.11). The bonds from each carbon point towards the corners of an equilateral triangle.

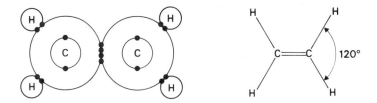

Figure 3.11. The ethene molecule

THE PERIODIC TABLE

A hydrogen atom consists only of a proton and an electron and so it is the element with the simplest atomic structure. The other elements can be built up from this by successive additions of a proton, an electron, and one or more neutrons. Now the electrons of an atom, as stated previously, are arranged in definite energy levels round the nucleus and it has been found that elements with similar electronic structures have similar properties. As a result of this, the elements have been arranged in a table, in order of their atomic numbers, in such a way that elements with similar electronic structures fall under one another. On going across the table, the number of electrons in the outer shell of the atoms increases steadily from one to eight. This arrangement is known as the *Periodic Table* and the position, electronic structure, and valency of the first 18 elements is shown in Figure 3.12 whilst the complete table is illustrated on page 280. The vertical columns of elements are known as *groups* whilst the horizontal lines are referred to as *periods*.

Group	I	II	III	IV	V	VI	VII	0
	H 1						H 1	He 2
	Li 2.1	Be 2.2	B 2.3	C 2.4	N 2.5	O 2.6	F 2.7	Ne 2.8
	Na 2.8.1	Mg 2.8.2	Al 2.8.3	Si 2.8.4	P 2.8.5	S 2.8.6	Cl 2.8.7	Ar 2.8.8
Valency of the group	1	2	3	4	3	2	1	0

Figure 3.12. The first eighteen elements in the Periodic Table

Difficulty has arisen in deciding on the most suitable place for hydrogen in the table. It has one electron in the outer shell, a valency of one, and it forms positive ions and so this suggests that it should head Group I. However, it also has similarities to the Group VII elements (F, Cl, Br, and I) in that it can form negative ions, it has a valency of one, and it exists as diatomic molecules, i.e. H_2. It is best considered in a class of its own but, in modern forms of the periodic table, hydrogen is generally placed at the tops of Group I and VII.

On passing from left to right across the periodic table, the elements change in character from metals to non-metals, the division in Figure 3.12 being indicated by the broken line. Elements to the left of the line are said to be *electropositive*. This means that they tend to lose electrons and form positive ions and so they form ionic compounds. However, the electropositivity decreases on going left to right since the greater the number of electrons that have to be lost, the more difficult it becomes (see page 30). On the other side of the line, the *electronegativity* increases on going from left to right, i.e. the elements have an increasing tendency to gain electrons to form negative ions. The elements in the centre groups, for example, boron, carbon, silicon, nitrogen, and phosphorus, form covalent compounds (although nitrogen and phosphorus do form a few ionic compounds as well), whereas the more electronegative elements can form covalent or electrovalent compounds (e.g. chlorine in CCl_4 and NaCl). The Group O elements are known as the noble or inert gases and they are unreactive because they have very stable electronic structures.

The elements in any particular group have similar properties but the first member in each group tends to have some peculiarities due to the small size of their atoms. The change in properties across a period is therefore best illustrated by reference to the series sodium, magnesium, aluminium, silicon, phosphorus, sulphur, and chlorine.

The sizes of the atoms in the above series decrease on going from left to right because the additional electrons enter the same shell whilst the charge on the nucleus increases. The increased nuclear charge pulls the outer electrons in closer and so a decrease in volume results. This explains why electropositivity decreases from left to right across the periodic table – the outer electrons are closer to the nucleus and so they are more firmly held.

Oxides of the series

Formula of oxide	Na_2O	MgO	Al_2O_3	SiO_2	P_2O_3 P_2O_5	SO_2 SO_3	Cl_2O
Type of oxide	Basic	Basic	Amphoteric	Acidic	Acidic	Acidic	Acidic
Type of bonding	Ionic	Ionic	Ionic	Covalent	Covalent	Covalent	Covalent

Other oxides of some of the elements are known but they are unimportant here.

Disodium oxide (sodium monoxide) reacts vigorously with water, giving a strongly alkaline solution of sodium hydroxide.

$$Na_2O(s) + H_2O(l) \rightarrow 2NaOH(aq)$$

Magnesium oxide is less basic than disodium oxide. It is insoluble in water but it reacts with acids to give salts (see page 73).

$$MgO(s) + H_2SO_4(aq) \rightarrow MgSO_4(aq) + H_2O(l)$$

Aluminium is in the region of the Periodic Table where metallic character decreases and non-metallic character increases and so it is only a weak metal. Its oxide is insoluble in water and is amphoteric, i.e. it can behave as a weak acid and as a weak base.

Acting as a base:

$$Al_2O_3(s) + 6HCl(aq) \rightarrow 2AlCl_3(aq) + 3H_2O(l)$$

Acting as an acid:

$$Al_2O_3(s) + 2NaOH(aq) + 3H_2O(l) \rightarrow 2NaAl(OH)_4(aq)$$

Silicon(IV) oxide (silicon dioxide or silica) is weakly acidic and dissolves in boiling solutions of alkalis giving silicates, e.g. with sodium hydroxide solution it gives sodium silicate.

$$SiO_2(s) + 2NaOH(aq) \rightarrow Na_2SiO_3(aq) + H_2O(l)$$

Phosphorus can form two oxides. When it burns in a limited supply of air phosphorus(III) oxide (phosphorus trioxide) is produced:

$$4P(s) + 3O_2(g) \rightarrow 2P_2O_3(s)$$

but phosphorus(V) oxide (phosphorus pentoxide) is the product when excess air or oxygen is present:

$$4P(s) + 5O_2(g) \rightarrow 2P_2O_5(s)$$

Both oxides dissolve in water giving acidic solutions:

$$P_2O_3(s) + 3H_2O(l) \rightarrow 2H_3PO_3(aq) \quad \text{phosphonic acid}$$
$$P_2O_5(s) + 3H_2O(l) \rightarrow 2H_3PO_4(aq) \quad \text{phosphoric acid}$$

Sulphur forms a dioxide and a trioxide. Sulphur dioxide reacts with water giving a weakly acidic solution, sulphurous acid:

$$H_2O(l) + SO_2(g) \rightarrow H_2SO_3(aq)$$

Sulphur(VI) oxide reacts vigorously with water, giving a very strong acid, sulphuric acid:

$$H_2O(l) + SO_3(s) \rightarrow H_2SO_4(aq)$$

The oxides of chlorine are relatively unimportant.

In summary, metals form basic oxides, weak metals form amphoteric oxides, and non-metals form acidic oxides.

Chlorides of the series

Formula of chloride	NaCl	MgCl$_2$	AlCl$_3$	SiCl$_4$	PCl$_3$ PCl$_5$	S$_2$Cl$_2$
Type of chloride	Ionic	Ionic	Covalent	Covalent	Covalent	Covalent
State at room temp.	Solid	Solid	Solid	Liquid	Liquid Solid	Liquid

The chlorides of sodium and magnesium dissolve in water without being hydrolysed, i.e. they are not decomposed by the water and the resultant solutions are neutral. However, chlorides of weak metals are hydrolysed and acidic solutions are obtained. The reaction between aluminium chloride and water is complicated but a simplified equation may be given as:

$$AlCl_3 + 3H_2O \rightleftharpoons Al(OH)_3 + 3HCl$$

the $\rightleftharpoons$ sign meaning that the reaction is reversible, i.e. it proceeds in both directions. A further indication that aluminium chloride is covalent is given by the fact that it sublimes on heating.

The non-metal chlorides are also generally hydrolysed by water. Silicon tetrachloride fumes in moist air due to hydrolysis.

$$SiCl_4(l) + 2H_2O(l) \rightarrow SiO_2(s) + 4HCl(aq)$$

Both phosphorus trichloride and the pentachloride fume in moist air. The former compound is instantly hydrolysed by cold water to give phosphonic acid (phosphorous acid):

$$PCl_3(l) + 3H_2O(l) \rightarrow H_3PO_3(aq) + 3HCl(aq)$$

Phosphorus pentachloride also reacts vigorously with water giving initially phosphorus trichloride oxide (phosphoryl chloride) and then phosphoric acid if excess water is present:

$$PCl_5(s) + H_2O(l) \rightarrow POCl_3(l) + 2HCl(aq)$$
$$POCl_3(l) + 3H_2O(l) \rightarrow H_3PO_4(aq) + 3HCl(aq)$$

The chloride of sulphur is unimportant.

The chlorides of strong metals are therefore seen to be ionic (electrovalent) and to dissolve in water, without reacting with it, giving neutral solutions. In contrast, the chlorides of weak metals and non-metals form covalent chlorides which are hydrolysed by water (there are some exceptions, e.g. tetrachloromethane, CCl_4) giving acidic solutions.

Hydrides of the series

Formula of hydride	NaH	Mg & Al hydrides	SiH_4	PH_3	H_2S	HCl
Type of hydride	Ionic	unimportant	Covalent	Covalent	Covalent	Covalent
State at room temp.	Solid		Gas	Gas	Gas	Gas

Sodium hydride is readily hydrolysed by cold water:

$$NaH(s) + H_2O(l) \rightarrow NaOH(aq) + H_2(g)$$

The hydrides of silicon, phosphorus, sulphur, and chlorine are all gases; covalent compounds generally have low boiling and melting points. Silane, SiH_4, inflames spontaneously in air but, like phosphine, PH_3, it is unaffected by water. Hydrogen sulphide is slightly soluble in water, giving a weakly acidic solution whereas hydrogen chloride is very soluble and gives a strongly acidic solution, hydrochloric acid:

$$HCl(g) + H_2O(l) \rightarrow H^+Cl^-(aq)$$

THE HALOGEN FAMILY

Elements in the same group have similar electronic structures and so their properties are basically similar. However, both their physical and chemical properties are modified as the group is descended. This can be demonstrated by the halogen family, i.e. Group VII.

Element	Electronic configuration	Boiling point	State	Colour
Fluorine	2.7	−188 °C	Gas	Pale green-yellow
Chlorine	2.8.7	−34 °C	Gas	Pale green-yellow
Bromine	2.8.18.7	58 °C	Liquid	Brown-red
Iodine	2.8.18.18.7	183 °C	Solid	Violet-black

The atoms increase in size and mass down the group as the number of protons, neutrons, and electrons increases. The increase in boiling point with increase in mass is in agreement with the general rule mentioned previously (page 25).

All the elements have a valency of one. They can achieve the noble gas electronic structure either by sharing electrons in the formation of covalent bonds, as in their molecules F_2, Cl_2, Br_2, and I_2, or by gaining electrons to give negative ions, i.e. F^-, Cl^-, Br^-, and I^-. Note that the ions are considerably larger than the corresponding atoms because they have more electrons than protons and the electrons repel one another.

The reactivity of the elements decreases sharply down the group, the reason being that with fluorine the outer electrons are close to the nucleus and there is a strong tendency to gain and keep an electron. On descending the group, the atoms become larger so that with iodine the outer electrons are a long way from the nucleus and there is little attraction for another electron. Iodine, in fact, prefers to share electrons and form covalent bonds rather than gain one and form ionic compounds. Fluorine is therefore strongly electronegative whilst iodine is only weakly so.

The decrease in reactivity down the group can be illustrated by their reaction with hydrogen.

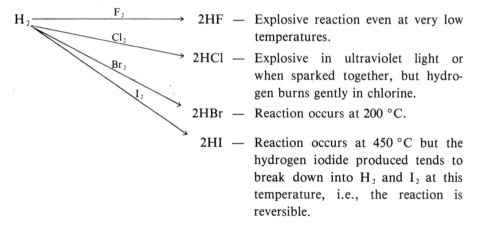

2HF — Explosive reaction even at very low temperatures.

2HCl — Explosive in ultraviolet light or when sparked together, but hydrogen burns gently in chlorine.

2HBr — Reaction occurs at 200 °C.

2HI — Reaction occurs at 450 °C but the hydrogen iodide produced tends to break down into H_2 and I_2 at this temperature, i.e., the reaction is reversible.

The decrease in reactivity is also illustrated by their reactions with other elements. Thus, fluorine is extremely reactive and it combines directly with every other element except oxygen and some of the noble gases. It reacts violently with many metals even at room temperature. Chlorine, bromine, and iodine react with many metals on heating but the reactions become progressively more sluggish as the atomic number of the halogen increases.

The halogens will displace those below them in the group from their halides. For example, chlorine will displace bromine from bromides:

$$Cl_2(g) + 2KBr(aq) \rightarrow 2KCl(aq) + Br_2(aq)$$

The reason for this is that chlorine is more electronegative than bromine. The chlorine molecule takes an electron from each of the Br^- ions giving $2Cl^-$ ions and two bromine atoms. The two bromine atoms then combine to give a bromine molecule, Br_2:

$$2Br^- + Cl_2 \rightarrow 2Cl^- + 2Br$$
$$2Br \rightarrow Br_2$$

Similarly, chlorine will displace iodine from iodides. Bromine can displace iodine from iodides but not chlorine from chlorides since chlorine is above it in the group. Fluorine is not normally encountered in the laboratory because its extreme reactivity makes it a difficult substance to handle.

SUMMARY

The elements are arranged in the Periodic Table in such a way that all those in a particular group have the same number of electrons in their outer shell. The strongly metallic elements are on the left hand side of the Table whilst the non-metals are on the right. On travelling from left to right across a period there is a steady transition from metallic to non-metallic character.

The reactivity of the metals increases as a group is descended since the outer electrons are further from the nucleus and more easily lost. The opposite is observed with the non-metals. Thus, in Group VII, the reactivity decreases from fluorine down to iodine as the attraction for an additional electron becomes less.

Questions

1. (a) State the main points of Dalton's atomic theory and explain how it has had to be amended in the light of subsequent discoveries.
   (b) Give the electronic structure of the elements with atomic numbers 14, 16, and 20 and hence predict their respective valencies.

2. By diagrams or equations, whichever is relevant, show the electronic changes which occur when the following pairs of elements combine:
   (i) silicon and fluorine;
   (ii) calcium and sulphur.
   In each case, state the properties you would expect to associate with the compound formed.

3. Explain briefly the following facts:
   (a) Not all atoms of a particular element have the same mass.
   (b) Oxygen (atomic number 8) has a valency of two.
   (c) Carbon exists as a solid with a melting point in excess of 3000 °C but nitrogen is a gas at room temperature.
   (d) Metals will readily conduct an electric current.

4. (a) The atomic number of potassium is 19. Give three deductions which can be made from this statement.
   (b) The following symbols represent atoms of elements showing their mass numbers and atomic numbers: $^{40}_{18}Ar$ $^{39}_{19}K$.
   (i) What is the electronic structure of the argon atom?
   (ii) How many neutrons are there in the nucleus of the argon atom?
   (iii) Explain why argon has a lower atomic number but a greater atomic mass than the potassium atom.
   (c) $^{32}_{16}S$ $^{35}_{17}Cl$ $^{40}_{18}Ar$ $^{39}_{19}K$ $^{40}_{20}Ca$
   Atoms and ions which have the same number of electrons are said to be isoelectronic. From the elements given above, write the formula of three ions which are isoelectronic with the argon atom.

   [J.M.B.]

5. By diagrams, show the electronic changes which occur when the covalent compounds silane, SiH_4 and phosphorus trifluoride, PF_3 are formed from the

elements. The atomic numbers of silicon, hydrogen, phosphorus, and fluorine are 14, 1, 15, and 9 respectively. Explain what shape you would expect the silane and phosphorus trifluoride molecules to adopt.

6   (a) "The type of bond present in a crystal can influence its chemical and physical properties."
Explain what is meant by the above statement in the case of crystals of:
(i) iodine,
(ii) diamond,
(iii) sodium chloride.
In your answer, in addition to the type of bond present, you should refer where relevant to the following:
the particles (i.e. atoms, ions, or molecules) present in the crystal,
the action of heat on the crystal,
whether the melting point is low or high,
the conditions under which the substance will conduct electricity.

(b) A compound, X, is a white crystalline solid. It has a melting point of 767 °C and is soluble in water. State, with a reason, what type of bonding is present in the crystal. Would you expect an aqueous solution of X to conduct electricity?
[J.M.B.]

7   This question is concerned with the following series of elements:
sodium, magnesium, aluminium, silicon, phosphorus, sulphur, and chlorine.

(a) State the physical state and type of bonding present in the hydrides of sodium and chlorine. Describe the action of each of these compounds with water, naming the products and giving an equation in each case.

(b) Write formulae for the chlorides of sodium and aluminium, and describe the effect of heat on each.

(c) Explain why the elements on the left of the series are metals while those on the right are non-metals.

(d) Explain how it is known that no new elements will ever be produced which fit between sodium and chlorine in this series. [J.M.B.]

8   Chlorine, bromine, and iodine are in Group VII of the Periodic Table and are members of the same family.

(a) Explain the characteristics of the family by reference to:
(i) the physical state of each element,
(ii) the similarities and differences in the electronic configuration of their atoms,
(iii) the gradation in the sizes of their atoms and ions,
(iv) the ease with which they combine with iron.
Give an equation for the reaction of iron with one of the elements.

(b) What is seen when an excess of chlorine is slowly bubbled through a solution of potassium iodide? Write an equation for the reaction.

(c) Dichlorodifluoromethane, CCl_2F_2, is used as a refrigerant called 'Freon'. Write its structural formula and suggest what shape its molecule might be. [J.M.B.]

9   The halogens have the following atomic numbers: fluorine 9, chlorine 17, bromine 35, iodine 53.

(a) Give the electronic configuration for:
(i) an atom of fluorine,
(ii) an atom of chlorine.

(b) How many valency electrons has an iodine atom?

(c) Write down the names of the four halogens and after each state whether it is a solid, a liquid, or a gas under ordinary laboratory conditions.

(d) How, and under what conditions, does chlorine react with:
(i) hydrogen,
(ii) iron,
(iii) potassium bromide?

(e) State whether iodine is more or less reactive than chlorine. Give an example to illustrate your answer. [J.M.B.]

4 Air and Water

EFFECT OF HEATING METALS IN AIR

Many common substances burn in air but they often have complex structures, and so an initial study is best confined to a few metals. For example, when a piece of copper is heated in air and then allowed to cool, it is found that it is covered with a black film. Lead readily melts on heating and the molten lead soon becomes covered with an orange powder. Magnesium, on the other hand, burns in air with an intense white flame and a white ash is left.

If known masses of the above metals are heated in air and the products are weighed, it is found in each case that there has been an increase in mass. The inference from this is that the metals have combined with air or something from the air. Analysis of each product shows it to be the oxide of the metal (with magnesium some nitride is also formed – see page 45). Air therefore contains oxygen.

COMPOSITION OF AIR

The experiments described below show that air is a mixture of gases. Its approximate composition, after the water vapour has been removed, is oxygen 21%, nitrogen 78%, noble gases (mainly argon) 1%, and carbon dioxide 0.03%.

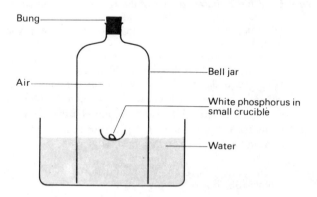

Figure 4.1. The combustion of white phosphorus in air

Since air is a mixture, its composition is variable, the various constituents can be separated by physical means (see page 113), and its properties are a combination of the properties of the individual constituents.

The approximate percentage of oxygen in air can be demonstrated with the apparatus shown in Figure 4.1. After noting the water level, the bung is removed, the phosphorus ignited by touching it with a hot wire, and then the bung is quickly replaced. The phosphorus burns and white clouds of phosphorus(V) oxide are formed.

$$4P(s) + 5O_2(g) \rightarrow 2P_2O_5(s)$$

When the reaction has finished and the apparatus has cooled, no white fumes are present since they dissolve in water. The trough is raised until its water level is the same as that in the bell-jar so that the residual gas is at atmospheric pressure. It is seen that the water level rises by about one fifth.

More accurate methods of determining the oxygen content of the air are possible. For example, a known volume of air can be repeatedly passed over heated copper (Figure 4.2) until there is no further change. The orange-red copper turns black as the oxide is formed:

$$2Cu(s) + O_2(g) \rightarrow 2CuO(s)$$

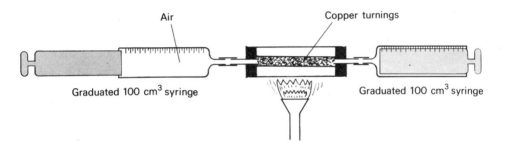

Figure 4.2. Determination of the percentage of oxygen in air

Alternatively, a known volume of air can be passed through an alkaline solution of benzene-1,2,3-triol (pyrogallol). In both cases about 79% of the air remains.

The presence of nitrogen in air may be demonstrated by strongly heating magnesium in the residual gas from above. Magnesium nitride is formed and addition of water to this produces ammonia which may be recognised by its smell and by its action of turning red litmus paper blue.

$$3Mg(s) + N_2(g) \rightarrow Mg_3N_2(s)$$
$$Mg_3N_2(s) + 6H_2O(l) \rightarrow 3Mg(OH)_2(s) + 2NH_3(g)$$

Limewater (a dilute solution of calcium hydroxide) turns milky when air is passed through it for some time. This is the standard test for carbon dioxide, the milkiness being due to the formation of a precipitate of calcium carbonate:

$$CO_2(g) + Ca(OH)_2(aq) \rightarrow CaCO_3(s) + H_2O(l)$$

Burning coal and oil, etc., produces large amounts of carbon dioxide but the carbon dioxide content of the atmosphere remains constant at about 0.03% due to the carbon cycle (see page 180).

The presence of water vapour in air is demonstrated by exposing calcium chloride to the atmosphere for a few days. The calcium chloride absorbs so much water vapour that it dissolves in it and a solution is formed; substances such as this are said to be *deliquescent*.

INDUSTRIAL USES OF AIR

Air is used in large quantities in the manufacture of sulphur(VI) oxide (page 209) and nitric acid (page 196), and in the blast furnace in the extraction of iron (page 163). Air is also used as a source of nitrogen and oxygen as outlined on page 113. Both nitrogen and oxygen are important industrial chemicals (see pages 130 and 117 respectively).

COMBUSTION

Whenever a substance burns in air, a chemical reaction is occurring in which the substance or its components are combining with oxygen from the air. The reaction is accompanied by the evolution of heat and light energy. For example, a candle will burn in a bell-jar of air over water as long as any oxygen is present. But, when all the oxygen has been used up the candle goes out. Candle wax is a hydrocarbon (a compound containing only carbon and hydrogen) and when it burns the carbon and hydrogen react with oxygen from the air to give carbon dioxide and water respectively.

$$C + O_2 \rightarrow CO_2$$
$$2H_2 + O_2 \rightarrow 2H_2O$$

When all the oxygen has been used up, these reactions cannot occur and so the flame goes out.

RESPIRATION

The apparatus shown in Figure 4.3 may be used to show that exhaled air contains more carbon dioxide than inhaled air.

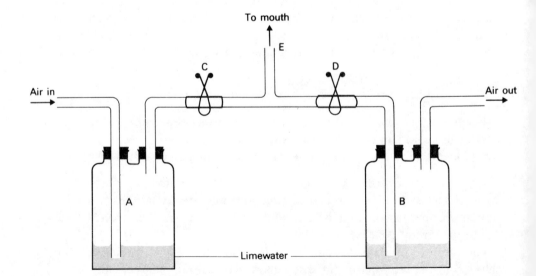

Figure 4.3. To show the presence of extra carbon dioxide in exhaled air

Air is inhaled through E with clip D closed and clip C open and then exhaled with C closed and D open. This process is repeated two or three times and it is seen that the limewater (dilute calcium hydroxide solution) in B turns milky whilst that in A remains clear. Since the same amount of air has passed through A and B, it is evident that exhaled air contains more carbon dioxide than inhaled air. In fact, exhaled air contains about 4% carbon dioxide.

Carbon dioxide is produced by respiration as follows. As air is drawn into the lungs, some of the oxygen is transferred to the red blood cells which then pass round the body. Food in the intestine is converted into water soluble products which are also absorbed by the blood. The water soluble products, e.g. sugar, and the oxygen react to give carbon dioxide and water, and energy is released.

$$C_6H_{12}O_6(aq) + 6O_2(g) \rightarrow 6CO_2(g) + 6H_2O(l)$$

When the blood returns to the lungs, the carbon dioxide is expelled and the process is repeated.

THE SOLUBILITY OF AIR IN WATER

The fact that air is slightly soluble in water may be demonstrated using the apparatus shown in Figure 4.4. The apparatus is completely filled with tap water through the tube fitted with the screw clip, A. On heating, bubbles appear in the flask and pass down the delivery tube to be collected in the graduated tube. Analysis of the gas collected shows it to be air with an enhanced oxygen content (about 30%). The high oxygen content is a consequence of this gas being more soluble in water than is nitrogen. The dissolved air is vital to marine fish and plant life.

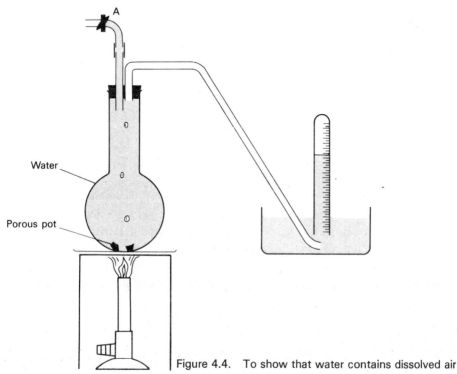

Figure 4.4. To show that water contains dissolved air

RUSTING

Iron will not rust when it is exposed to dry air nor when exposed to air-free water, i.e. water which has been boiled until all the dissolved air has been removed. However, it rusts rapidly in the presence of air and water. These facts may be demonstrated using the apparatus shown in Figure 4.5. Rusting is a complicated process which is accelerated by the presence of carbon dioxide. During rusting, iron combines with oxygen from the air and water to give hydrated iron(III) oxide, $Fe_2O_3 \cdot xH_2O$.

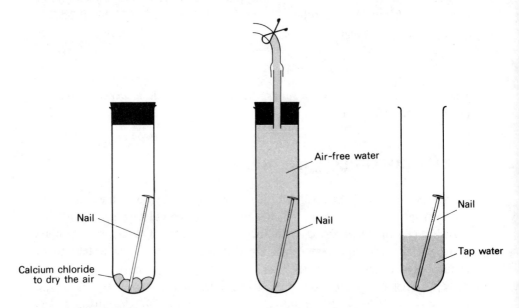

Figure 4.5. To find the conditions for rusting to occur

WATER

Water is the most abundant and most important liquid. It is essential to life and is a constituent of all living organisms. Pure water freezes at 0 °C and boils at 100 °C when under a pressure of 760 mm mercury (1 atmosphere). It is neutral to litmus and is colourless, odourless, and tasteless.

DETECTION OF WATER

When copper(II) sulphate crystallises from water, it forms blue crystals of the pentahydrate, $CuSO_4 \cdot 5H_2O$, i.e. five molecules of water are loosely combined with the copper(II) sulphate. The copper(II) sulphate is said to have five molecules of water of crystallisation. On gentle heating, these crystals break down, the water is driven off, and a white powder of anhydrous copper(II) sulphate is left. When water is added to the powder this process is reversed:

$$CuSO_4 + 5H_2O \rightleftharpoons CuSO_4 \cdot 5H_2O$$
$$\text{White} \hspace{4em} \text{Blue}$$

Thus, water may be detected by its property of turning anhydrous copper(II) sulphate blue.

Water may also be detected by the use of cobalt chloride paper. Filter paper impregnated with cobalt(II) chloride is blue when dry but pink when moist. Cobalt(II) chloride is often added to silica gel – a common drying agent – in order to show when it has absorbed so much water that its effectiveness is impaired. The exhausted silica gel is heated in an oven until the blue colour, due to the anhydrous cobalt(II) chloride, is restored.

It should be noted that both anhydrous copper(II) sulphate and cobalt(II) chloride only indicate the presence of water, they do not indicate that the water is pure.

VOLUMETRIC COMPOSITION OF WATER

The composition by volume of water may be demonstrated using Hofmann's voltameter (Figure 4.6) in which water is decomposed by passing an electric current through it. Since water is a very poor conductor of electricity, a little dilute sulphuric acid is added to help to carry the current. Bubbles of gas form at each electrode when the current is switched on. The reaction is allowed to proceed for a few minutes so that the dilute sulphuric acid becomes saturated with the gases. The evolved gases are discarded and then the current is passed again. It is seen that twice as much gas is produced at the cathode (the negative electrode) than at the anode (the positive electrode). If the gas produced at the cathode is held to a flame it 'pops' thus indicating that it is hydrogen. The gas formed at the anode relights a glowing splint and so it is oxygen. All the sulphuric acid is still present at the end of the reaction and so it is apparent that some of the water has decomposed into its elements. Water therefore consists of hydrogen and oxygen combined in the ratio of two volumes of hydrogen to one of oxygen.

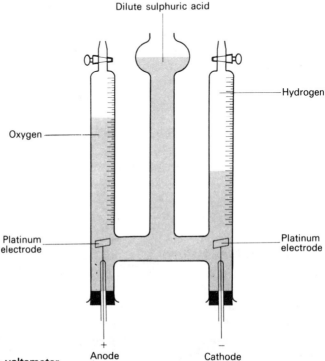

Figure 4.6. Hofmann's voltameter

SYNTHESIS OF WATER

Water may be synthesised from its elements using the apparatus illustrated in Figure 4.7. Hydrogen is produced from the reaction between zinc and cold, dilute sulphuric acid.

$$Zn(s) + H_2SO_4(aq) \rightarrow ZnSO_4(aq) + H_2(g)$$

The hydrogen is dried by passing it through calcium chloride and then it escapes through the jet. When the hydrogen has had time to displace all the air from the apparatus (oxygen–hydrogen mixtures are explosive), it is lit at the jet and the flame is allowed to play on a retort cooled by running water. The burning hydrogen reacts with oxygen from the air to form water vapour and this condenses on the retort and is collected in an evaporating basin. The identity of the liquid in the evaporating basin is confirmed by its action on anhydrous copper sulphate.

$$2H_2(g) + O_2(g) \rightarrow 2H_2O(l)$$

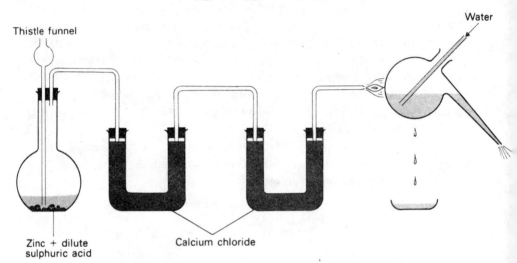

Figure 4.7. Synthesis of water

THE REACTION OF WATER WITH METALS

A number of metals react with water; in fact, the reaction with water may be used to help to arrange the metals in order of their reactivity (see page 135). A few of the reactions will be discussed here.

a. With sodium

Sodium is a very reactive metal and it has to be kept under paraffin or solvent naphtha, etc., in order to retard attack on it by moisture, oxygen, and carbon dioxide in the air. Even so, it is usually covered with a whitish coating of its hydroxide and carbonate. If a freshly cut piece of sodium is added to water, a vigorous reaction occurs in which the sodium melts and darts round the surface of the water, and a gas is evolved. The gas can be shown to be hydrogen by the 'pop' it gives when held to a flame (see page 112).

$$2Na(s) + 2H_2O(l) \rightarrow 2NaOH(aq) + H_2(g)$$

b. With calcium

Calcium is a silvery white metal which soon tarnishes in the atmosphere due to the formation of a layer of its oxide. When the metal is added to water it sinks and, once the surface layer of oxide has dissolved, a fairly brisk stream of hydrogen is given off. A white precipitate may be produced because the product, calcium hydroxide, is not very soluble in water.

$$Ca(s) + 2H_2O(l) \rightarrow Ca(OH)_2(aq) + H_2(g)$$

c. With magnesium

Magnesium undergoes negligible reaction with cold water but it reacts with boiling water or steam and hydrogen is liberated (see page 145).

$$Mg(s) + H_2O(g) \rightarrow MgO(s) + H_2(g)$$

d. With iron

It has already been mentioned (page 48) that iron rusts in the presence of air and water but is not attacked by pure water at room temperature. However, if steam is passed over red-hot iron (Figure 4.8), hydrogen is produced along with iron(II) diiron(III) oxide, Fe_3O_4.

$$3Fe(s) + 4H_2O(g) \rightarrow Fe_3O_4(s) + 4H_2(g)$$

This reaction has industrial importance.

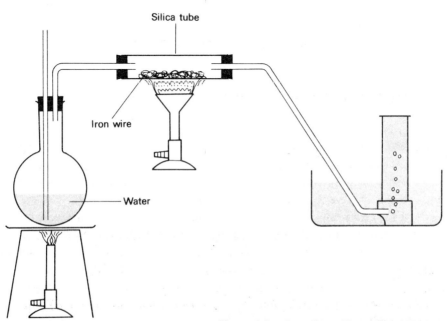

Figure 4.8. Reaction of iron wire with steam

THE BOILING POINT OF WATER

A water molecule consists of two hydrogen atoms joined by covalent bonds to an oxygen atom. Now oxygen atoms are strongly electronegative (see page 38) and so the oxygen atom takes more than its fair share of the electrons in the O—H bonds. Since electrons are negative, the oxygen becomes slightly negatively charged whilst the hydrogens, which do not have their fair share of the electrons, become slightly positively charged. Water molecules are said to be polarised and the polarisation may be represented as

$$\overset{\delta+}{H}-\overset{\delta-}{\underset{\underset{H^{\delta+}}{\diagdown}}{O}}$$

The δ sign is the small Greek letter delta and it is used in chemistry to indicate a small value of some property. A consequence of the polarisation is that water molecules attract one another and so move in clusters.

$$\overset{\delta+}{H}\diagdown\underset{\underset{\overset{|}{H}}{\overset{|}{O}}}{}\overset{\delta-}{\diagup}\overset{\delta+}{H}\diagdown\underset{\underset{\overset{|}{H}}{\overset{|}{O}}}{}\overset{\delta-}{\diagup}\overset{\delta+}{H}\diagdown\underset{\underset{\overset{|}{H}}{\overset{|}{O}}}{}\overset{\delta-}{\diagup}\overset{\delta+}{H}\diagdown\underset{\underset{\overset{|}{H}}{\overset{|}{O}}}{}\overset{\delta-}{\diagup}$$

The water molecules are said to undergo *hydrogen bonding or association*. Energy is required to overcome the hydrogen bonding before water molecules can escape from the liquid to become vapour. Water therefore has a higher boiling point than would be expected from the mass of its molecules. The effect of the hydrogen bonding may be seen from the fact that hydrogen sulphide, H_2S, with molecules nearly twice as heavy as water molecules, has a boiling point of $-61\,°C$. Hydrogen bonding is virtually non-existent in hydrogen sulphide because sulphur is much less electronegative than oxygen.

WATER AS A SOLVENT

Water is an excellent solvent and this may be ascribed to the polarisation of its molecules. Consider an ionic substance such as sodium chloride. The sodium chloride lattice consists of Na^+ and Cl^- ions arranged alternately in a three-dimensional network (Figure 3.1). If sodium chloride is added to water, Na^+ ions at the surface of the crystal are attracted by the slightly negatively charged oxygen atoms of the water molecules. Similarly, the surface Cl^- ions are attracted by the slightly positively charged hydrogen atoms in water molecules. The crystal slowly breaks down and the ions go into solution surrounded by several water molecules (Figure 4.9); the ions are said to be *hydrated*. Ionic solids will be appreciably soluble in water if the force of attraction between the water and ions is greater than the forces holding the crystal together.

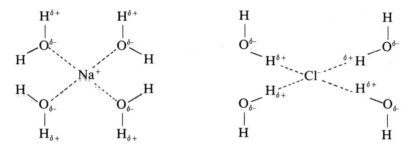

Figure 4.9. Hydration of ions

Some covalent substances also will dissolve in water. For example methanol (an alcohol), CH_3OH, is soluble in water in all proportions because it is polarised in the same way as water, i.e., $CH_3 - \overset{\delta-}{O} - \overset{\delta+}{H}$. The methanol and water molecules mutually attract one another.

Solubility of solids in water

If a solid is added to a liquid and it dissolves, a *solution* is formed. The substance which dissolves is known as the *solute* whilst the substance which does the dissolving is called the *solvent*.

There is a limit to the amount of solute that will dissolve in a solvent at a given temperature. If, for example, small amounts of potassium chloride are added to a given volume of water, it is found that initially all the solid dissolves. However, the stage is eventually reached where no more of the solute will dissolve at that temperature and the solution is then said to be *saturated*. Sodium nitrate is found to be more soluble than potassium chloride, but again the solution becomes saturated at a given temperature if enough of the solute is added. *A saturated solution is defined as one which, in the presence of excess solute, will not dissolve any more of the solute at a particular temperature.*

The solubility of a solute is the number of grams of it required to saturate 1000 grams of the solvent at a given temperature. As a general rule, it is found that the solubility of a solute increases as the temperature increases. However, there are exceptions to the rule: for example, the solubility of calcium hydroxide decreases as the temperature is raised.

The solubility of a solid at a given temperature may be determined by evaporating a known mass of the saturated solution to dryness and then weighing the residue. The mass of solute which will dissolve in 1000 g of solvent is calculated from the results by simple proportion.

When the solubility of a solute has been determined at a number of temperatures, a graph of the results may be plotted. Some typical solubility curves are illustrated in Figure 4.10. It is seen that the curve for sodium sulphate differs from the rest in that the solubility rises to a peak at 32.4 °C and then falls. The reason for this is that at 32.4 °C the decahydrate, $Na_2SO_4 \cdot 10H_2O$, loses its water of crystallisation and so the latter part of the graph is the solubility curve for the anhydrous salt.

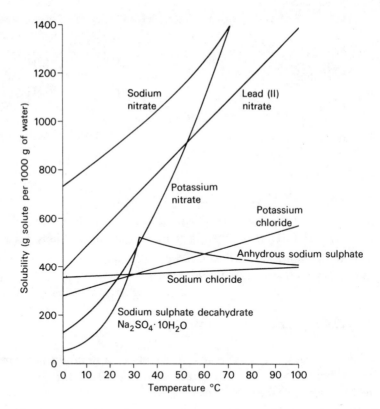

Figure 4.10. Solubility curves for salts in water

The value of a solubility curve lies in the fact that the solubility of the substance can be read off at any temperature covered by the graph. Thus, the solubility of potassium chloride at 48 °C is found by drawing a vertical line from this temperature to the curve and reading off the solubility opposite the point of intersection. Figure 4.10 shows the solubility of potassium chloride at this temperature to be 420 grams per litre of water.

Supersaturated solutions

Sodium thiosulphate crystals, $Na_2S_2O_3 \cdot 5H_2O$, have a solubility of 1099 g in 1000 g water at 20 °C. Hence, 5 g water will dissolve about 5.5 g of the crystals at 20 °C. However, if say 10 g of sodium thiosulphate crystals are dissolved in 5 g of hot water and the resulting solution is kept perfectly still as it cools, no crystals are deposited. The cold solution is then said to be *supersaturated*. Crystallisation immediately occurs if a small crystal is added to the solution, this process being known as '*seeding*'. The phenomenon of supersaturation occurs with other compounds, particularly with those whose solubility increases rapidly with temperature. Crystallisation from supersaturated solutions may sometimes be induced by shaking the solution, adding a particle of dust, or by scratching the glass below the solution with a glass rod. Once started, the crystallisation continues until a truly saturated solution results.

A supersaturated solution may be defined as one which contains more solute than the saturated solution at that temperature. It should be noted that the definition of a saturated solution stipulates that excess solute must be present; supersaturated solutions cannot exist when this condition prevails.

SOLUBILITY OF GASES IN WATER

Gases differ from most solids in that their solubility decreases as the temperature is raised. This may be readily demonstrated by the experiment described on page 47 in which the dissolved air is expelled from tap water. A further important difference is that the solubility of gases in a solvent is greatly affected by pressure; the greater the pressure the greater the solubility. A common illustration of this is the streams of bubbles produced when the stopper is removed from a bottle of mineral water.

Obviously, the solubility of gases can only be effectively compared if they are measured under the same conditions of temperature and pressure. The solubilities of a few common gases at s.t.p. are given in the table below.

Gas	Formula	Solubility (vol/vol) at s.t.p.
Ammonia	NH_3	1200
Hydrogen chloride	HCl	500
Sulphur dioxide	SO_2	80
Carbon dioxide	CO_2	1.8
Oxygen	O_2	0.05
Nitrogen	N_2	0.023
Hydrogen	H_2	0.019

The high solubilities of ammonia, hydrogen chloride, and sulphur dioxide may be attributed to their reaction with the water.

$$NH_3(g) + H_2O(l) \rightarrow NH_3(aq)$$
$$HCl(g) + H_2O(l) \rightarrow HCl(aq) \quad \text{(i.e. } H^+(aq) + Cl^-(aq))$$
$$SO_2(g) + H_2O(l) \rightarrow H_2SO_3(aq)$$

The high solubility of ammonia and hydrogen chloride is utilised in the fountain experiment (page 192).

HARD WATER

It has already been mentioned that water is an excellent solvent and so it is not surprising that water from all natural sources contains impurities. The purest natural source is rainwater; this contains dissolved gases from the air along with dust particles, etc. Water from springs contains dissolved solids due to its passage through the ground, whilst river water contains further impurities as a result of effluents discharged into it. The sea is the sink for rivers and so sea-water is particularly impure. The dissolved solid content of sea-water is generally in the region of three per cent.

Tap water is normally obtained from rivers and springs. The water is allowed to stand so that solid particles settle, the process often being accelerated by addition of a little aluminium sulphate. The water is then filtered by passing it through sand and gravel, and finally, about one part per million of chlorine is added in order to kill the bacteria. Dissolved salts are left in the water because they are beneficial to health.

Water which will not readily form a lather with soap is said to be 'hard'. The hardness of tap water varies considerably throughout the country but it always has the same cause, i.e. dissolved calcium and magnesium salts. Hardness can be determined by titrating water with a standard soap solution. Alternatively, the magnesium and calcium ion concentration can be determined directly with a reagent commonly known as edta. Hardness can be divided into two types: temporary and permanent.

(a) *Temporary hardness* is caused by the hydrogencarbonates of calcium and magnesium, the former being the main culprit. Calcium hydrogencarbonate is formed when water containing dissolved carbon dioxide comes into contact with limestone or chalk.

$$CaCO_3(s) + H_2O(l) + CO_2(g) \rightarrow Ca(HCO_3)_2(aq)$$

This hardness is said to be temporary because it is destroyed when the water is boiled.

$$Ca(HCO_3)_2(aq) \rightarrow CaCO_3(s) + CO_2(g) + H_2O(l)$$

A fur or scale of insoluble carbonate is deposited as the hydrogencarbonates decompose.

(b) *Permanent hardness* is unaffected by boiling and is caused by calcium and magnesium sulphates and chlorides.

The formation of a scum when soap is added to hard water may be explained as follows. Soap is the sodium salt of a long chain fatty acid and is made by refluxing vegetable oils or animal fats with sodium hydroxide solution. It has the formula $C_{17}H_{35} \cdot COONa$ but for simplicity it may be referred to as sodium soap. When soap is added to hard water, a scum is formed because the calcium and magnesium ions attract the anion from the soap to give insoluble products, e.g.

sodium soap (aq) + calcium sulphate (aq) $\rightarrow$ sodium sulphate (aq) + calcium soap (s)

No lather will form until all the calcium and magnesium ions have been removed from solution.

Detergents are more effective than soap at bridging the interface between water and grease and hence releasing dirt. They are compounds such as $C_{12}H_{25} \cdot O \cdot SO_2 \cdot ONa$ and they do not give a scum with hard water because the calcium and magnesium salts of the detergent are water soluble. Note that both soap and detergents have a long carbon chain which will dissolve fats and a water soluble group, i.e. $-COO^-Na^+$ or $-O \cdot SO_2 \cdot O^-Na^+$.

REMOVAL OF HARDNESS

Both types of hardness can be removed by distillation of the water, but this is uneconomical both in terms of time and cost. Temporary hardness may be removed by adding calcium hydroxide to the water:

$$Ca(HCO_3)_2(aq) + Ca(OH)_2(aq) \rightarrow 2CaCO_3(s) + 2H_2O(l)$$

i.e., the acid salt is neutralised by the alkali. However, the amount of calcium hydroxide added must be very carefully controlled since any excess would just dissolve in the water and give rise to permanent hardness.

Methods which remove permanent hardness simultaneously remove temporary hardness. Two important methods are discussed.

(a) *Addition of sodium carbonate.* This will result in the precipitation of calcium and magnesium carbonates, e.g.

$$Ca(HCO_3)_2(aq) + Na_2CO_3(aq) \rightarrow CaCO_2(s) + 2NaHCO_3(aq)$$
$$CaSO_4(aq) + Na_2CO_3(aq) \rightarrow CaCO_3(s) + Na_2SO_4(aq)$$

Note that the sodium salts remain in solution but this is immaterial since they do not cause hardness.

(b) *The permutit process.* Permutit is a complex sodium aluminium silicate and when water is passed over it, the sodium ions interchange with the calcium and magnesium ions in the water. This may be represented as

$$\underset{\substack{\text{In solution causing}\\\text{hardness}}}{Ca^{2+}} + \underset{\text{In permutit}}{2Na^+} \longrightarrow \underset{\text{In solution}}{2Na^+} + \underset{\substack{\text{In permutit, i.e.}\\\text{hardness removed}}}{Ca^{2+}}$$

Obviously, the permutit will eventually become exhausted. However, permutit may be regenerated by slowly passing a concentrated solution of sodium chloride over it: the Ca^{2+} and Mg^{2+} ions are washed away as their chlorides.

WATER OF CRYSTALLISATION

It has already been mentioned (page 48) that, when copper sulphate crystallises from water, the pentahydrate, $CuSO_4 \cdot 5H_2O$, is obtained. The crystals can be obtained perfectly dry, the five water molecules being loosely combined with the copper sulphate. Many other compounds crystallise with water of crystallisation, some common examples being $Na_2SO_4 \cdot 10H_2O$, $Na_2CO_3 \cdot 10H_2O$, and $FeSO_4 \cdot 7H_2O$. On the other hand, some salts, such as sodium chloride and potassium nitrate, never contain water of crystallisation.

When compounds containing water of crystallisation are gently heated, the water is lost and this fact may be utilised to determine the formula of the crystals. The practical details are given on page 253 but the calculation based on a set of results obtained with barium chloride crystals, $BaCl_2 \cdot xH_2O$ is illustrated over.

Mass of crucible + lid = 21.3827 g
Mass of crucible + lid + barium chloride crystals = 25.9316 g
Mass of crucible + lid + anhydrous barium chloride = 25.2600 g
Mass of water driven off = 0.6716 g
Mass of anhydrous barium chloride = 3.8773 g

∴ The mass ratio in the crystals is 3.8773 g $BaCl_2$: 0.6716 g H_2O

Now dividing the mass of each substance by its relative molecular mass (page 63) will give the ratio of the number of molecules, i.e.

$$\frac{3.8733}{208.3} BaCl_2 \quad : \quad \frac{0.6716}{18} H_2O$$

$$= 0.01861 \ BaCl_2 \quad : \quad 0.0373 \ H_2O$$

Dividing through by the small value will give the simple whole number ratio, i.e.

$$\frac{0.01861}{0.01861} BaCl_2 \quad : \quad \frac{0.0373}{0.01861} H_2O$$

$$= 1 \ BaCl_2 \quad : \quad 2H_2O$$

The formula of the crystals is therefore $BaCl_2 \cdot 2H_2O$.

Questions

1 Describe the experiments that you would carry out to confirm the following observations:
   (a) Air contains both water vapour and carbon dioxide.
   (b) Air contains approximately 20% oxygen.
   (c) Air diffuses faster than carbon dioxide under the same conditions of temperature and pressure.
   [J.M.B.]

2 In order to determine the proportion of one of the major components of air, the air was passed in turn through sodium hydroxide solution, through concentrated sulphuric acid and into a glass syringe. The volume of remaining gas was measured and the gas was passed repeatedly over red hot copper until no further contraction occurred. The gas was then allowed to cool and its volume was measured.
Volume of gas before passing over hot copper = 90.0 cm³
Volume of gas after passing over hot copper = 70.2 cm³
   (a) Why was the air passed through the sodium hydroxide solution?
   (b) Why was it passed through concentrated sulphuric acid?
   (c) How would the appearance of the copper before heating and after cooling differ?
   (d) Which gas was removed by the copper?
   (e) Name the main gas remaining in the syringe at the end of the experiment and calculate the approximate percentage of this gas in the air from the data provided.
   (f) Give the name of another element which would still be present in the residual gas.
   [J.M.B.]

Air and Water 59

3 (a) Name one compound in each case which can cause:
    (i) temporary hardness,
    (ii) permanent hardness in water.

(b) Show, by equations, how both temporary and permanent hard water react with soap and state what you would see during the process. Contrast this with the action of a soapless detergent on a sample of hard water.

(c) Describe how these two types of hardness arise in nature.

(d) Why is it necessary, in hard water areas, to check domestic hot water and central heating pipes regularly?

[J.M.B.]

4 (a) Describe how you would prepare a solution of potassium nitrate that is saturated at room temperature. How would you use this solution to determine the solubility of potassium nitrate (in g/100 g water) at room temperature? Mention the precautions you would take to ensure accuracy, and show how you would calculate the result.

(b) Explain the following experiments which were carried out with a single sample of hard water.
    (i) When boiled, a small amount of a white solid was deposited.
    (ii) Some hardness remained even after the water had been boiled for a considerable time.
    (iii) The water could be softened completely by adding sodium carbonate.

[A.E.B. June 1975]

5 (a) Explain what is meant by:
    (i) a saturated solution,
    (ii) a supersaturated solution.

(b) A mass of 4.133 g of sodium carbonate crystals, $Na_2CO_3 \cdot xH_2O$, was heated gently and 3.533 g of the anhydrous compound remained. Calculate the value of x.

5 Quantitative Relationships

THE LAWS OF CHEMICAL COMBINATION

1. **Law of constant composition:** *a given compound always contains the same elements combined in the same proportions by mass.* The law is sometimes known as the law of definite proportions.

    Verification of the law may be achieved by making a compound in a number of ways and analysing the various samples. For example, copper(II) oxide may be made by the three methods below (see page 251 for full practical details).

    (a) By heating copper(II) carbonate to constant mass.

    $$CuCO_3(s) \rightarrow CuO(s) + CO_2(g)$$

    (b) By dissolving copper in concentrated nitric acid, evaporating the resultant copper(II) nitrate solution to dryness, and then heating to constant mass.

    $$Cu(s) + 4HNO_3(aq) \rightarrow Cu(NO_3)_2(aq) + 2H_2O(l) + 2NO_2(g)$$
    $$2Cu(NO_3)_2(s) \rightarrow 2CuO(s) + 4NO_2(g) + O_2(g)$$

    (c) By adding excess sodium hydroxide solution to copper(II) sulphate solution, and then boiling the mixture formed to convert the copper(II) hydroxide to the oxide. The copper(II) oxide is filtered, washed thoroughly with water, and then dried.

    $$CuSO_4(aq) + 2NaOH(aq) \rightarrow Cu(OH)_2(s) + Na_2SO_4(aq)$$
    $$Cu(OH)_2(s) \rightarrow CuO(s) + H_2O(l)$$

    Each sample of copper(II) oxide is analysed by heating a known mass of it in a stream of hydrogen. The hydrogen combines with the oxygen of the oxide to give water, which escapes as steam, and the copper remains.

    $$CuO(s) + H_2(g) \rightarrow Cu(s) + H_2O(g)$$

    The residual copper is weighed and the percentage composition of the oxide is calculated. The calculation is illustrated using the results given below.

Mass of porcelain boat	= 22.4792 g
Mass of boat + copper(II) oxide	= 23.9556 g
Mass of copper(II) oxide	= 1.4764 g
Mass of boat + copper	= 23.6603 g
Mass of copper	= 1.1811 g

$$\% \text{ copper in copper(II) oxide} = \frac{\text{mass of copper}}{\text{mass of oxide}} \times 100$$

$$= \frac{1.1811}{1.4764} \times 100$$

$$= 80.01\%$$

2. **Law of multiple proportions:** *when two elements A and B combine to form more than one compound, the masses of A which combine with a fixed mass of B are in a simple ratio to one another.*

   The law may be illustrated by the two oxides of copper, one of which is black and the other red. Heating known masses of each oxide in a current of hydrogen shows that:

   (i) in the black oxide, each gram of copper combines with 0.252 g oxygen,

   (ii) in the red oxide, each gram of copper combines with 0.126 g oxygen.

   In this example, the fixed mass of B is 1 g of copper and so, according to the law, the two masses of oxygen should be in a simple ratio. Since 0.252 : 0.126 is 2 : 1, the results are in agreement with the law. The two oxides are, in fact, copper(II) oxide, CuO, and copper(I) oxide, Cu_2O (only half as much oxygen as the other oxide).

3. **Gay-Lussac's law of gaseous volumes:** *when gases combine, the volumes of the reactants, and of the products if gaseous, are in a simple ratio to each other provided that they are all measured under the same conditions of temperature and pressure.*

   For example, under a given set of conditions, it is found that 1 volume of nitrogen combines with 3 of hydrogen to give 2 volumes of ammonia, or 2 volumes of hydrogen react with 1 of oxygen to give 2 volumes of steam.

4. **Avogadro's law:** *equal volumes of gases at the same temperature and pressure contain equal numbers of molecules.*

   The equation

   $$N_2(g) + 3H_2(g) \rightarrow 2NH_3(g)$$

   shows that 1 molecule of nitrogen reacts with 3 molecules of hydrogen to give 2 molecules of ammonia. However, it was seen above (Gay-Lussac's law) that the volumes involved are in the same ratio. The same relationship between the number of molecules and the volumes is found in all gaseous reactions and so it is apparent that Avogadro's law is correct.

   The application of Gay-Lussac's and Avogadro's laws is illustrated by the following example.

   Example Calculate the volume of oxygen required to completely burn 40 cm³ of butane, C_4H_{10}, all measurements being made at 20 °C and 760 mm pressure.

   Butane burns in air or oxygen to give carbon dioxide and water and so the equation for the reaction is

   $$2C_4H_{10} + 13O_2 \rightarrow 8CO_2 + 10H_2O$$

Applying Avogadro's and Gay-Lussac's laws, it is apparent that

2 volumes of butane react with 13 volumes of oxygen

$\therefore$ 40 cm³ of butane react with $\frac{13}{2} \times 40 = 260$ cm³ of oxygen

RELATIVE ATOMIC MASSES

Hydrogen atoms have a mass of 1.66×10^{-24} g, i.e. 1.66 million million million millionths of a gram. Now all other elements have heavier atoms than this but, nevertheless, the absolute masses are far too small for convenient use. For this reason, relative atomic masses are used instead of absolute masses. Initially, the masses of the atoms of all other elements were compared to that of the hydrogen atom which was said to have a mass of one unit. Thus, the fact that sodium has a relative atomic mass of 23 meant that a sodium atom weighed 23 times as much as a hydrogen atom. For various reasons, the standard was changed from hydrogen to oxygen and finally to the $^{12}_{6}C$ isotope. Relative atomic masses are normally used correct to one decimal place and changing the standard did not affect these figures.

The relative atomic mass of an element may now be defined as the number of times one atom of the element is heavier than $\frac{1}{12}$ of the mass of the $^{12}_{6}C$ isotope. Relative atomic mass may be represented by the symbol A_r. A list of relative atomic masses is given on page 279.

Many elements have relative atomic masses which are not whole numbers and the main reason for this is that they exist as mixtures of isotopes (see page 27). For example, the relative atomic mass of chlorine is 35.5 because it exists as a mixture of the isotopes $^{35}_{17}Cl$ and $^{37}_{17}Cl$ in the ratio of 3 : 1.

Relative atomic masses have no units since they are just a ratio. However, the relative atomic mass of an element expressed in grams is known as a *mole* of atoms of the element. Hence, a mole of hydrogen atoms has a mass of 1 gram whilst a mole of oxygen atoms has a mass of 16 grams.

Since the absolute mass of a hydrogen atom is 1.66×10^{-24} g, and 1 mole of hydrogen atoms has a mass of 1 g, then the number of hydrogen atoms in 1 mole

$$= \frac{1}{1.66 \times 10^{-24}} = 6.023 \times 10^{23}$$

Now the absolute mass of the atoms of any element and the mass of 1 mole of its atoms differ from the figures for hydrogen in the same ratio. For example, the absolute mass of an oxygen atom is $16 \times 1.66 \times 10^{-24}$ g and 1 mole of oxygen atoms has a mass of 16 g. Therefore, the number of oxygen atoms in 1 mole

$$= \frac{16}{16 \times 1.66 \times 10^{-24}} = 6.023 \times 10^{23}$$

The same number is found for all elements and this is known as the *Avogadro constant*. Thus, *1 mole of atoms of any element consists of 6.023×10^{23} atoms.*

RELATIVE MOLECULAR MASSES

Molecules, like atoms, have extremely small absolute masses and so they are measured on the same scale as atoms. *Relative molecular mass may be defined as the sum of the relative atomic masses of all the atoms in the molecule.* For example, the relative atomic masses of hydrogen, sulphur, and oxygen are 1, 32, and 16 respectively and so the relative molecular mass of sulphuric acid $H_2SO_4 = (2 \times 1) + 32 + (4 \times 16) = 98$.

Relative molecular masses, like relative atomic masses, have no units. Relative molecular mass is merely a number, i.e. the number of times the molecule is heavier than $\frac{1}{12}$ of the $^{12}_{6}C$ isotope. However, if the relative molecular mass is expressed in grams, then that mass is known as 1 mole of the substance. Now, the absolute mass of a hydrogen molecule, H_2, is $2 \times 1.66 \times 10^{-24}$ g and 1 mole of hydrogen molecules weighs 2 g and so the number of molecules in 1 mole of hydrogen

$$= \frac{2}{2 \times 1.66 \times 10^{-24}} = 6.023 \times 10^{23}$$

It will be recalled that this is the same as the number of atoms in 1 mole of atoms. The same result will be obtained for the number of molecules in 1 mole of any substance. *The Avogadro constant may therefore be defined as the number of particles in 1 mole of any substance.*

RELATIONSHIP BETWEEN RELATIVE MOLECULAR MASS AND RELATIVE DENSITY

Relative molecular masses can be determined with great accuracy using an instrument known as a mass spectrometer. However, these instruments are very expensive. A simple alternative but less accurate method, for gases and vapours, is to determine their relative densities (vapour densities) since relative molecular mass and relative density are related as shown below.

$$\text{Relative density} = \frac{\text{mass of 1 volume of gas}}{\text{mass of 1 volume of hydrogen}}$$

both gases being at the same temperature and pressure.

Now, according to Avogadro's law, equal volumes of gases at the same temperature and pressure contain equal numbers of molecules.

$$\therefore \quad \text{Relative density} = \frac{\text{mass of } n \text{ molecules of gas}}{\text{mass of } n \text{ molecules of hydrogen}}$$

$$= \frac{\text{mass of 1 molecule of gas}}{\text{mass of 1 molecule of hydrogen}}$$

$$= \frac{\text{mass of 1 molecule of gas}}{\text{mass of 2 atoms of hydrogen}}$$

But relative molecular mass $= \dfrac{\text{mass of 1 molecule of gas}}{\text{mass of 1 atom of hydrogen}}$

$\therefore$ Relative density $= \frac{1}{2}$ relative molecular mass

or relative molecular mass $= 2 \times$ relative density.

This expression is based on the old relative atomic mass scale of H = 1 and not on the present scale of $^{12}_{6}C = 12.0000$. However, this is unimportant since it makes negligible difference to the answer.

QUANTITATIVE INTERPRETATION OF EQUATIONS

The relative molecular masses of substances in an equation show the quantitative relationships, i.e. they show the masses of the substances which react together and of the products. Consider, for example, the reaction between sodium carbonate and hydrochloric acid. The equation for the reaction is:

$$Na_2CO_3 + 2HCl \rightarrow 2NaCl + H_2O + CO_2$$

The masses of each substance involved can be found given the following relative atomic masses: Na = 23, C = 12, O = 16, H = 1, and Cl = 35.5.

$$\begin{aligned}
\text{Mass of } Na_2CO_3 &= (2 \times 23) + 12 + (3 \times 16) = 106 \\
2HCl &= 2(1 + 35.5) &= 73 \\
2NaCl &= 2(23 + 35.5) &= 117 \\
H_2O &= (2 \times 1) + 16 &= 18 \\
CO_2 &= 12 + (2 \times 16) &= 44
\end{aligned}$$

Thus, 106 g of sodium carbonate react with 73 g of hydrochloric acid to give 117 g of sodium chloride, 18 g of water, and 44 g of carbon dioxide. Note that the total mass of products must be the same as the total mass of reactants.

The mass of products formed by using different masses of reactants can now be calculated.

Example Calculate how much sodium chloride will be formed if excess hydrochloric acid is added to 12 g of sodium carbonate.

$$Na_2CO_3 + 2HCl \rightarrow 2NaCl + H_2O + CO_2$$
$$106 \qquad 73 \qquad 117 \qquad 18 \qquad 44$$

106 g of Na_2CO_3 produce 117 g NaCl

1 g of Na_2CO_3 produces $\dfrac{117}{106}$ g NaCl

∴ 12 g of Na_2CO_3 produce $\dfrac{117}{106} \times 12 = 13.24$ g NaCl

Note: The procedure in all calculations of this type should be as follows.

(a) Write down the equation and the masses of the substances concerned.

(b) Pick out the two substances involved in the calculation.

(c) Write down the information which the *equation* gives concerning the theoretical masses of the two substances involved. Make sure that the first substance written about is the one whose actual mass to be used in the experiment is known – in the above case, sodium carbonate.

(d) Proceed with the actual calculation.

The mass of a reactant required to produce a given mass of one of the products may also be calculated.

Example Sulphuric acid reacts with copper(II) oxide to give copper(II) sulphate and water. Calculate how much copper(II) oxide is required to make 30 g of copper(II) sulphate. Relative atomic masses: Cu = 63.5, O = 16, H = 1, and S = 32.

$$CuO + H_2SO_4 \rightarrow CuSO_4 + H_2O$$
$$79.5 \quad\; 98 \quad\;\; 159.5 \quad\; 18$$

159.5 g of $CuSO_4$ are made from 79.5 g CuO

1 g of $CuSO_4$ is made from $\dfrac{79.5}{159.5}$ g CuO

∴ 30 g of $CuSO_4$ are made from $\dfrac{79.5}{159.5} \times 30 = 14.95$ g CuO

MOLAR VOLUME

The relative atomic mass of hydrogen is 1.008 and so its relative molecular mass is 2.016 (there are two atoms in the hydrogen molecule). Further, experiments show that 1 litre of hydrogen at standard temperature and pressure has a mass of 0.09 g. These facts can be utilised to calculate the volume occupied at s.t.p. by 1 mole of hydrogen:

0.09 g hydrogen occupies 1 litre at s.t.p.

∴ 2.016 g hydrogen occupies $\dfrac{1}{0.09} \times 2.016 = 22.4$ litres at s.t.p.

This volume is known as the *molar volume* of hydrogen.

Now according to Avogadro's law, equal volumes of gases at the same temperature and pressure contain equal numbers of molecules. Therefore, since a mole of all substances contains the same number of molecules (see the Avogadro constant), it follows that a mole of any gas will occupy 22.4 litres at standard temperature and pressure.

Molar volume is defined as the volume occupied by one mole of any gas and at s.t.p. this volume is 22.4 litres. Molar volume is very useful because gases are normally measured by volume, not by mass. The volumes of gases evolved in reactions can now be calculated as illustrated in the following examples.

Example Calculate the volume of hydrogen evolved at s.t.p. when excess hydrochloric acid is added to 4 g of zinc. Molar volume is 22.4 litres at s.t.p. and the relative atomic masses of zinc, hydrogen, and chlorine are 65.4, 1, and 35.5 respectively.

The equation for the reaction is

$$Zn(s) + 2HCl(aq) \rightarrow ZnCl_2(aq) + H_2(g)$$
$$65.4 \quad\;\; 73 \quad\quad\;\; 136.4 \quad\;\; 2$$

and so 65.4 g of zinc produce 2 g of hydrogen. However 2 g of hydrogen is 1 mole and so this has a volume of 22.4 litres at s.t.p.

i.e. 65.4 g Zn give 22.4 litres H_2 at s.t.p.

∴ 4 g Zn give $\dfrac{22.4}{65.4} \times 4 = 1.37$ litres H_2 at s.t.p.

Example Potassium chlorate, $KClO_3$, decomposes on heating to give potassium chloride and oxygen. Calculate what volume of oxygen at s.t.p. will be obtained when 5 g of potassium chlorate is heated. $K = 39.1$, $Cl = 35.5$, $O = 16$, and molar volume = 22.4 litres at s.t.p.

The equation involved is

$$2KClO_3 \rightarrow 2KCl + 3O_2$$
$$245.2 \quad 149.2 \quad 96$$

and so 245.2 g of potassium chlorate produce 96 g or 3 moles of oxygen which will occupy 3×22.4 litres at s.t.p.

i.e.   245.2 g $KClO_3$ produce 3×22.4 litres O_2 at s.t.p.

∴   5 g $KClO_3$ produce $\dfrac{3 \times 22.4}{245.2} \times 5 = 1.37$ litres O_2 at s.t.p.

Example Ammonium chloride reacts with sodium hydroxide to give sodium chloride, water, and ammonia. Calculate how much ammonium chloride will be required to produce 448 cm³ of ammonia measured at s.t.p. $N = 14$, $Cl = 35.5$, $H = 1$, $O = 16$ and $Na = 23$.

The equation is

$$NH_4Cl + NaOH \rightarrow NaCl + H_2O + NH_3$$
$$53.5 \quad\quad 40 \quad\quad 58.5 \quad 18 \quad 17$$

and so 53.5 g of ammonium chloride produce 1 mole (17 g) of ammonia which will have a volume of 22.4 litres at s.t.p.

i.e.   22 400 cm³ of NH_3 at s.t.p. are made from 53.5 g NH_4Cl

∴   448 cm³ of NH_3 at s.t.p. are made from $\dfrac{53.5 \times 448}{22\,400} = 1.07$ g NH_4Cl

Note that, in the layout of this calculation, the ammonia figures are given first because this is the substance about which all the information is known.

Example Iron reacts with hydrochloric acid to give iron(II) chloride and hydrogen. Calculate the volume of hydrogen measured at 20 °C and 750 mm pressure which will be produced if excess hydrochloric acid is added to 7 g of iron. $Fe = 56$, $H = 1$, and $Cl = 35.5$. Molar volume = 22.4 litres at s.t.p.

The equation is

$$Fe + 2HCl \rightarrow FeCl_2 + H_2$$
$$56 \quad 73 \quad\quad 127 \quad 2$$

Hence 56 g of iron produces 1 mole or 22.4 litres of hydrogen at s.t.p.

i.e.   56 g of Fe produce 22.4 litres H_2 at s.t.p.

∴   7 g of Fe produce $\dfrac{22.4}{56} \times 7 = 2.8$ litres of H_2 at s.t.p.

Quantitative Relationships

Now the question asks for the volume of hydrogen at 20 °C and 750 mm pressure and so this entails the use of the equation

$$\frac{P_1 V_1}{T_1} = \frac{P_2 V_2}{T_2}$$

i.e.
$$\frac{760 \times 2.8}{273} = \frac{750 \times V_2}{293}$$

$$\therefore \quad V_2 = \frac{760 \times 2.8 \times 293}{273 \times 750} = 3.045 \text{ litres}$$

Example Iron(II) sulphide reacts with hydrochloric acid giving iron(II) chloride and hydrogen sulphide. What mass of iron(II) sulphide will be required to produce 1.5 litres of hydrogen sulphide measured at 18 °C and 745 mm pressure? Fe = 56, S = 32, H = 1, and Cl = 35.5.

The equation for the reaction is

$$\text{FeS} + 2\text{HCl} \rightarrow \text{FeCl}_2 + \text{H}_2\text{S}$$
$$88 \qquad 73 \qquad 127 \qquad 34$$

Therefore, 22.4 litres of hydrogen sulphide at s.t.p. (i.e. 1 mole or 34 g) are made from 88 g of iron(II) sulphide. Now, before the actual mass of iron(II) sulphide required in the experiment can be calculated, it is necessary to know the volume of hydrogen sulphide required under the conditions of standard temperature and pressure. Hence, using the equation

$$\frac{P_1 V_1}{T_1} = \frac{P_2 V_2}{T_2}$$

$$\frac{745 \times 1.5}{291} = \frac{760 \times V_2}{273}$$

$$\therefore \quad V_2 = \frac{745 \times 1.5 \times 273}{291 \times 760} = 1.379 \text{ litres}$$

But, 22.4 litres of H_2S at s.t.p. are made from 88 g FeS

$\therefore$ 1.377 litres of H_2S at s.t.p. are made from $\frac{88}{22.4} \times 1.377 = 5.41$ g FeS

Questions

1 (a) State the law of constant composition. Briefly outline how the law may be verified, including relevant equations in your answer.

   (b) Two oxides of chromium, Cr, were analysed: 1·169 g of oxide A were found to contain 0·800 g of chromium whilst 1·154 g of oxide B contained 0·600 g of chromium. Show how these results may be used to illustrate the law of multiple proportions.

2 An oxide of lead was weighed in a porcelain boat and it was then reduced to lead by heating it in a stream of hydrogen. The boat with the lead in it was then allowed to cool with the hydrogen still passing over it, and it was then weighed. It was reheated in hydrogen, recooled and reweighed until a constant weight was attained for the boat and the lead.

The following weighings were obtained:
Weight of boat = 10.20 g
Weight of boat + lead oxide = 17.37 g
Final weight of boat + lead = 16.41 g
(Relative atomic masses: Pb = 207.0, O = 16.0)

(a) Name a drying agent which could be used to dry the hydrogen used in the experiment.

(b) Why was:
  (i) the boat cooled with the hydrogen still passing over it,
  (ii) the experiment repeated until a constant weight was obtained?

(c) (i) What weight of lead was produced in the experiment?
  (ii) What weight of oxygen was originally combined with this weight of lead?
  (iii) Calculate the weight of oxygen which combines with 1 mole of lead.
  (iv) How many moles of oxygen combine with 1 mole of lead?
  (v) Write the formula and name of the oxide of lead used in the experiment.

(d) In a second experiment 4.14 g of lead was obtained from 4.46 g of another oxide of lead.
  (i) Calculate how many moles of oxygen combine with 1 mole of lead to form this oxide.
  (ii) Give the name and formula of this oxide.

[J.M.B.]

3 Sodium chloride and silver nitrate react according to the equation

$$NaCl(aq) + AgNO_3(aq) \rightarrow AgCl(s) + NaNO_3(aq)$$

Calculate how much silver chloride will be precipitated if excess silver nitrate solution is added to an aqueous solution containing 2.56 g of sodium chloride.

4 The action of heat on copper(II) carbonate gives copper(II) oxide and carbon dioxide. Calculate how much copper(II) carbonate is required to make 10 g of copper(II) oxide.

5 Hydrochloric acid reacts with sodium sulphite to give sodium chloride, water, and sulphur dioxide. Calculate the volume of sulphur dioxide, measured at s.t.p., which is obtained when excess of the acid is added to 5 g of sodium sulphite.

6 Sulphuric acid reacts with zinc to give zinc sulphate and hydrogen. Calculate how much zinc would be required to produce 1120 cm³ of hydrogen measured at s.t.p.

7 The reaction between calcium carbonate and hydrochloric acid gives calcium chloride, water, and carbon dioxide. Calculate the volume of carbon dioxide, measured at 18 °C and 750 mm, which will be produced if excess of the acid is added to 10 g of calcium carbonate.

6 Acids, Bases, and Salts

ACIDS

Acids are compounds which, when dissolved in water, give hydrogen ions, H^+, as the only positive ion.

The equations for the dissociation of some common acids in water are as follows:

$$HCl \xrightarrow{H_2O} H^+ + Cl^-$$
$$HNO_3 \xrightarrow{H_2O} H^+ + NO_3^-$$
$$H_2SO_4 \xrightarrow{H_2O} 2H^+ + SO_4^{2-}$$

For simplicity, the hydration of the ions has not been shown. In fact, the H^+ ion (proton) is attracted by a lone pair of electrons on the oxygen atom of a water molecule. A co-ordinate bond is formed between the oxygen and hydrogen and an oxonium ion is produced:

$$H_2\ddot{O} + H^+ \rightarrow H-\overset{+}{\underset{\downarrow}{O}}-H$$
$$\phantom{H_2\ddot{O} + H^+ \rightarrow H-\overset{+}{\underset{\downarrow}{O}}}H$$

In this way, the H^+ ion achieves the electronic structure of helium. Hydration of the ions occurs in a similar manner to that described on page 53.

Hydrochloric, nitric, and sulphuric acids are said to be strong acids because their dissociation into ions is virtually complete, i.e., the equilibrium lies almost entirely to the right. However, only partial dissociation occurs with many acids. For example, if ethanoic acid is dissolved in water so that the solution contains 0.1 mole of the acid per litre, only about 130 molecules in every million are dissociated.

$$CH_3-C\underset{O-H}{\overset{O}{\diagup}} \xrightleftharpoons{H_2O} CH_3-C\underset{O^-}{\overset{O}{\diagup}} + H^+$$

Acids such as this are described as weak acids. *It should be noted that the terms weak or strong, as applied to acids, have nothing to do with the concentration of the solution but refer only to the degree of dissociation.*

Acids may be classified according to the number of H^+ ions one molecule of them can produce. For example, nitric acid, hydrochloric acid, and ethanoic acid ($CH_3 \cdot COOH$) are all said to be *monobasic* or to have a basicity of one, because each molecule of them is capable of providing only one H^+ ion. Note that three atoms of hydrogen in ethanoic acid are firmly bound to the carbon atom and it is only the one attached to the oxygen that can escape as H^+.

$$CH_3-C\underset{O-H}{\overset{O}{\diagup}} \xrightleftharpoons{H_2O} CH_3-C\underset{O^-}{\overset{O}{\diagup}} + H^+$$

Sulphuric acid is said to be *dibasic* or to have a basicity of two, because each molecule can provide two H$^+$ ions, whilst phosphoric acid is *tribasic*, i.e.,

$$H_3PO_4 \xrightleftharpoons{H_2O} 3H^+ + PO_4^{3-}$$

The basicity of an acid may be defined as the number of H$^+$ ions that one molecule of the acid can produce when it is dissolved in water. Alternatively, basicity can be regarded as being the number of hydrogen atoms in a molecule of an acid that can be replaced by a metal.

PREPARATION OF ACIDS

There are two main ways of making acids.

1. Acids are formed when acidic oxides react with water, e.g.

    $H_2O(l) + SO_2(g) \rightarrow H_2SO_3(aq)$ sulphurous acid
    $3H_2O(l) + P_2O_5(s) \rightarrow 2H_3PO_4(aq)$ phosphoric acid
    $H_2O(l) + CO_2(g) \rightarrow H_2CO_3(aq)$ carbonic acid

    Oxides which react with water to give acids are known as acid anhydrides. Thus, sulphur dioxide is the anhydride of sulphurous acid.

2. Weak acids are displaced from their salts by stronger ones, for example, sulphuric acid displaces ethanoic acid from its salts.

    $2CH_3 \cdot COONa(s) + H_2SO_4(aq) \rightarrow Na_2SO_4(aq) + 2CH_3 \cdot COOH(aq)$

    Also, strong acids are displaced from their salts by heating them with less volatile strong acids. Thus, nitric acid may be prepared by heating sodium nitrate with concentrated sulphuric acid.

    $2NaNO_3(s) + H_2SO_4(conc) \rightarrow Na_2SO_4(s) + 2HNO_3(l)$

GENERAL PROPERTIES OF ACIDS

Acids have a number of characteristic properties as detailed below.

1. Many acids have a sour taste. The taste of dilute ethanoic acid in vinegar, 2-hydroxypropane-1,2,3-tricarboxylic acid (citric acid) in lemons, and ethanedioic acid (oxalic acid) in rhubarb will be familiar, but dilute hydrochloric acid, etc., have similar tastes.

2. Most of them turn blue litmus paper red but a few very weak acids, such as carbonic acid, H_2CO_3, just turn it pink. Substances which change colour according to the acidity or basicity of a solution are known as indicators.

3. Most acids liberate carbon dioxide from metal carbonates, e.g.

    $Na_2CO_3(s) + 2HNO_3(aq) \rightarrow 2NaNO_3(aq) + H_2O(l) + CO_2(g)$

    This reaction is used as a test for acids. The sodium carbonate may be used either in the solid state or in solution.

Acids, Bases, and Salts

BASES

It was seen above that acids are proton donors and so *bases may be regarded as proton acceptors*. Alternatively, bases may be regarded as being substances which react with acids to give a salt and water only. (For the definition of a salt, see below.)

Most metal oxides and hydroxides are bases. Some typical acid-base reactions are:

$$ZnO(s) + H_2SO_4(aq) \rightarrow ZnSO_4(aq) + H_2O(l)$$
$$Fe_2O_3(s) + 6HCl(aq) \rightarrow 2FeCl_3(aq) + 3H_2O(l)$$
$$Cu(OH)_2(s) + 2HNO_3(aq) \rightarrow Cu(NO_3)_2(aq) + 2H_2O(l)$$
$$KOH(aq) + HCl(aq) \rightarrow KCl(aq) + H_2O(l)$$

It is instructive to have a look at the above reactions written as ionic equations, i.e.,

$$Zn^{2+} + O^{2-} + 2H^+ + SO_4^{2-} \rightarrow Zn^{2+} + SO_4^{2-} + H_2O$$
$$2Fe^{3+} + 3O^{2-} + 6H^+ + 6Cl^- \rightarrow 2Fe^{3+} + 6Cl^- + 3H_2O$$
$$Cu^{2+} + 2HO^- + 2H^+ + 2NO_3^- \rightarrow Cu^{2+} + 2NO_3^- + 2H_2O$$
$$K^+ + HO^- + H^+ + Cl^- \rightarrow K^+ + Cl^- + H_2O$$

Thus, in any reaction between an acid and an oxide, the net ionic change is

$$2H^+ + O^{2-} \rightarrow H_2O$$

whilst in reactions between acids and hydroxides, the net change is

$$H^+ + HO^- \rightarrow H_2O$$

The reason for the definition of bases as proton acceptors is therefore apparent.

Ammonia is also classed as a base because the lone pair of electrons on the nitrogen atom attracts a H^+ ion. Thus, the reaction between ammonia and hydrochloric acid takes place as follows:

$$NH_3 + H^+ \longrightarrow NH_4^+ \xrightarrow{Cl^-} NH_4^+Cl^-$$

It should be noted that lead(IV) oxide is not classed as a base because it does not react with acids to give a salt and water only. For example, it reacts with hydrochloric acid to give chlorine as well as a salt and water:

$$PbO_2(s) + 4HCl(conc) \rightarrow PbCl_2(aq) + 2H_2O(l) + Cl_2(g)$$

Most bases are insoluble in water. However, a few bases are water soluble and these are known as alkalis. *Thus, alkalis are defined as water soluble bases.* The common alkalis are sodium hydroxide, potassium hydroxide, calcium hydroxide, and aqueous ammonia solution. Only dilute solutions of calcium hydroxide can be obtained because it is not very soluble in water.

The hydroxides of sodium, potassium, and calcium are strong bases because they are fully dissociated into ions. However, aqueous ammonia is a weak base because it only contains low concentrations of HO^- (and NH_4^+) ions.

PREPARATION OF BASES

1. Oxides may be prepared by the following methods.

    (a) By burning or heating metals in air or oxygen, e.g.
    $$2Mg(s) + O_2(g) \rightarrow 2MgO(s)$$

    (b) By thermal decomposition of hydroxides, carbonates, and nitrates other than those of sodium and potassium, e.g.
    $$Cu(OH)_2(s) \rightarrow CuO(s) + H_2O(g)$$
    $$ZnCO_3(s) \rightarrow ZnO(s) + CO_2(g)$$
    $$2Pb(NO_3)_2(s) \rightarrow 2PbO(s) + 4NO_2(g) + O_2(g)$$

2. Hydroxides are prepared as follows.

    (a) From the reaction between a metal or a metal oxide and water. This method is virtually limited to sodium, potassium, calcium, and their oxides, e.g.
    $$2Na(s) + 2H_2O(l) \rightarrow 2NaOH(aq) + H_2(g)$$
    $$CaO(s) + H_2O(l) \rightarrow Ca(OH)_2(aq)$$

    (b) By addition of an alkali to a solution of a salt, e.g.
    $$MgSO_4(aq) + 2NaOH(aq) \rightarrow Mg(OH)_2(s) + Na_2SO_4(aq)$$
    $$FeCl_3(aq) + 3NaOH(aq) \rightarrow Fe(OH)_3(s) + 3NaCl(aq)$$

GENERAL PROPERTIES OF ALKALIS

1. They have a bitter taste.
2. Solutions of alkalis turn red litmus paper blue.
3. The solutions feel soapy. Concentrated solutions of sodium and potassium hydroxide are highly corrosive and should on no account be brought into contact with the skin.
4. They neutralise acids to give a salt and water, e.g.
    $$NaOH(aq) + HCl(aq) \rightarrow NaCl(aq) + H_2O(l)$$
    $$2KOH(aq) + H_2SO_4(aq) \rightarrow K_2SO_4(aq) + 2H_2O(l)$$

    In all reactions of this type, the net ionic change is
    $$H^+ + HO^- \rightarrow H_2O$$
    and this process is known as *neutralisation*.

5. Alkalis liberate ammonia when warmed with ammonium salts, e.g.
    $$Ca(OH)_2(s) + (NH_4)_2SO_4(s) \rightarrow CaSO_4(s) + 2H_2O(l) + 2NH_3(g)$$

    These reactions take place because the hydroxide ions abstract H^+ ions from the ammonium ions:
    $$NH_4^+ + HO^- \rightarrow NH_3 + H_2O$$

6. Solutions of alkalis precipitate the hydroxides of other metals from solutions of their salts, e.g.

$$FeCl_3(aq) + 3NaOH(aq) \rightarrow Fe(OH)_3(s) + 3NaCl(aq)$$
$$MgSO_4(aq) + 2KOH(aq) \rightarrow Mg(OH)_2(s) + K_2SO_4(aq)$$

SALTS

Salts are the substances formed when some or all of the hydrogen ions of acids are replaced by a metal or ammonium ion. Some examples of acids and salts obtained from them are given below.

Acid	Examples of salts	General name of salts
H^+Cl^-	Na^+Cl^-, $NH_4^+Cl^-$, $Ca^{2+}(Cl^-)_2$	Chlorides
$H^+NO_3^-$	$K^+NO_3^-$, $NH_4^+NO_3^-$, $Pb^{2+}(NO_3^-)_2$	Nitrates
$(H^+)_2SO_4^{2-}$	$(Na^+)_2SO_4^{2-}$, $Cu^{2+}SO_4^{2-}$, $Fe^{2+}SO_4^{2-}$	Sulphates
$(H^+)_2SO_3^{2-}$	$(Na^+)_2SO_3^{2-}$, $Ca^{2+}SO_3^{2-}$	Sulphites
$(H^+)_2CO_3^{2-}$	$(NH_4^+)_2CO_3^{2-}$, $Pb^{2+}CO_3^{2-}$, $Ca^{2+}CO_3^{2-}$	Carbonates
$(H^+)_3PO_4^{3-}$	$(Na^+)_3PO_4^{3-}$, $(Ca^{2+})_3(PO_4^{3-})_2$	Phosphates

Two points should be noted.

(a) Brackets are only used in formula such as $Ca^{2+}(Cl^-)_2$ because the charge on the ion is being shown. This is done because $Ca^{2+}Cl_2^-$ could be erroneously interpreted as meaning that the two chlorines have only one negative charge between them.

(b) The number of charges on each ion is the same as its valency.

ACID SALTS

Consider the addition of a solution containing one mole of sulphuric acid to one containing a mole of sodium hydroxide. The equation for the reaction would be:

$$H_2SO_4(aq) + NaOH(aq) \rightarrow NaHSO_4(aq) + H_2O(l)$$

The resultant solution of sodium hydrogensulphate is found to be acidic because this salt is dissociated into Na^+, H^+, and SO_4^{2-} ions. Sodium hydrogensulphate is therefore known as an acid salt: it contains a hydrogen that can be replaced by a metal. The acid salt may be neutralised by adding a further mole of sodium hydroxide:

$$NaHSO_4(aq) + NaOH(aq) \rightarrow Na_2SO_4(aq) + H_2O(l)$$

The product, sodium sulphate, has no replaceable hydrogens and so it is known as a *normal salt*. Any di- or tribasic acid can form acid salts; further examples are given below.

Acid	Acid salt	Normal salt
H_2SO_3	Calcium hydrogensulphite, $Ca(HSO_3)_2$	Calcium sulphite, $CaSO_3$
H_2CO_3	Magnesium hydrogencarbonate, $Mg(HCO_3)_2$	Magnesium carbonate, $MgCO_3$
H_2S	Sodium hydrogensulphide, $NaHS$	Sodium sulphide, Na_2S

BASIC SALTS

Some bases, like acids, may be neutralised in stages and, if the neutralisation is incomplete, a basic salt is formed. For example, stepwise neutralisation of zinc hydroxide gives a basic and then a normal salt.

$$Zn(OH)_2 + HCl \rightarrow Zn(OH)Cl + H_2O$$
base   basic salt, zinc hydroxide chloride

$$Zn(OH)Cl + HCl \rightarrow ZnCl_2 + H_2O$$
                         normal salt, zinc chloride

PREPARATION OF SALTS

There are several ways of preparing salts and the method chosen is largely determined by the solubility of the salt concerned. Hence, some general knowledge of salt solubilities is necessary.

The following salts are soluble in water at room temperature.

1. All the common sodium, potassium, and ammonium salts.
2. All the common nitrates.
3. Most chlorides with the exception of silver, mercury(I), and lead(II) chlorides. Since lead(II) chloride is quite soluble in hot water but not in cold, it may be purified by crystallisation.
4. Most sulphates with the exception of lead(II), calcium, and barium sulphates.

The salts given below are insoluble in water.

1. Most carbonates except sodium, potassium, and ammonium carbonates.
2. The hydroxides except ammonium, sodium, and potassium hydroxides. Calcium hydroxide is sparingly soluble (1.13 g l^{-1} at 25 °C).
3. The sulphides except ammonium, sodium, and potassium sulphides.

Soluble salts are made by methods (a) to (c) below, the salt in each case being isolated by crystallisation. Insoluble salts are made by precipitation as outlined in method (d), whilst a few compounds are made by direct combination of the elements – method (e).

a. Action of an acid on a metal

The metal is added, a little at a time, to the acid until reaction ceases, i.e. until no more gas is evolved and excess metal is present. This ensures that all the acid is used up and so contamination of the salt is prevented. The excess metal is removed by filtration, the solution is concentrated by evaporation, and then allowed to cool. The resultant crystals are filtered and then dried between sheets of filter paper.

Zinc and magnesium sulphates are examples of salts made by this method:

$$Zn(s) + H_2SO_4(aq) \rightarrow ZnSO_4(aq) + H_2(g)$$
$$Mg(s) + H_2SO_4(aq) \rightarrow MgSO_4(aq) + H_2(g)$$

Both salts are obtained as the heptahydrate, i.e. $ZnSO_4 \cdot 7H_2O$ and $MgSO_4 \cdot 7H_2O$.

b. Action of an acid on a base

If a soluble base, i.e. an alkali, is being used, then a known volume of it is titrated with dilute acid from a burette. (The practical details involved are outlined on pages 78–79). An indicator is used to show when the base has been neutralised. In a separate experiment, the equivalent volumes of acid and base are mixed in the absence of the indicator. The resultant solution is concentrated and allowed to cool. The crystals are filtered and dried as above. Sodium, potassium, and ammonium salts are made by this method, e.g.

$$KOH(aq) + HNO_3(aq) \rightarrow KNO_3(aq) + H_2O(l)$$

In experiments involving insoluble bases, heat may be necessary to speed up the reaction. Excess of the base is added to the acid and, when reaction is complete, the mixture is filtered and the solution is concentrated, and allowed to cool, etc. as above. Salts such as lead(II) nitrate and copper(II) sulphate may be made by this method.

$$PbO(s) + 2HNO_3(aq) \rightarrow Pb(NO_3)_2(aq) + H_2O(l)$$
$$CuO(s) + H_2SO_4(aq) \rightarrow CuSO_4(aq) + H_2O(l)$$

The lead nitrate has no water of crystallisation but the copper sulphate is obtained as the pentahydrate, $CuSO_4 \cdot 5H_2O$.

c. Action of an acid on a carbonate

The equivalent volumes of an acid solution and a solution of a carbonate may be found by titration, using methyl orange as indicator. The same volumes are then mixed in the absence of the indicator and the resultant solution is concentrated, cooled, etc. as above. Ammonium, sodium, and potassium salts are made by this method, e.g.

$$K_2CO_3(aq) + H_2SO_4(aq) \rightarrow K_2SO_4(aq) + H_2O(l) + CO_2(g)$$

Most carbonates are insoluble in water, but they may be converted into other salts by adding excess of them to a dilute solution of an acid. When reaction is complete, i.e. when no more carbon dioxide is evolved, the excess carbonate is removed by filtration. The solution of the salt is concentrated and then allowed to cool. The resultant crystals are filtered and dried. Salts such as zinc chloride and copper(II) nitrate may be made by this method.

$$ZnCO_3(s) + 2HCl(aq) \rightarrow ZnCl_2(aq) + H_2O(l) + CO_2(g)$$
$$CuCO_3(s) + 2HNO_3(aq) \rightarrow Cu(NO_3)_2(aq) + H_2O(l) + CO_2(g)$$

d. Precipitation

This method entails mixing two solutions, the solutions being chosen so that one of the products is soluble and the other insoluble. The precipitate is filtered, washed, and dried. Many lead and silver salts are made by this method, e.g.

$$Pb(NO_3)_2(aq) + 2KI(aq) \rightarrow PbI_2(s) + 2KNO_3(aq)$$
$$AgNO_3(aq) + HCl(aq) \rightarrow AgCl(s) + HNO_3(aq)$$

Note that, in these reactions, one of the products remains in solution and is removed by the filtration process.

This type of reaction is often referred to as double decomposition. However, the term 'double decomposition' is somewhat misleading because, normally, both reactants are fully dissociated in solution and so no decomposition occurs. Reaction takes place because two of the ions in solution form an insoluble product. Thus, the reaction between lead(II) nitrate and potassium iodide may be represented as:

$$Pb^{2+}(aq) + 2NO_3^-(aq) + 2K^+(aq) + 2I^-(aq) \rightarrow Pb^{2+}(I^-)_2(s) + 2K^+(aq) + 2NO_3^-(aq)$$

Hence the effective change is:

$$Pb^{2+}(aq) + 2I^-(aq) \rightarrow Pb^{2+}(I^-)_2(s)$$

e. Direct combination of the elements

This method has limited application. However, anhydrous iron(III) chloride is prepared by heating iron in a current of chlorine:

$$2Fe(s) + 3Cl_2(g) \rightarrow 2FeCl_3(s)$$

Also, iron(II) sulphide is made by heating a mixture of iron and sulphur:

$$Fe(s) + S(s) \rightarrow FeS(s)$$

VOLUMETRIC ANALYSIS

The process of volumetric analysis involves running a measured volume of a solution from a burette into a known volume of a second solution until reaction is complete as shown by an indicator. The concentration of one of the solutions is accurately known and, from this and the reacting volumes, the concentration of the other solution is calculated.

The apparatus used in volumetric analysis is illustrated in Figure 6.1.

Figure 6.1. Apparatus for volumetric analysis

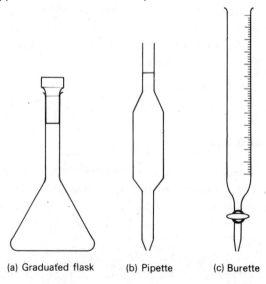

(a) Graduated flask   (b) Pipette   (c) Burette

Graduated flasks (Figure 6.1(a)) hold a stated volume, e.g. 100, 250, or 1000 cm^3, when filled, with solution at 20 °C, to the mark on the neck of the flask. The bottom of the meniscus should just touch the mark. The pipettes (Figure 6.1(b)) normally employed in simple volumetric analysis, hold 25 cm^3 when filled, to the mark on the stem, with solution at 20 °C. Burettes (Figure 6.1(c)) are calibrated in 0.1 cm^3 divisions; 50 cm^3 ones are generally used and they are calibrated between 0 and 50 cm^3, the 50 cm^3 mark being a few cm above the tap.

As stated above, the concentration of one of the solutions in volumetric analysis is accurately known. This solution is made up from, or standardised against, a substance known as a primary standard. *Primary standards* are substances which can be obtained in a pure state and which are stable to the atmosphere. Anhydrous sodium carbonate is an important primary standard, since any traces of moisture it picks up, or any hydrogencarbonate it forms from absorption of moisture and carbon dioxide, is readily removed by heating the sample in an oven at 105 °C for half an hour. The salt is allowed to cool in a desiccator (page 11). Sodium hydroxide is unsuitable as a primary standard because of its deliquescent nature and its irreversible reaction with carbon dioxide.

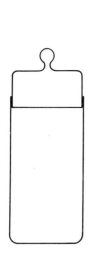

Figure 6.2.   A weighing bottle   Figure 6.3.   A wash bottle

Standard solutions are those in which the mass of solute in a given volume of solution is known. A standard solution of sodium carbonate is made by accurately weighing out a sample of the dried salt in a weighing bottle (Figure 6.2). Weighing bottles are often made of glass and they have a ground glass stopper. The known mass of sodium carbonate is carefully tipped into some distilled water contained in a beaker. Any of the salt remaining in the weighing bottle is washed out into the beaker by means of a plastic wash bottle (Figure 6.3). Squeezing the bottle forces out a jet of water. The sodium carbonate is stirred with the water until all the solid has dissolved. The solution is then transferred to a graduated flask with the aid of a filter funnel and wash bottle (Figure 6.4). The beaker, glass rod, and funnel are thoroughly washed with distilled water to make sure that all the carbonate solution ends up in the flask. Next, the funnel is removed and the solution is made up to the mark with distilled water. Finally, the stopper is fitted to the flask which is then shaken until the solution is homogeneous.

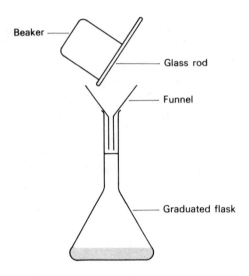

Figure 6.4. Transferring a solution to a graduated flask

The concentration of a solution of an acid may be found by titrating the acid with the standard sodium carbonate solution as outlined below.

A burette is washed out with distilled water and then with two small portions of the acid. The burette is then filled above the zero mark with the acid with the aid of a funnel. The tap is opened fully so that all the air is displaced from the jet and then acid is run out carefully until the bottom of the meniscus is just level with the zero mark. Note that in all the measurements with graduated flasks, pipettes, and burettes, the eyes must be level with the graduation mark in order to obtain accurate results.

The tip of a clean pipette is now placed well below the surface of the sodium carbonate solution and a little of the solution is sucked up. A finger is then placed over the top of the pipette, the pipette withdrawn, and the solution swirled round the pipette to wash it. The wash solution is discarded. Now, solution is sucked up above the mark of the pipette and a finger again placed in position. The pipette is withdrawn from the solution and the finger is raised slightly so that the solution can run out until the bottom of the meniscus is level with the mark. The finger is then pressed tightly to stop the flow and any drop of liquid adhering to the tip is removed by touching it on the side of the flask. The pipette is now held vertically over a clean conical flask and the solution allowed to run out completely unassisted, i.e. no blowing, etc. When the solution has run out, the pipette is allowed to drain for 15 seconds and then the tip is touched on the surface of the solution. Any solution still remaining in the pipette is meant to be there and should not be expelled.

Acids, Bases, and Salts

The conical flask is placed on a white tile under the burette (Figure 6.5) and two drops of methyl orange indicator are added. Acid is allowed to run, about 1 cm³ at a time, from the burette into the carbonate solution whilst the flask is swirled. The approximate volume of acid required to turn the indicator in the solution from yellow to orange is noted. If excess acid is added the indicator turns red. The conical flask is then thoroughly washed out with distilled water and the titration is repeated. This time, the acid is allowed to run in fairly rapidly to within 1 cm³ of the volume used in the previous rough titration. Acid is then added dropwise until the solution is neutral as shown by the indicator turning orange. Further titrations are done until two agree within 0.1 cm³. The average of the two accurate titrations is used to calculate the molarity (concentration in terms of moles per litre) of the acid. The calculation is illustrated by Example 1 on page 81.

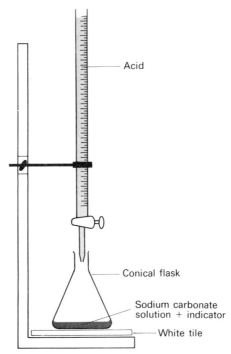

Figure 6.5. An acid-carbonate titration

The neutral or equivalence point in acid-alkali or acid-carbonate titrations is shown by an indicator. However, it should be stressed that the indicator has to be chosen to suit the particular conditions encountered in the reaction. The common indicators are methyl orange, screened methyl orange, litmus, and phenolphthalein. Their colour changes are as follows.

Indicator	Colour		
	Acidic solution	Neutral solution	Alkaline solution
Methyl orange	Red	Orange	Yellow
Screened methyl orange	Red	Blue-grey	Green
Litmus	Red	Purple	Blue
Phenolphthalein	Colourless	Colourless	Red

The indicators that can be used in the various types of titration are given below.

Titration	Indicator
Strong acid–strong alkali	Methyl orange, litmus, or phenolphthalein
Strong acid–weak alkali	Methyl orange
Weak acid–strong alkali	Phenolphthalein
Weak acid–weak alkali	No suitable indicator
Strong acid–carbonate	Methyl orange

CONCENTRATION OF SOLUTIONS IN TERMS OF MOLARITY

A molar solution is defined as one which contains one mole of solute in one litre of solution. Note that it is one litre of solution, not solvent, that is specified. Thus, since sulphuric acid has a relative molecular mass of 98, molar sulphuric acid contains 98 g of sulphuric acid in one litre of the solution. Similarly, molar sodium hydroxide solution contains 40 g of sodium hydroxide (NaOH = 40) per litre of solution. A 0.5 molar solution of sodium hydroxide will obviously contain 20 g of the solute per litre of solution.

Molarity is generally represented by the symbol, M, e.g. 2 M sulphuric acid, 0.5 M sodium hydroxide. The one is often omitted if the solution is molar.

The equation for the reaction between sulphuric acid and sodium hydroxide is

$$H_2SO_4 + 2NaOH \rightarrow Na_2SO_4 + 2H_2O$$

This shows that one mole of sulphuric acid reacts with two moles of sodium hydroxide and so it follows that

    1 litre of M sulphuric acid reacts with 2 litres of M sodium hydroxide

or

    1 litre of M sulphuric acid reacts with 1 litre of 2 M sodium hydroxide

or

    1 litre of M sulphuric acid reacts with 4 litres of 0.5 M sodium hydroxide

Consider now the general reaction

$$aA + bB \rightarrow \text{products}$$

in which compounds A and B react in the ratio of a moles of A to b moles of B. Suppose that in a titration V_A cm³ of solution A which has molarity M_A reacts with V_B cm³ of solution B whose molarity is M_B.

Since   1000 cm³ of solution A contain M_A moles of A

    V_A cm³ of solution A contain $\dfrac{M_A \times V_A}{1000}$ moles of A

Similarly, since 1000 cm³ of solution B contains M_B moles of B

    V_B cm³ of solution B contain $\dfrac{M_B \times V_B}{1000}$ moles of B

Therefore, the ratio of the moles of A and B reacting is

$$\frac{a}{b} = \frac{\frac{M_A \times V_A}{1000}}{\frac{M_B \times V_B}{1000}}$$

which on rearranging and cancelling gives:
$$aM_B V_B = bM_A V_A$$

Now, in any titration, only one of these variables is unknown and it may be calculated using this relationship. A number of calculations involving the molarity of solutions are illustrated below.

Example 1 2.832 g of anhydrous sodium carbonate were dissolved in water to give 250 cm^3 of solution. 25.0 cm^3 of this solution required 24.3 cm^3 of a sulphuric acid solution to neutralise it. Calculate the molarity of the acid solution. Na = 23, C = 12, O = 16.

250 cm^3 of sodium carbonate solution contains 2.832 g Na$_2$CO$_3$

∴ 1000 cm^3 of sodium carbonate solution contains $\frac{2.832 \times 1000}{250}$ = 11.328 g

106 gl^{-1} of sodium carbonate gives a M solution

∴ 11.328 gl^{-1} of sodium carbonate gives a $\frac{1 \times 11.328}{106}$ = 0.1069 M solution

Now, using the relationship
$$aM_B V_B = bM_A V_A$$

let A be sulphuric acid and B be sodium carbonate. The equation

$$H_2SO_4 + Na_2CO_3 \rightarrow Na_2SO_4 + H_2O + CO_2$$

shows that 1 mole of H$_2$SO$_4$ reacts with 1 mole of Na$_2$CO$_3$ and so

$$a = b = 1.$$

Therefore $\qquad 1 \times 0.1069 \times 25 = 1 \times M_A \times 24.3$

or $\qquad M_A = \frac{0.1069 \times 25}{24.3} = 0.1100$ M

Example 2 25.0 cm^3 of a sodium carbonate solution required 23.4 cm^3 of 0.5120 M hydrochloric acid solution to neutralise it. Calculate the molarity of the sodium carbonate solution and hence its concentration in gl^{-1}. Na = 23, C = 12, O = 16.

$$Na_2CO_3 + 2HCl \rightarrow 2NaCl + H_2O + CO_2$$

Using the equation
$aM_B V_B = bM_A V_A$, let A = Na$_2$CO$_3$ and B = HCl

$1 \times 0.5120 \times 23.4 = 2 \times M_A \times 25.0$

∴ $\qquad M_A = \frac{0.5120 \times 23.4}{2 \times 25.0} = 0.2396$ M

Now M Na$_2$CO$_3$ contains 106 gl^{-1}

∴ 0.2396 M Na$_2$CO$_3$ contains $106 \times 0.2396 = 25.40$ gl^{-1}

Example 3 25.0 cm^3 of a solution containing 0.196 g of a metal hydroxide, XOH, were just neutralised by 35.0 cm^3 of 0.1 M hydrochloric acid. Calculate (a) the relative molecular mass of the metal hydroxide, (b) the relative atomic mass of X. O = 16, H = 1.

(a) The equation for the reaction is

$$X\text{OH} + \text{HCl} \rightarrow X\text{Cl} + \text{H}_2\text{O}$$

Using the relationship $aM_B V_B = bM_A V_A$, let A = HCl and $B = X$OH then

$$1 \times M_B \times 25.0 = 1 \times 0.1 \times 35.0$$

$$M_B = \frac{0.1 \times 35.0}{25.0} = 0.140$$

Now, since 25.0 cm^3 of XOH solution contains 0.196 g XOH

$$1000 \text{ cm}^3 \text{ of } X\text{OH solution contains } \frac{0.196 \times 1000}{25.0} = 7.84 \text{ g } X\text{OH}$$

Hence   0.140 M XOH solution contains 7.84 g l^{-1}

∴ 1 M XOH solution contains $\frac{7.84 \times 1}{0.140}$ = 56 g XOH l^{-1}

The relative molecular mass of XOH is therefore 56.

(b) Since XOH has a relative molecular mass of 56 and O = 16 and H = 1, the relative atomic mass of X = 56−(16+1)
$$= 39$$

Example 4 The equation for the reaction between iron(III) hydroxide and hydrochloric acid is:

$$\text{Fe(OH)}_3 + 3\text{HCl} \rightarrow \text{FeCl}_3 + 3\text{H}_2\text{O}$$

Calculate the volume of molar hydrochloric acid which will just dissolve 5.35 g of iron(III) hydroxide. Fe = 56, O = 16, H = 1, Cl = 35.5.
From the equation and the relative molecular masses:

$$\text{Fe(OH)}_3 + 3\text{HCl} \rightarrow \text{FeCl}_3 + 3\text{H}_2\text{O}$$
$$107 \quad\quad 109.5 \quad\quad 162.5 \quad\quad 54$$

it is seen that 107 g of Fe(OH)$_3$ react with 109.5 g of HCl. However, the question is concerned with the volume of acid used. Now 109.5 g (3 moles) of hydrochloric acid will make 1 litre of 3 M solution or 3 litres of M solution, i.e.

107 g Fe(OH)$_3$ are dissolved by 3000 cm^3 of M HCl

∴ 5.35 g Fe(OH)$_3$ are dissolved by $\frac{3000 \times 5.35}{107}$ = 150 cm^3 M HCl

Example 5 An excess of copper(II) carbonate was added to 25.0 cm^3 of 2M nitric acid. Calculate (a) the mass of copper(II) carbonate that would react and (b) the volume of carbon dioxide, at s.t.p., that would be evolved. Cu = 63.5, C = 12, O = 16, H = 1, N = 14, and molar volume = 22.4 litres at s.t.p.

(a) $\quad CuCO_3 + 2HNO_3 \rightarrow Cu(NO_3)_2 + H_2O + CO_2$
$\qquad$ 123.5 $\quad$ 1 litre of 2M $\qquad\qquad\qquad$ 22.4 litre at s.t.p.

$\qquad$ 1000 cm^3 of 2M HNO_3 dissolve 123.5 g of $CuCO_3$

$\therefore\;$ 25.0 cm^3 of 2M HNO_3 dissolve $\dfrac{123.5 \times 25.0}{1000}$ = 3.087 g $CuCO_3$

(b) $\quad$ 1000 cm^3 of 2M HNO_3 react with $CuCO_3$ to give 22 400 cm^3 CO_2 at s.t.p.

$\therefore\;$ 25.0 cm^3 of 2M HNO_3 react with $CuCO_3$ to give $\dfrac{22\,400 \times 25.0}{1000}$

$\qquad\qquad\qquad\qquad\qquad\qquad\qquad\qquad\qquad$ = 560 cm^3 CO_2 at s.t.p.

THE pH SCALE OF ACIDITY AND ALKALINITY

Tap water is a weak conductor of electricity due mainly to dissolved salts and carbon dioxide. However, water still conducts an electric current to a small extent even after the most careful purification. The conductivity of pure water is, in fact, a consequence of slight dissociation:

$$H_2O \rightleftharpoons H^+ + HO^-$$

It is found that pure water, at 25 °C, has a hydrogen ion, and hydroxide ion, concentration of 10^{-7} mole litre^{-1}. The degree of dissociation increases as the temperature is raised because bonds are weakened if they are given more energy.

In neutral solutions, the hydrogen ion and hydroxide ion concentrations are the same. However, acidic solutions contain more H^+ than HO^- ions. Thus, molar solutions of strong monobasic acids, for example, hydrochloric and nitric acids, contain 1 mole of H^+ and 10^{-14} moles HO^- ions per litre. On the other hand, alkaline solutions contain more HO^- than H^+: a molar solution of sodium hydroxide contains 1 mole of HO^- and 10^{-14} mole of H^+ per litre.

Now, comparing the acidity or alkalinity of solutions by their hydrogen ion concentrations is not very convenient and so the pH scale is used where pH = $-\log H^+$ concentration. This gives a scale where:

$\qquad$ molar solutions of strong monobasic acids have pH = 0,
$\qquad$ neutral solutions have pH = 7,
and $\quad$ molar solutions of strong monoacidic bases have pH = 14.

Thus, as a solution becomes progressively more acidic its pH falls from 7 to 0 whilst as it becomes more alkaline the pH rises from 7 to 14. The pH scale is normally used over the range 0 to 14, although more concentrated solutions of strong acids and bases can give pH values outside this range.

It should be noted that the pH of a molar solution of ethanoic acid is between 2 and 3 because dissociation is incomplete. Similarly, molar aqueous ammonia solution has a pH of between 11 and 12.

The approximate pH of a solution may be determined by adding a few drops of Universal Indicator to it. Universal Indicator is a mixture of indicators which appears pale green in neutral solutions. As solutions become more acidic, the indicator changes through yellow to orange to pink to red. Increasing alkalinity is shown by the indicator changing through dark green to blue to violet. A colour chart is supplied for comparison purposes, the chart showing the colours obtained with a series of solutions with known pH (Figure 6.6).

pH	2	4	6	7	8	10	12
Colour	Deep pink	Orange	Yellow	Pale Green	Green	Turquoise	Violet

Figure 6.6. Variation of colour of Universal Indicator with pH

Questions

1. 25 cm^3 of a solution containing 6.0 g of sodium hydroxide per litre, exactly neutralise 30 cm^3 of a solution of nitric acid.
    (a) Calculate the molarity (normality) of the sodium hydroxide solution.
    (b) Using the above results, calculate the molarity (normality) of the nitric acid and its concentration in grams per litre.
    (c) If you were supplied with 200 cm^3 of each of the solutions describe carefully how, by titration, you would arrive at the results given in the first statement.
    (d) Having done the titration, describe how you would obtain a pure, dry crystalline sample of sodium nitrate.
    [J.M.B.]

2. (a) It is required to determine the concentration (in terms of either molarity or normality) of a sodium hydroxide solution by titration against 0.115 M (0.115 N) hydrochloric acid. A suitable indicator is provided, but it is one whose colour change you do not know. Describe carefully how you would investigate the colour change of the indicator, perform the titration and use the results to determine the concentration of the alkali solution.
    (a) Suppose that you found that 25.00 cm^3 of alkali were neutralised by 20.50 cm^3 of the hydrochloric acid. Calculate the concentration (in terms of either molarity or normality) of the alkali solution.
    [J.M.B.]

3. (a) In a reaction between a solution of a metallic hydroxide (of formula MOH) and dilute hydrochloric acid, 20.0 cm^3 of the alkali reacted with 25.0 cm^3 of the acid. The acid concentration was 4.00 g per litre and that of the alkali was 7.67 g per litre. Calculate the relative molecular mass of the alkali and hence the relative atomic mass of the metal.
    (b) Describe in detail how you would determine the volume of acid and alkali which exactly neutralise each other.
    [J.M.B.]

4. (a) Describe in detail how you would determine experimentally the volume of a solution of sodium hydroxide which would exactly neutralise 25 cm^3 of a solution of sulphuric acid containing 4.9 g of pure acid, in 1 litre of solution.
    (b) Calculate the molarity (or normality) of the sodium hydroxide solution if 30.0 cm^3 were required for neutralisation.
    (c) Using 25 cm^3 of the sulphuric acid, how would you obtain a solution of sodium hydrogensulphate? Write an equation for the reaction.
    (d) How would you distinguish between solutions of sodium sulphate and sodium hydrogensulphate? Describe the test and state what would happen in each case.
    (e) What would you see on adding a solution of barium nitrate to sodium sulphate solution?
    [J.M.B.]

Acids, Bases, and Salts

5 (a) Which ion causes the acidic character shown by aqueous solutions of hydrochloric acid and ethanoic acid, $CH_3 \cdot COOH$? Explain briefly why one of these solutions behaves as a 'strong' acid, the other as a 'weak' acid.

   (b) You are provided with solutions of hydrochloric acid and sodium hydroxide. Describe how you would determine, as accurately as possible, the volume of hydrochloric acid required to react exactly with one litre of sodium hydroxide solution.

   (c) A water-soluble compound $MHCO_3$ reacts with hydrochloric acid according to the equation

   $$MHCO_3 + HCl \rightarrow MCl + CO_2 + H_2O$$

   It was found that 25.0 cm³ of a solution containing 21.6 grams per litre of $MHCO_3$ reacted with 27.0 cm³ of a solution of hydrochloric acid containing 7.30 grams per litre. Calculate the following and justify your working:
   (i) the molarity of the hydrochloric acid,
   (ii) the molarity of the solution containing $MHCO_3$,
   (iii) the mass of one mole of $MHCO_3$,
   (iv) the relative atomic mass of M.

   [A.E.B. June 1974]

7 Electrolysis

INTRODUCTION

Many substances may be decomposed by energy in the form of heat. For example, calcium carbonate decomposes in the region of 800 °C into calcium oxide and carbon dioxide:

$$CaCO_3(s) \rightarrow CaO(s) + CO_2(g)$$

whilst potassium nitrate is thermally decomposed into potassium nitrite and oxygen at about 400 °C:

$$2KNO_3(s) \rightarrow 2KNO_2(s) + O_2(g)$$

It should come as no great surprise, therefore, that chemical changes can also be brought about in some cases by energy in the form of electricity.

A convenient way of determining whether a substance will conduct an electric current is to use the apparatus illustrated in Figure 7.1.

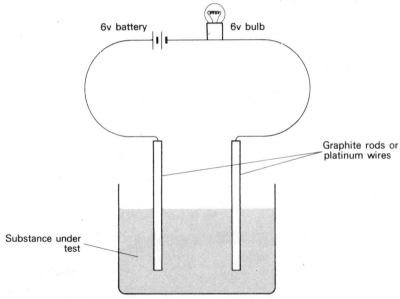

Figure 7.1. Apparatus to determine whether a substance will conduct an electric current

If the substance under test completes the circuit, i.e. conducts an electric current, the bulb will glow. It is found that all metals conduct an electric current whilst non-metals, with the exception of graphite, do not. Dry salts, e.g. sodium chloride, and covalent compounds, e.g. sugar, ethanol, and tetrachloromethane do not conduct. An electric current is passed by fused (molten) salts and by aqueous solutions of salts, acids, and alkalis, and changes are apparent at the graphite rods or the platinum wires whichever are used.

Before an attempt is made to explain the observations above, it is necessary to be familiar with the definition of a number of terms encountered in this work.

An *electrolyte* is a solution or fused (molten) compound which will conduct an electric current and is decomposed by it. Note that metals are not electrolytes.

A *non-electrolyte* is a compound which cannot be decomposed by an electric current.

Electrolysis is the decomposition of a compound in solution or in the molten state by passing an electric current through it.

An *electrode* is the metal or graphite rod by which the current enters or leaves an electrolyte. The *cathode* is the negatively charged electrode whilst the *anode* is the positively charged electrode.

NATURE OF ELECTRIC CURRENTS

An electric current is a flow of electrons, the electrons moving from a position of high to one of low electron density. By convention, the electric current is regarded as flowing in the opposite direction to the electrons, i.e. from positive to negative.

There are two types of electric current: alternating current, a.c., and direct current, d.c. The electrical supply to houses is alternating current, so called because it changes direction one hundred times a second. Direct current flows in one direction only and it is this current which must be used in electrolysis experiments.

FLOW OF ELECTRICITY IN METALS

The reason why metals can conduct an electric current has been discussed previously (page 34). The only detectable change which takes place when a metal conducts electricity is that the metal becomes hot; practical use of this is made in electric light bulbs.

FLOW OF ELECTRICITY IN ELECTROLYTES

When an electric current is passed through an electrolyte, electrons flow from the negative pole of the battery or other source to the cathode. The positive ions in the electrolyte are attracted to the cathode where they accept electrons and are discharged, whereas the negative ions migrate to the anode where they give up electrons and are liberated. The electrons produced at the anode flow to the positive pole of the battery.

A visual demonstration of the migration of ions to the electrodes may be obtained with the apparatus shown in Figure 7.2.

A crystal of a coloured salt such as potassium manganate(VII) is placed in the centre of the moist oblong strip of filter paper and a d.c. current of at least 50 volts is switched on. The purple colour, due to the MnO_4^- ion, is seen to move towards the anode. No colour is apparent at the cathode because K^+ ions are colourless. If a crystal of copper(II) sulphate is used, the blue colour of the hydrated Cu^{2+} ion is seen to move towards the cathode. The SO_4^{2-} ion is colourless and so no colour is observed moving to the anode.

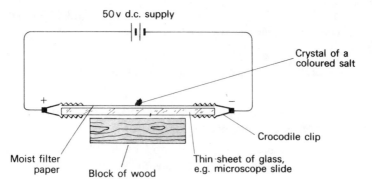

Figure 7.2. Visual demonstration of ion migration

The movement of ions and electrons during electrolysis is summarised by Figure 7.3.

Dry salts cannot conduct an electric current because the ions are not free to move to the electrodes – they are firmly held in position in the crystal lattice.

A few important specific cases of electrolysis will now be considered. It should be noted that, in every case, ions gain electrons at the cathode, i.e. reduction occurs (see page 98). Conversely, negative ions lose electrons at the anode and so oxidation occurs.

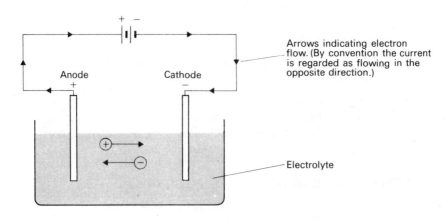

Figure 7.3. Migration of ions during electrolysis

ELECTROLYSIS OF FUSED (i.e. MOLTEN) SODIUM CHLORIDE

Sodium is manufactured by the electrolysis of fused sodium chloride using a steel cathode and graphite anode. Sodium ions migrate to the cathode where they accept electrons and are discharged as sodium atoms. The chlorine ions move to the anode where they lose electrons and chlorine is liberated.

$$\text{Cathode} \quad Na^+ + e^- \rightarrow Na$$
$$\text{Anode} \quad Cl^- - e^- \rightarrow Cl; \quad 2Cl \rightarrow Cl_2$$

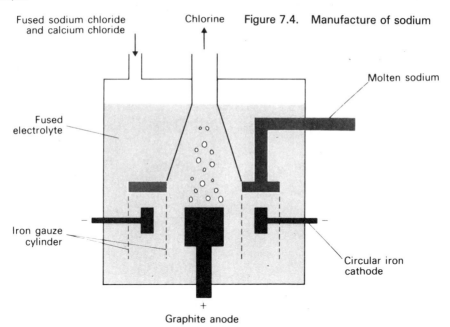

Figure 7.4. Manufacture of sodium

Precautions have to be taken to keep the liberated chlorine and sodium apart so that they do not recombine; the cell is illustrated in Figure 7.4. The calcium chloride is added to lower the operating temperature from 800 °C to 600 °C.

ELECTROLYSIS OF SATURATED SODIUM CHLORIDE SOLUTION (BRINE)

The solution will contain Na^+ and Cl^- ions from the sodium chloride and H^+ and HO^- ions from slight dissociation of the water (see page 83). On electrolysis, the cations, i.e. the Na^+ and H^+ ions, move to the cathode. However, hydrogen ions accept electrons more readily than sodium ions (see the electrochemical series, pages 103 to 105) and so it is hydrogen that is liberated. This using up of the hydrogen ions results in more water dissociating to restore the equilibrium:

$$H_2O \rightleftharpoons H^+ + HO^-$$

The anions, i.e. the HO^- and Cl^-, migrate to the anode. Now, in solutions where their concentrations are similar, HO^- ions will lose electrons more readily than Cl^- ions. However, in brine, the concentration of Cl^- is very high compared to that of HO^- ions and this proves to be the decisive factor. Chlorine gas is therefore evolved at the anode.

$$\text{Cathode} \quad H^+ + e^- \rightarrow H; \quad 2H \rightarrow H_2$$
$$\text{Anode} \quad Cl^- - e^- \rightarrow Cl; \quad 2Cl \rightarrow Cl_2$$

Eventually, all the hydrogen and chlorine ions will be liberated and only sodium ions and hydroxide ions will remain. Sodium hydroxide is in fact manufactured by this method; the hydrogen and chlorine are useful by-products from the process.

ELECTROLYSIS OF COPPER(II) SULPHATE SOLUTION USING PLATINUM ELECTRODES

The solution will contain Cu^{2+} and SO_4^{2-} ions from the copper(II) sulphate and H^+ and HO^- ions from the water. The Cu^{2+} and H^+ ions migrate to the cathode but copper ions accept electrons more readily than hydrogen ions (see the electrochemical series, pages 103 to 105) and so copper is deposited on the cathode.

$$\text{Cathode} \quad Cu^{2+} + 2e^- \rightarrow Cu$$

The SO_4^{2-} and HO^- ions travel to the anode but the HO^- ions lose electrons the easiest and so they are discharged.

$$\text{Anode} \quad HO^- - e^- \rightarrow HO \quad \text{hydroxyl radical}$$
$$2HO \rightarrow H_2O + O$$
$$2O \rightarrow O_2$$

i.e.
$$4HO^- - 4e^- \rightarrow 2H_2O + O_2$$

The changes taking place at the anode may be clearer if the outer electrons of the various species are shown:

$$H-\ddot{O}\!:^- - e^- \rightarrow H-\ddot{O}\cdot$$
$$H-\ddot{O}\cdot + H-\ddot{O}\cdot \rightarrow H-\ddot{O}-H + \cdot\ddot{O}\cdot$$
$$2\cdot\ddot{O}\cdot \rightarrow \ddot{O}=\ddot{O}$$

ELECTROLYSIS OF COPPER(II) SULPHATE SOLUTION USING COPPER ELECTRODES

As in the previous example, the ions present are Cu^{2+}, SO_4^{2-}, H^+, and HO^- and copper ions are discharged at the cathode.

$$\text{Cathode} \quad Cu^{2+} + 2e^- \rightarrow Cu$$

The SO_4^{2-} and HO^- ions migrate to the anode where there are now three possibilities: either the SO_4^{2-} or HO^- ions could be discharged or copper atoms in the anode could lose electrons and go into solution as copper ions. In fact, the last process occurs.

$$\text{Anode} \quad Cu - 2e^- \rightarrow Cu^{2+}$$

The net result of the electrolysis, therefore, is the loss of copper from the anode and deposition of copper at the cathode. Copper is purified industrially in this manner. Thus, a large piece of crude copper is used as the anode and a small rod of pure copper as the cathode.

Metallic impurities which are more reactive than copper, e.g. iron, are not discharged, they merely dissolve due to reactions of the type:

$$Fe(s) + Cu^{2+}(aq) \rightarrow Fe^{2+}(aq) + Cu(s) \qquad \text{(see page 105)}$$

Less reactive impurities, such as silver and gold, sink to the bottom of the cell as a sludge and are recovered later.

ELECTROLYSIS OF SODIUM HYDROXIDE SOLUTION WITH PLATINUM ELECTRODES

The sodium hydroxide solution contains Na^+, HO^-, and H^+ ions. The Na^+ and H^+ ions both migrate to the cathode but hydrogen is liberated since it is lower in the electrochemical series. The hydroxide ions are discharged at the anode.

$$\text{Cathode} \quad H^+ + e^- \rightarrow H; \quad 2H \rightarrow H_2$$
$$\text{Anode} \quad 4HO^- - 4e^- \rightarrow 2H_2O + O_2$$

Note that H^+ and HO^- ions are discharged in the same proportion and so two moles of hydrogen will be liberated to every one of oxygen.

ELECTROLYSIS OF DILUTE SULPHURIC ACID USING PLATINUM ELECTRODES

Sulphuric acid solution contains H^+, SO_4^{2-}, and HO^- ions. The H^+ and HO^- ions are discharged at the cathode and anode respectively.

$$\text{Cathode} \quad H^+ + e^- \rightarrow H; \quad 2H \rightarrow H_2$$
$$\text{Anode} \quad 4HO^- - 4e^- \rightarrow 2H_2O + O_2$$

The electrolysis of dilute sulphuric acid in Hofmann's voltameter is described on page 49.

APPLICATIONS OF ELECTROLYSIS

Electrolysis has a number of important industrial applications as outlined below.

1. Manufacture of chemicals

Sodium hydroxide is manufactured by the electrolysis of brine (page 89), hydrogen and chlorine being important by-products.

2. Extraction and purification of metals

The very electropositive metals such as sodium, magnesium, and calcium have to be extracted by electrolysis of their fused chlorides since their oxides cannot be reduced by carbon. In some cases the oxide is reduced but a carbide is then formed. Aluminium is obtained by electrolysis of the fused oxide because the chloride is covalent. Copper is purified as described on page 90.

3. Electroplating

Metals may be electroplated by making them the cathode of a cell. Car bumpers and hub caps are coated with chromium in this way. Another familiar example is the silver plating of cutlery, etc.

4. Anodising

Aluminium is fairly resistant to corrosion because it readily forms a thin protective layer of the oxide on exposure to the atmosphere. A stronger and thicker layer of the oxide may be formed by a process known as anodising. In this process, the aluminium article is made the anode in the electrolysis of dilute sulphuric acid – the liberated oxygen oxidises the aluminium.

FARADAY'S LAWS OF ELECTROLYSIS

There are two laws of electrolysis, both being due to Faraday in the early nineteenth century.

1. The first law states that: *the mass of a substance liberated during electrolysis is proportional to the quantity of electricity passed*. Now, quantity of electricity is measured in coulombs where a coulomb is the passage of a current of one amp for one second. The law may be represented mathematically as:

    $m \propto I \times t$ where m = mass of substance liberated
    I = current in amps
    t = time in seconds

    or $m = e \times I \times t$

    where e is a proportionality constant known as the electrochemical equivalent of the substance. *The electrochemical equivalent is the mass of a substance liberated by the passage of one coulomb of electricity*.

    The law may be explained as follows. During electrolysis, the number of electrons being used up at the cathode by the positive ions must equal the number of electrons released at the anode by the negative ions. Therefore if the current, i.e. the number of electrons flowing in a given time, is increased, the number of ions discharged must be increased proportionally. The same applies if the current is passed for a longer time. The mass of a substance liberated is obviously directly proportional to the number of ions discharged.

2. The second law may be expressed as: *the quantities of electricity required to liberate one mole of different elements are in simple whole number ratios*.

    It is found experimentally that 96 500 coulombs are required to discharge one mole of monovalent ions such as H^+, Na^+, Ag^+, Cl^-, and HO^-. This quantity of electricity is often known as the *Faraday constant and it is defined as the quantity of electricity required to liberate one mole of silver during electrolysis*. A Faraday represents one mole of electrons since this is the quantity of electrons required to convert one mole of Ag^+ ions into one mole of silver.

    The truth of Faraday's second law will be seen from the fact that:

1 Faraday is required to liberate one mole of H^+, Na^+, or Ag^+ ions,
2 Faradays are required to liberate one mole of Cu^{2+}, Pb^{2+}, or Mg^{2+} ions,
3 Faradays are required to liberate one mole of Al^{3+} ions.

Electrolysis

These figures may be obtained using the apparatus shown in Figure 7.5. Two or more voltameters are used, a voltameter being the apparatus in which an electrolysis is performed. The voltameters are connected in series so that the same current passes through each. The copper voltameter consists of copper electrodes in copper(II) sulphate solution whilst the silver voltameter has platinum electrodes in silver nitrate solution. The two cathodes are cleaned, dried, and weighed. A suitable current is then passed for an accurately measured time. Copper is deposited on the cathode of the copper voltameter and silver on the cathode of the silver voltameter. After the passage of the current, the cathodes are carefully washed, dried, and reweighed. The increase in mass represents the mass of copper or silver deposited. The results may be used to calculate the quantity of electricity required to deposit one mole of each element as shown in Example 1 below.

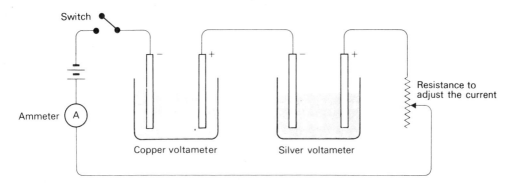

Figure 7.5. Apparatus for confirmation of Faraday's second law

The following examples illustrate calculations involving Faraday's laws.

Example 1 A current of 0.1 amp, flowing for 45 minutes through a silver and a copper voltameter connected in series, was found to deposit 0.302 g of silver and 0.089 g of copper on the respective cathodes. Calculate the quantity of electricity which would be required to deposit one mole of each of the elements. $Ag = 107.9$, $Cu = 63.5$.

The quantity of electricity passed = amps × seconds
$$= 0.1 \times 45 \times 60$$
$$= 270 \text{ coulombs}$$

0.302 g of silver are deposited by 270 coulombs

∴ 107.9 g of silver are deposited by $\dfrac{270 \times 107.9}{0.302}$ coulombs

$$= 96\,467 \text{ coulombs (1 Faraday)}$$

0.089 g of copper are deposited by 270 coulombs

∴ 63.5 g of copper are deposited by $\dfrac{270 \times 63.5}{0.089}$ coulombs

$$= 192\,640 \text{ coulombs (2 Faradays)}$$

It is seen that the quantities of electricity required to deposit one mole of each element are in a simple ratio as predicted by Faraday's second law.

Example 2 0.03 Faraday of electricity were passed through sodium hydroxide solution using platinum electrodes.

(a) Represent the reactions taking place at the electrodes by ionic equations.

(b) Calculate the number of moles of each gas produced and also the volume each gas would occupy at s.t.p.

(c) Calculate the time required to complete the passage of 0.03 Faraday if a current of 2.5 amps were passed through the solution.

Molar volume = 22.4 litres at s.t.p., 1 Faraday = 96 500 coulombs.

(a) At the cathode $H^+ + e^- \rightarrow H$; $2H \rightarrow H_2$
    At the anode $4HO^- - 4e^- \rightarrow 2H_2O + O_2$

(b) 1 Faraday liberates 1 mole of H^+ ions, i.e. 0.5 mole of H_2 molecules
    ∴ 0.03 Faraday liberates $0.5 \times 0.03 = 0.015$ mole of H_2
    0.015 mole of H_2 occupies $0.015 \times 22.4 = 0.336$ litres at s.t.p.

1 Faraday liberates 0.5 mole of O^{2-} ions, i.e. 0.25 mole of O_2 molecules
∴ 0.03 Faraday liberates $0.25 \times 0.03 = 0.0075$ mole of O_2
0.0075 mole of O_2 occupies $0.0075 \times 22.4 = 0.168$ litres at s.t.p.

(c) 1 coulomb = 1 amp for 1 second
    96 500 coulombs = 1 amp for 96 500 seconds
    = 2.5 amps for $\frac{96\,500}{2.5}$ seconds

If a current of 2.5 amps were passed, 1 Faraday would take $\frac{96\,500}{2.5}$ seconds

∴ 0.03 Faraday would take $\frac{96\,500}{2.5} \times 0.03 = 1158$ seconds

Questions

The following circuit was set up.

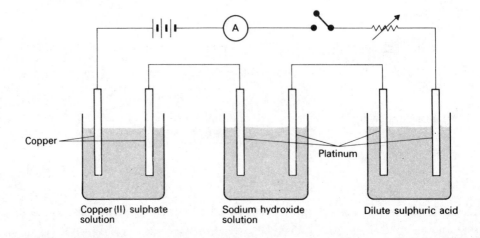

Electrolysis 95

The switch was closed and a current of 0.5 A was allowed to flow for 16 min 5 sec.

(a) Describe how the electricity is conducted in:
 (i) the connecting wires,
 (ii) the copper(II) sulphate solution.

(b) Name the products formed during the electrolysis of sodium hydroxide solution at:
 (i) the cathode,
 (ii) the anode.

(c) State whether the masses of
 (i) the copper cathode, and,
 (ii) the copper anode will increase, decrease or remain constant during the electrolysis.

(d) Write equations for the reactions taking place in dilute sulphuric acid at:
 (i) the cathode, and
 (ii) the anode.

(e) Calculate the volume of hydrogen which would be released from the dilute sulphuric acid during the electrolysis.
 (1 g of hydrogen at room temperature and pressure occupies 12 litres. The Faraday constant = 96500 coulombs (mole of electrons)$^{-1}$. Relative atomic mass: H = 1)

(f) What would happen to the concentration of each solution during the passage of the current?

[J.M.B.]

2   0.02 Faradays of electricity were passed through a solution of sodium hydroxide using platinum electrodes.

(a) Give the names of the gases evolved, and the names or signs of the electrodes at which they were produced.

(b) Draw a labelled diagram of a suitable apparatus for this electrolysis and for the collection of the products.

(c) Represent the reactions taking place at the electrodes by ionic equations.

(d) Calculate the number of moles of each gas produced and also the volume which each gas would occupy at s.t.p.

(e) Calculate the time required to complete the passage of 0.02 Faradays if a current of 2 amps were passed through the solution.

(f) Write an equation to represent the reaction which would take place if the volumes of gases mentioned in (d) were mixed and ignited. State the number of moles of the product which would be formed.
 (The molar volume of a gas at s.t.p. is 22.4 litres. The Faraday constant = 96500 coulombs.)

[J.M.B.]

3 (a) By reference to the structures of the substances mentioned, briefly explain the following:
 (i) solid sodium conducts electricity,
 (ii) solid sodium chloride is a non-conductor of electricity but it will conduct when molten.

(b) Scandium (Sc) is a trivalent metal of relative atomic mass 45. The same quantity of electricity was passed through two cells, liberating scandium in the first cell and 0.3 mole of sodium in the second. Calculate the mass of scandium liberated.

(c) Describe how crude copper is purified by an electrolytic method. Indicate the reaction that takes place at each electrode and explain briefly what happens to other metals present in the crude copper.

[A.E.B. June 1976]

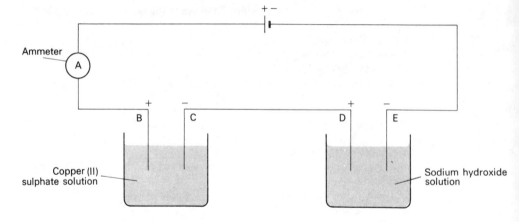

4 A current of 0.5 amp is passed in turn through solutions of:
(i) copper(II) sulphate by means of copper electrodes B and C,
(ii) sodium hydroxide by means of platinum electrodes D and E, as illustrated in the above diagram.

(a) Write ionic equations for the reactions occurring at the electrodes B, C, D, and E and name the products formed at electrodes D and E. (Make sure that you state clearly in your answer the letter of each electrode.)

(b) What other measurements, in addition to the current, would you need to take if you were asked to find the mass (w) of copper deposited by 0.1 Faraday? State briefly how you would use all your results to calculate 'w'. (The Faraday constant = 96 500 coulombs.)

(c) If 0.1 Faraday were passed through copper(II) sulphate solution, how many moles of copper would be deposited?

(d) Given solutions of copper(II) sulphate and sodium hydroxide, describe briefly how you would obtain a sample of dry copper(II) oxide.

[J.M.B.]

5 (a) State what particles (atoms, molecules, or ions) are present in the crystal lattices of:
(i) sugar,
(ii) sodium chloride,
(iii) copper.

(b) Explain the following in terms of the structures of the substances involved.
(i) Copper will conduct an electric current but sugar will not.
(ii) Solid sodium chloride is a non-conductor but the molten salt conducts and is decomposed by the electric current.

(c) Suppose you are asked to perform an experiment to determine the quantity of electricity necessary to deposit one mole of zinc (relative atomic mass 65.4) by electrolysis of zinc sulphate solution.
(i) Sketch the circuit you would use.
(ii) State the measurements you would make.
(iii) Outline how you would use your results to calculate the required answer.

(d) Explain why the quantity of electricity needed to deposit one mole of zinc, from zinc sulphate solution, is twice that required to deposit one mole of sodium from molten sodium chloride.

6 (a) Sodium is manufactured from sodium chloride. For this process:
(i) state whether the electrolyte is liquid, molten or in solution,
(ii) name the materials of which the electrodes are made,
(iii) write an equation for the reaction in which the sodium is formed.

(b) How is the process in (a) modified to obtain sodium hydroxide as the product? Explain how the sodium hydroxide is formed.

Electrolysis

(c) Name the other product common to both processes and outline the reactions of this substance with:
(i) water,
(ii) hydrogen. In each case, give the conditions and either an equation or the name(s) of the product(s).

[J.M.B.]

7 (a) Copper(II) sulphate solution is electrolysed using copper electrodes.
(i) What ions are present in the solution prior to electrolysis?
(ii) State and explain the changes which occur at the electrodes.

(b) Experiments show that 1.08 g of silver is liberated by the passage of 965 coulombs through silver nitrate solution. If the same quantity of electricity is passed through copper(II) sulphate solution, 0.32 g of copper is deposited. Explain these results.
(Relative atomic masses: Cu = 64, Ag = 108; the Faraday constant = 96 500 coulombs.)

8 Oxidation and Reduction

Originally, oxidation was regarded as any process in which oxygen was added to a substance or hydrogen was removed from a substance. Thus, the following reactions involved the oxidation of magnesium and hydrogen sulphide respectively.

$$2Mg + O_2 \rightarrow 2MgO$$
$$H_2S + Cl_2 \rightarrow S + 2HCl$$

Conversely, reduction was defined as the addition of hydrogen to a substance or the removal of oxygen from a substance, e.g.

$$CH_2{=}CH_2 + H_2 \rightarrow CH_3 \cdot CH_3 \quad \text{(ethene reduced)}$$
$$S + H_2 \rightarrow H_2S \quad \text{(sulphur reduced)}$$
$$PbO + H_2 \rightarrow Pb + H_2O \quad \text{(lead(II) oxide reduced)}$$

However, the above definitions proved very limiting. For example, heating copper in oxygen:

$$2Cu + O_2 \rightarrow 2CuO$$

was classed as an oxidation but heating copper with sulphur:

$$Cu + S \rightarrow CuS$$

was not. Now oxygen and sulphur are the first and second elements respectively in Group VI of the Periodic Table and so their reactions are very similar. The common factor in these reactions is that the copper atoms lose electrons to give copper ions:

$$Cu - 2e^- \rightarrow Cu^{2+}$$

It is apparent that the oxygen and sulphur gain the electrons which the copper loses:

$$O_2 + 4e^- \rightarrow 2O^{2-}$$
$$S + 2e^- \rightarrow S^{2-}$$

Appreciation of facts such as these has led to more general and useful definitions of oxidation and reduction. Thus, *oxidation is now defined as the loss of electrons by an element, compound, or ion* whilst *reduction is the gain of electrons by an element, compound or ion*. It follows that every oxidation must be accompanied by a corresponding reduction and vice versa, because if one substance loses electrons then another must gain them. In the reaction between copper and oxygen, therefore, the copper is oxidised whilst the oxygen is reduced. Similarly, in the reaction between copper and sulphur, the copper is oxidised and the sulphur is reduced.

Reactions in which simultaneous reduction and oxidation are occurring are often known as *redox reactions*. Further examples of redox reactions are given below.

1. Iron(II) chloride reacts with chlorine to give iron(III) chloride:

$$2FeCl_2 + Cl_2 \rightarrow 2FeCl_3$$

The electronic changes occurring are given by the two half-equations:

$$2Fe^{2+} - 2e^- \rightarrow 2Fe^{3+} \quad (Fe^{2+} \text{ oxidised})$$
$$Cl_2 + 2e^- \rightarrow 2Cl^- \quad (Cl_2 \text{ reduced})$$

Addition of the two half-equations gives the full ionic equation:

$$2Fe^{2+} + Cl_2 \rightarrow 2Fe^{3+} + 2Cl^-$$

In ionic equations, any ions appearing on both sides are omitted.

The spontaneous reaction between iron(II) oxide and oxygen:

$$4FeO + O_2 \rightarrow 2Fe_2O_3$$

is very similar, the half-equations being:

$$4Fe^{2+} - 4e^- \rightarrow 4Fe^{3+} \quad (Fe^{2+} \text{ oxidised})$$
$$O_2 + 4e^- \rightarrow 2O^{2-} \quad (O_2 \text{ reduced})$$

This gives the ionic equation:

$$4Fe^{2+} + O_2 \rightarrow 4Fe^{3+} + 2O^{2-}$$

2. Zinc displaces copper from copper(II) sulphate solution:

$$Zn + CuSO_4 \rightarrow ZnSO_4 + Cu$$

The half-equations are:

$$Zn - 2e^- \rightarrow Zn^{2+} \quad (Zn \text{ oxidised})$$
$$Cu^{2+} + 2e^- \rightarrow Cu \quad (Cu^{2+} \text{ reduced})$$

The ionic equation is therefore:

$$Zn + Cu^{2+} \rightarrow Zn^{2+} + Cu$$

Note that the sulphate ion, SO_4^{2-}, does not occur in the ionic equation since it undergoes no change in the reaction.

3. Hydrogen sulphide reacts with bromine to give hydrogen bromide and sulphur:

$$H_2S + Br_2 \rightarrow 2HBr + S$$

The half-equations are:

$$S^{2-} - 2e^- \rightarrow S \quad (S^{2-} \text{ oxidised})$$
$$Br_2 + 2e^- \rightarrow 2Br^- \quad (Br_2 \text{ reduced})$$

and so the ionic equation is:

$$S^{2-} + Br_2 \rightarrow S + 2Br^-$$

It should be noted that hydrogen bromide is, in fact, covalent and not ionic as the above equations imply. The attraction of the H^+ ion for electrons of the Br^- ion is so great that they are shared, i.e.

$$H^+ + Br^- \rightarrow H\text{—}Br$$

4. Copper may be obtained from copper(II) oxide by heating it in a current of hydrogen:

$$CuO + H_2 \rightarrow Cu + H_2O$$

The half-equations are:

$$Cu^{2+} + 2e^- \rightarrow Cu \quad (Cu^{2+} \text{ reduced})$$
$$H_2 - 2e^- \rightarrow 2H^+ \quad (H_2 \text{ oxidised})$$

and so the ionic equation is:

$$Cu^{2+} + H_2 \rightarrow Cu + 2H^+$$

This is another example where the ionic equation does not give the complete picture. The O^{2-} ion is strongly attracted by the H^+ ions and so water is almost completely covalent.

5. Electrolyses are redox reactions – see pages 88 to 91.

OXIDISING AGENTS

Since oxidation involves loss of electrons from a substance, an *oxidising agent may be defined as an electron acceptor*. A number of important oxidising agents and examples of their use are given below. The half-equations of some of the oxidising agents are not required at this level but the relevant equation is given in brackets for the sake of completeness.

1. Oxygen $\qquad O_2 + 4e^- \rightarrow 2O^{2-}$

   Oxygen oxidises most metals and non-metals to their oxides when the elements are heated together, e.g:

   $$2Mg(s) + O_2(g) \rightarrow 2MgO(s)$$
   $$S(s) + O_2(g) \rightarrow SO_2(g)$$

2. Hydrogen peroxide $\quad (H_2O_2 + 2H^+ + 2e^- \rightarrow 2H_2O)$

   (a) Solutions of iron(II) salts are oxidised to iron(III) salts by warming them with hydrogen peroxide solution.

   $$2FeSO_4(aq) + H_2O_2(aq) + H_2SO_4(aq) \rightarrow Fe_2(SO_4)_3(aq) + 2H_2O(l)$$

   (b) Hydrogen peroxide oxidises acidified solutions of potassium iodide, iodine being liberated. The acid liberates hydrogen iodide from the potassium iodide and it is the hydrogen iodide which is oxidised, i.e.

   $$KI + HCl \rightarrow KCl + HI$$
   $$2HI + H_2O_2 \rightarrow I_2 + 2H_2O$$

   The full reaction is therefore:

   $$2KI(aq) + 2HCl(aq) + H_2O_2(aq) \rightarrow I_2(aq \text{ or } s) + 2KCl(aq) + 2H_2O(l)$$

3. Chlorine $\qquad Cl_2 + 2e^- \rightarrow 2Cl^-$

   (a) Chlorine oxidises iron, on heating, to iron(III) chloride:

   $$2Fe(s) + 3Cl_2(g) \rightarrow 2FeCl_3(s)$$

(b) Chlorine oxidises hydrogen sulphide to sulphur:

$$H_2S(g) + Cl_2(g) \rightarrow S(s) + 2HCl(g)$$

4. Manganese(IV) oxide oxidises concentrated hydrochloric acid to chlorine on heating:

$$MnO_2(s) + 4HCl(conc) \rightarrow MnCl_2(aq) + 2H_2O(l) + Cl_2(g)$$

5. Concentrated nitric acid $(2HNO_3 + e^- \rightarrow H_2O + NO_3^- + NO_2)$

This is often boiled with solutions of iron(II) salts to oxidise them to iron(III) salts.

6. Concentrated sulphuric acid $(2H_2SO_4 + 2e^- \rightarrow 2H_2O + SO_4^{2-} + SO_2)$

Hot concentrated sulphuric acid oxidises copper to copper(II) sulphate:

$$Cu(s) + 2H_2SO_4(conc) \rightarrow CuSO_4(s) + 2H_2O(l) + SO_2(g)$$

7. Potassium manganate(VII) $(MnO_4^- + 8H^+ + 5e^- \rightarrow Mn^{2+} + 4H_2O)$

Acidified solutions of potassium manganate(VII) oxidise iron(II) salts to iron(III) salts, sulphites to sulphates, and nitrites to nitrates; in each case the purple solution turns colourless.

8. Potassium dichromate(VI) $(Cr_2O_7^{2-} + 14H^+ + 6e^- \rightarrow 2Cr^{3+} + 7H_2O)$

Acidified solutions of potassium dichromate(VI) oxidise iron(II) salts to iron(III) salts, and sulphur dioxide to sulphuric acid, the orange solution turning green in the process.

REDUCING AGENTS

Reduction is the gain of electrons by a substance and so *reducing agents may be defined as electron donors*. Some important reducing agents, and examples of their use, are given below.

1. Hydrogen in the reduction of metal oxides $H_2 + O^{2-} \rightarrow H_2O + 2e^-$.

   The oxides of the less reactive metals may be reduced by heating them in a current of hydrogen, e.g:

$$CuO(s) + H_2(g) \rightarrow Cu(s) + H_2O(l)$$
$$PbO(s) + H_2(g) \rightarrow Pb(s) + H_2O(l)$$

2. Carbon in the reduction of metal oxides $C + O^{2-} \rightarrow CO + 2e^-$

   Zinc and lead are extracted by heating their oxides with carbon:

$$ZnO(s) + C(s) \rightarrow Zn(s) + CO(g)$$
$$PbO(s) + C(s) \rightarrow Pb(s) + CO(g)$$

3. Carbon monoxide in the reduction of metal oxides

$$CO + O^{2-} \rightarrow CO_2 + 2e^-$$

Iron(III) oxide is reduced in the blast furnace by carbon monoxide (and by carbon):

$$Fe_2O_3(s) + 3CO(g) \rightarrow 2Fe(s) + 3CO_2(g)$$

4. Hydrogen sulphide $S^{2-} \rightarrow S + 2e^-$

Hydrogen sulphide reduces sulphur dioxide to sulphur provided that a trace of water is present.

$$2H_2S(g) + SO_2(g) \rightarrow 3S(s) + 2H_2O(l)$$

5. Potassium iodide in acid solution $2I^- \rightarrow I_2 + 2e^-$

Acidified potassium iodide solution is used to detect oxidising agents; iodine is liberated and this gives a brown solution or black precipitate depending on the concentrations of the solutions used. With hydrogen peroxide the equation is:

$$2KI(aq) + H_2SO_4(aq) + H_2O_2(aq) \rightarrow K_2SO_4(aq) + I_2(aq\ or\ s) + 2H_2O(l)$$

6. Sulphurous acid $(SO_3^{2-} + H_2O \rightarrow SO_4^{2-} + 2H^+ + 2e^-)$

Sulphurous acid is made by passing sulphur dioxide into water.

$$H_2O(l) + SO_2(g) \rightarrow H_2SO_3(aq)$$

It turns acidified potassium dichromate(VI) solution from orange to green and acidified potassium manganate(VII) solution from purple to colourless. In both cases the sulphurous acid is oxidised to sulphuric acid.

It is important to note that some oxidising and reducing agents are more powerful than others and so they will not all undergo the same reactions. Further, some substances can behave both as oxidising and reducing agents. For example, hydrogen peroxide will oxidise iron(II) to iron(III) salts but it is itself oxidised by a more powerful oxidising agent such as potassium manganate(VII). Similarly, sulphur dioxide reduces iron(III) to iron(II) salts but it oxidises hydrogen sulphide to sulphur.

TESTS FOR OXIDISING AGENTS

There is no single test which can be used to detect all oxidising agents. A number of tests may be necessary but the following will generally provide the answer.

1. Add the substance to an acidified solution of potassium iodide. Most oxidising agents will liberate iodine which will turn starch-iodide paper (filter paper soaked in potassium iodide and starch solution) blue.

2. Boil the substance with a little freshly prepared iron(II) sulphate solution acidified with dilute sulphuric acid, cool, and then make alkaline with sodium hydroxide solution. A green precipitate indicates iron(II) hydroxide whilst a brown precipitate shows iron(III) hydroxide. Most oxidising agents will oxidise iron(II) to iron(III) salts and so a brown precipitate would be expected in this test.

3. Pass hydrogen sulphide through a solution of the substance: a pale yellow precipitate indicates that the hydrogen sulphide has been oxidised to sulphur.
4. Warm the substance with concentrated hydrochloric acid. Some oxidising agents oxidise the acid to chlorine which may be recognised by its colour and its bleaching effect on moist litmus.

TESTS FOR REDUCING AGENTS

Again there is no specific test to cover all cases. However, the following tests are generally adequate.
1. Acidify a dilute solution of potassium manganate(VII) with dilute sulphuric acid. Dropwise addition of this purple solution to most reducing agents will result in its decolourisation.
2. Acidify a dilute solution of potassium dichromate(VI) with dilute sulphuric acid. Most reducing agents, if present in excess, will turn this solution from orange to green.
3. Add the substance to a little iron(III) chloride solution. Many reducing agents will convert iron(III) to iron(II) salts. Addition of sodium hydroxide solution will give a green precipitate of iron(II) hydroxide if reduction has taken place but a brown precipitate or iron(III) hydroxide if it has not.

SIMPLE CELLS AND THE ELECTROCHEMICAL SERIES

A simple cell, the Daniell cell, can be made by placing a sheet of zinc and a sheet of copper in dilute sulphuric acid and connecting them externally by a copper wire. A one volt bulb in the circuit (Figure 8.1) will glow, thus showing that a current is flowing.

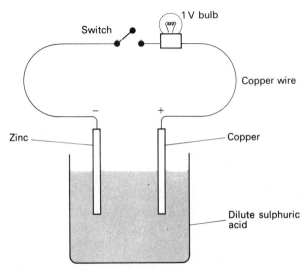

Figure 8.1. The Daniell cell

In electrolysis, electrical energy is used to bring about chemical changes but, in a cell, chemical actions generate electrical energy. In the above cell, the current arises due to zinc going into solution as zinc ions,

$$Zn \rightarrow Zn^{2+} + 2e^-$$

the electrons released flowing through the wire from the zinc to the copper. The copper then attracts the H^+ ions from the acid and they are discharged:

$$H^+ + e^- \rightarrow H; \quad 2H \rightarrow H_2$$

If the cell is left on, it is found that the current rapidly falls due to the surface of the copper being covered with bubbles of hydrogen (this reduces the number of hydrogen ions reaching the copper). If the bubbles are removed, the cell temporarily revives. This defect in the cell is known as *polarisation* and it may be overcome by continual brushing of the copper. However, a more convenient method is to add a depolarising agent such as potassium dichromate(VI) – this oxidises the hydrogen atoms to water as they are formed.

Cells can be made by placing any two metals in an electrolyte and connecting them externally. The greater the difference in reactivity between the two metals, the greater the voltage or e.m.f. (electromotive force) of the cell.

If one electrode is adopted as a standard electrode and the voltages obtained using a different second electrode each time are determined, the metals can be arranged in a reactivity series. The standard electrode used is the hydrogen electrode, the details of which are beyond the scope of this text. However, the results its use produces are important. The series obtained by placing the metals in order of decreasing negative voltage is known as the *electrochemical series*. A high negative voltage indicates that the metal has a strong tendency to go into solution as positive ions, the electrons being left on the electrode.

A reactivity series may also be derived by studying the action of air, water, and acids on the metals and the ease of reduction of their oxides (see page 136). The electrochemical and reactivity series are similar but not identical. A short version of the electrochemical series is given in Table 8.1.

Table 8.1. The Electrochemical Series

Element	Standard electrode potential (volts)	Electropositivity
Potassium	−2.92	
Calcium	−2.87	
Sodium	−2.71	
Magnesium	−2.39	
Aluminium	−1.66	
Zinc	−0.76	Steadily decreases
Iron	−0.44	
Tin	−0.14	
Lead	−0.13	
Hydrogen	0.0	
Copper	+0.34	
Silver	+0.80	

The electrochemical series has a number of important uses.

1. It shows the general order of reactivity of the elements. Thus, the reactivity decreases as the series is descended.
2. It shows which elements will displace each other from solutions of their salts. For example, it is possible to predict that if iron is placed in a solution of copper(II) sulphate, some of the iron will go into solution as ions whilst copper will be deposited on the iron. The reason for this is that iron is higher up the electrochemical series than copper and so it has a greater tendency to exist as positive ions in solution.

i.e.
$$Fe(s) + CuSO_4(aq) \rightarrow FeSO_4(aq) + Cu(s)$$
$$Fe + Cu^{2+} \rightarrow Fe^{2+} + Cu$$

Similarly, it may be predicted that metals above hydrogen in the electrochemical series will liberate hydrogen from dilute acids but the metals below it will not, e.g.

$$Mg(s) + H_2SO_4(aq) \rightarrow MgSO_4(aq) + H_2(g)$$
$$Cu(s) + H_2SO_4(aq) \rightarrow \text{no reaction}$$

3. It is possible to predict which metals will be liberated preferentially when solutions containing their ions are electrolysed. However, complications may arise here because the nature of the electrodes, the concentration of the ions, the current density, and the temperature can all influence the discharge of ions.
4. It can explain the effects of galvanising and tin plating. Galvanising involves dipping iron or steel into molten zinc. The resultant thin coating of zinc protects the iron because, even if the zinc is scratched to expose the iron, the latter does not readily rust because zinc is above it in the electrochemical series and so corrodes preferentially. The zinc reacts with moisture and carbon dioxide in the atmosphere to give zinc hydroxide and basic zinc carbonate. Thus, a white coating dries on the iron and this soon inhibits further attack.

Iron plated with tin is used for canning food. In this case, however, if the tin is scratched to expose the iron, it is the iron which corrodes because it is higher up the electrochemical series. The tin plating is necessary because tin is more resistant than iron to the acids in fruit juices, etc.

Questions

1   (a) Define the term 'reduction' in terms of:
      (i) oxygen,
      (ii) hydrogen, and
      (iii) electrons.
   (b) In each of the following reactions, state which species (i.e. which atom, molecule or ion) is responsible for the observation in italics and say whether the formation of each product is by a process of oxidation or by a process of reduction.
      (i) When hydrogen sulphide is bubbled into a *yellow solution* of iron(III) chloride a *pale green solution* is obtained together with a *yellow precipitate*.

(ii) When carbon monoxide is passed over heated lead(II) oxide, a *silver-coloured substance* is formed and a *colourless gas* may be collected. Write an equation for the reaction.

(iii) A solution of chlorine is added in excess to potassium iodide solution. A *black solid* is formed. Write an ionic equation for this reaction.

[J.M.B.]

2 (a) Define the terms 'oxidation' and 'oxidising agent'.

(b) For each of the following reactions, state which substance is being oxidised and give the relevant half-equations in support of your answers.
(i) $PbO_2 + 4HCl \rightarrow PbCl_2 + 2H_2O + Cl_2$
(ii) $2FeCl_2 + 2HCl + H_2O_2 \rightarrow 2FeCl_3 + 2H_2O$
(iii) $H_2S + Br_2 \rightarrow S + 2HBr$
(iv) $Cu + S \rightarrow CuS$
(v) $H_2SO_4 + 2HBr \rightarrow 2H_2O + Br_2 + SO_2$

3 State whether each of the following reactions involves
(i) oxidation only,
(ii) reduction only,
(iii) both oxidation and reduction, or
(iv) neither oxidation nor reduction.

(a) $Fe^{3+} + 3e^- \rightarrow Fe$ (e) $CuO + H_2SO_4 \rightarrow CuSO_4 + H_2O$
(b) $H_2S + Cl_2 \rightarrow S + 2HCl$ (f) $2Br^- - 2e^- \rightarrow Br_2$
(c) $PbO + C \rightarrow Pb + CO$ (g) $Fe^{3+} + Al \rightarrow Fe + Al^{3+}$
(d) $NaOH + HNO_3 \rightarrow NaNO_3 + H_2O$ (h) $2H_2O + 2Cl_2 \rightarrow 4HCl + O_2$

Give the reasoning behind your answers to reactions (b) and (e).

4 Predict, by use of the electrochemical series, what if anything would happen when:
(i) copper is placed in dilute hydrochloric acid,
(ii) tin is placed in magnesium sulphate solution,
(iii) aluminium is placed in iron(II) sulphate solution,
(iv) iron is added to zinc sulphate solution,
(v) copper is added to silver nitrate solution.

Explain your answers and give equations where relevant.

9 Hydrogen, Oxygen, and Hydrogen Peroxide

HYDROGEN

Occurrence and manufacture

The atmosphere contains a negligible amount of the gas. However, large quantities of hydrogen occur in combination with other elements as, for example, in water, natural gas, and crude petroleum.

Hydrogen is an important by-product from the manufacture of sodium hydroxide by the electrolysis of brine (page 89). Also, very large quantities are obtained from the petroleum industry. Thus, in the cracking process (page 238) long chain alkanes are thermally decomposed to give hydrogen and short chain alkanes and alkenes. On the other hand, the catalytic reforming process converts alkanes containing six or more carbon atoms into aromatic hydrocarbons and hydrogen, e.g.:

$$\underset{\text{Hexane}}{C_6H_{14}} \xrightarrow[500\,°C]{\text{Molybdenum and aluminium oxides}} \underset{\text{Benzene}}{C_6H_6} + 4H_2$$

The manufacture of hydrogen in conjunction with carbon monoxide to give a mixture known as water gas is discussed on page 180.

Laboratory preparations

1. *From water*
   (a) The more reactive metals such as potassium, sodium, and calcium react with cold water, with decreasing vigour, giving hydrogen and the metal hydroxide, e.g:

   $$2Na(s) + 2H_2O(l) \rightarrow 2NaOH(aq) + H_2(g)$$
   $$Ca(s) + 2H_2O(l) \rightarrow Ca(OH)_2(aq) + H_2(g)$$

   These reactions have little practical importance but the apparatus used in the case of calcium is illustrated in Figure 9.1.

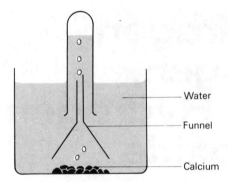

Figure 9.1. Preparation of hydrogen from calcium and water

(b) Steam reacts with hot magnesium or iron (Figure 9.2), the products being the metal oxide and hydrogen; the magnesium burns vigorously:

$$Mg(s) + H_2O(g) \rightarrow MgO(s) + H_2(g)$$
$$3Fe(s) + 4H_2O(g) \rightarrow Fe_3O_4(s) + 4H_2(g)$$

Note that a safety valve is always included in preparations such as this. Hence, if a blockage occurred in the apparatus, water would be forced up the tube out of the flask until the pressure was released. If the valve was not included, a blockage could result in the apparatus blowing up.

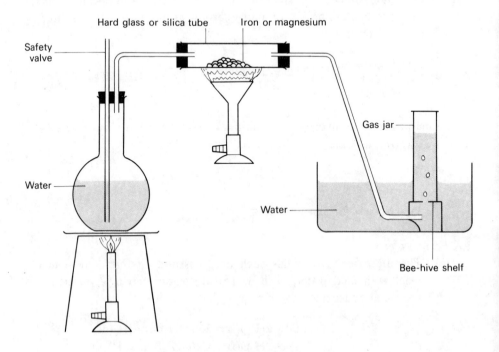

Figure 9.2. Preparation of hydrogen from steam and hot magnesium or iron

2. *From acids*

A number of metals, e.g. zinc, iron, magnesium, and aluminium react with dilute acids giving hydrogen and a salt. The usual laboratory preparation uses slightly impure zinc and dilute sulphuric acid (Figure 9.3):

$$Zn(s) + H_2SO_4(aq) \rightarrow ZnSO_4(aq) + H_2(g)$$

i.e. $$Zn(s) + 2H^+(aq) \rightarrow Zn^{2+}(aq) + H_2(g)$$

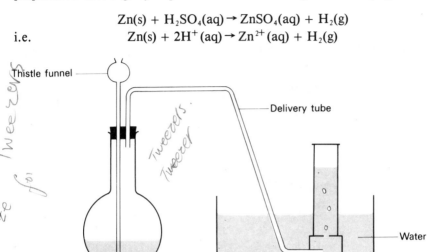

Figure 9.3. Laboratory preparation of hydrogen

Acid is added as required down the thistle funnel. The end of the thistle funnel must be below the level of the acid otherwise the hydrogen would escape to the atmosphere.

The hydrogen from above is slightly impure (impure zinc is used) and it contains water vapour. Pure dry hydrogen can be obtained by bubbling the gas through an acidified solution of potassium manganate(VII) to remove traces of hydrogen sulphide, phosphine, etc. The gas is then passed through concentrated sulphuric acid to dry it and is finally collected by downward displacement of air (Figure 9.4).

Figure 9.4. Preparation of pure dry hydrogen

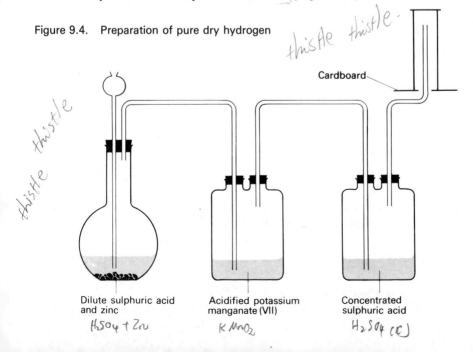

A ready supply of hydrogen is conveniently obtained in the laboratory by the use of a Kipp's apparatus (Figure 9.5).

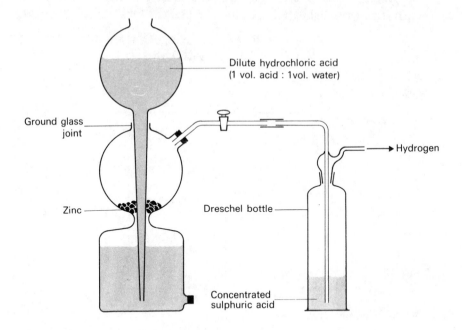

Figure 9.5. Use of Kipp's apparatus to prepare hydrogen

When the tap is opened, the pressure in the middle compartment falls and acid runs into it from the top through the bottom. Reaction then occurs between the acid and zinc to give hydrogen. When the tap is closed, the pressure in the centre compartment builds up and forces the acid back into the top and so reaction ceases. The Dreschel bottle containing a little concentrated sulphuric acid serves not only to dry the gas but also to show the rate at which hydrogen is leaving the apparatus. Note that a Kipp's apparatus could not be used if heat was required in the reaction.

If pure zinc is used in the preparation of hydrogen, the reaction is very slow and it is necessary to add a little copper(II) sulphate as catalyst.

Hydrogen may also be obtained by electrolysis of dilute sulphuric acid or sodium hydroxide using platinum electrodes (see page 91).

Properties and reactions of hydrogen

Hydrogen is a colourless odourless gas which is only slightly soluble in water, i.e. about 2% by volume at room temperature. It is the lightest gas known, its density being approximately one-fourteenth that of air. The low density may be readily demonstrated by passing a stream of hydrogen through a solution of liquid detergent (Figure 9.6). Hydrogen bubbles rise quickly from the solution.

As indicated previously (page 37), it is difficult to classify hydrogen since it has some properties characteristic of the Group I metals and others typical of

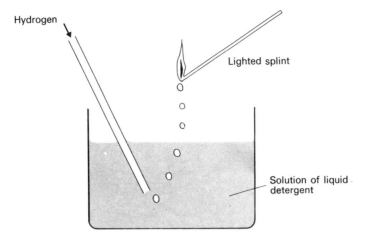

Figure 9.6. The production and ignition of bubbles of hydrogen

the Group VII elements, i.e. the halogens, which are non-metals. Its atoms consist simply of one proton and one electron and it always has a valency of one. The atoms may attain the electronic structure of helium by sharing their electron with one from another atom, i.e. forming a covalent bond as in the H_2 or HCl molecules. Alternatively, they may gain an electron to give a H^- ion (see below). Loss of an electron from a hydrogen atom gives a proton or H^+ ion.

Hydrogen combines with sodium, potassium, and calcium on heating to give ionic (electrovalent) hydrides, e.g:

$$Ca(s) + H_2(g) \rightarrow CaH_2(s)$$

The structure of calcium hydride is $Ca^{2+}(H^-)_2$.

Hydrogen combines directly with many non-metals on heating to give covalent hydrides, e.g:

$$H_2(g) + S(s) \rightarrow H_2S(g)$$

The reactions with nitrogen and the halogens are described on pages 130 and 41 respectively.

Hydrogen burns in air but does not support combustion. This may be demonstrated by plunging a lighted splint into an inverted gas jar of hydrogen (Figure 9.7). The splint is extinguished but the hydrogen burns, with a faint blue flame, at the mouth of the jar and traces of water appear on the sides of the jar:

$$2H_2(g) + O_2(g) \rightarrow 2H_2O(l)$$

The synthesis of water by burning hydrogen in air was described on page 50. Hydrogen and oxygen mixtures, in the ratio of two volumes to one, explode violently on ignition.

Hydrogen reduces the oxides of some of the less reactive metals on heating, e.g:

$$PbO(s) + H_2(g) \rightarrow Pb(s) + H_2O(g)$$

The reduction of copper(II) oxide by hydrogen is described on page 252.

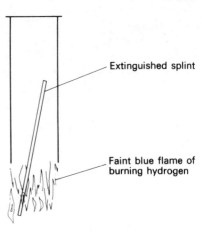

Figure 9.7. Demonstration that hydrogen burns but does not support combustion

Test for hydrogen

If a flame is applied to the mouth of a test-tube of hydrogen, there is a squeak or pop as the gas explodes with oxygen from the air to give water. A similar result is obtained when a light is applied to bubbles of hydrogen (Figure 9.6).

Uses of hydrogen

1. Large quantities of hydrogen are used in the Haber process (page 130) for manufacturing ammonia. Ammonium compounds are important fertilizers.

2. Heating hydrogen with some animal or vegetable oils, in the presence of a nickel catalyst, gives solid fats. Margarine is made in this way.

3. The very exothermic reaction between hydrogen and oxygen is utilised in oxy-hydrogen flames. These flames reach 2000 °C and are used in welding and cutting metals.

4. In the past, hydrogen was used extensively for lifting balloons and airships. However, the flammability was a severe disadvantage and so it has been largely replaced by helium.

OXYGEN

Occurrence and extraction

Oxygen is the most abundant element (page 3). In the free state, it accounts for 21% by volume of the atmosphere (see page 44). Vast quantities also occur in combination with other elements as, for example, in water (H_2O), sand (SiO_2), and rocks ($CaSiO_3$, etc.).

Oxygen is obtained on a commercial scale by fractional distillation of liquid air. Initially, air is cooled under pressure to remove carbon dioxide and water because these compounds readily solidify and would block the apparatus. Next, the air is cooled further by liquid nitrogen, compressed to about 200 atmospheres, and then allowed to expand rapidly. The rapid expansion causes the temperature to fall (the Joule-Thomson effect) as energy is used up in overcoming the attractions which exist between molecules when they are very close together. Repeated compression and expansion eventually causes the air to liquefy. Liquid air is pale blue and contains nitrogen (b.p. −196 °C), oxygen (b.p. −183 °C), and small amounts of the noble gases which may be neglected for many purposes. If the liquid air is allowed to warm up very slowly, the colourless nitrogen evaporates first and the blue oxygen remains. Careful fractionation can also lead to the isolation of some of the noble gases.

Laboratory preparations

1. The usual laboratory preparation of oxygen involves the catalytic decomposition of hydrogen peroxide by manganese(IV) oxide (Figure 9.8).

$$2H_2O_2(aq) \xrightarrow{MnO_2 \text{ catalyst}} 2H_2O(l) + O_2(g)$$

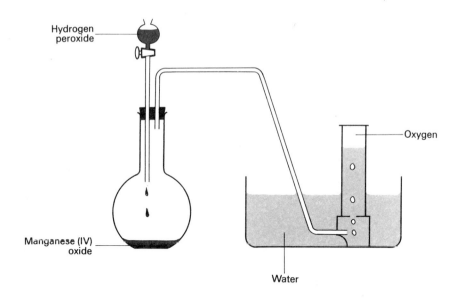

Figure 9.8. Preparation of oxygen from hydrogen peroxide

The reaction is very rapid at first but it slows down as the hydrogen peroxide is used up − the rate at which a reaction occurs is proportional to the concentration of the reactants (see page 127).

2. Oxygen may be prepared by the thermal decomposition of the compounds below using the apparatus illustrated in Figure 9.9. Generally, however, these methods are no longer used as laboratory preparations of oxygen in view of various hazards associated with them.

(a) *From potassium chlorate, $KClO_3$.* If potassium chlorate is heated, it decomposes slowly at about 400 °C:

$$2KClO_3(s) \rightarrow 2KCl(s) + 3O_2(g)$$

When mixed with about one-quarter of its mass of manganese(IV) oxide, however, oxygen is given off much more rapidly and at a lower temperature. Manganese(IV) oxide, MnO_2, does not evolve oxygen when heated alone and it may be recovered fully at the end of the experiment and, so in this reaction, it is acting as a catalyst (see page 128).

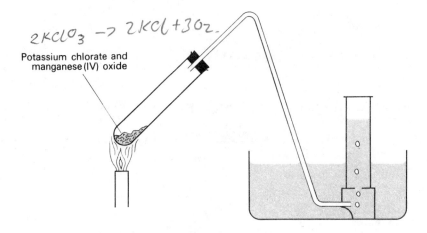

Figure 9.9. Preparation of oxygen

The reaction is potentially explosive if traces of impurity are present.

(b) *From dilead(II) lead(IV) oxide, Pb_3O_4.* Dilead(II) lead(IV) oxide (red lead) decomposes on heating to give lead(II) oxide, the colour changing from red to yellow in the process:

$$2Pb_3O_4(s) \rightarrow 6PbO(s) + O_2(g)$$

Lead compounds are now known to be very toxic.

(c) *From lead(IV) oxide, PbO_2.* Lead(IV) oxide is a purple-brown solid which decomposes into yellow lead(II) oxide on heating:

$$2PbO_2(s) \rightarrow 2PbO(s) + O_2(g)$$

Again the toxicity of lead compounds is a drawback.

(d) *From mercury(II) oxide, HgO.* Mercury(II) oxide is a red solid which decomposes on heating into mercury and oxygen, the mercury condensing as silver globules in the cooler parts of the apparatus.

$$2HgO(s) \rightarrow 2Hg(l) + O_2(g)$$

Mercury vapour is very toxic.

(e) *From sodium or potassium nitrates, $NaNO_3$ or KNO_3.* Most nitrates decompose on heating to give oxygen and the metal oxide (see page

198). Sodium and potassium nitrates, however, are exceptions in that they give the corresponding nitrite and oxygen, e.g.:

$$2NaNO_3(s) \rightarrow 2NaNO_2(s) + O_2(g)$$

Fairly strong heating is necessary and in both cases the white solid melts before oxygen is evolved.

Properties and reactions of oxygen

Oxygen is a colourless, odourless gas, a little denser than air. It is slightly soluble in water (100 cm^3 of water at room temperature dissolves about 4 cm^3 of oxygen) but this is sufficient for the needs of fish and other marine life.

Oxygen is the first element in Group VI and has the electronic structure 2.6. It can attain the noble gas electronic structure either by gaining two electrons to give an O^{2-} ion, as for example in $Mg^{2+}O^{2-}$, or by sharing two electrons in covalent bond formation, e.g. H—O—H.

Oxygen does not burn but it supports combustion. It combines vigorously with many metals and non-metals, some examples being described below. The combustions are normally performed using a deflagrating spoon in a gas jar of oxygen (Figure 9.10).

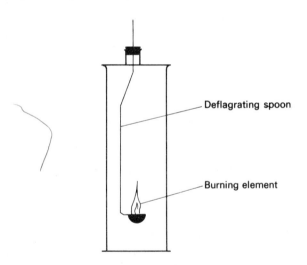

Figure 9.10. Combustion of elements in oxygen

Sodium burns with an intense yellow flame to give sodium peroxide:

$$2Na(s) + O_2(g) \rightarrow Na_2O_2(s) \quad (Na^{+\,-}O—O^-Na^+)$$

The resultant yellowish solid dissolves in water to give an alkaline solution, i.e. it turns red litmus blue.

Calcium burns with a red flame to give a white residue of calcium oxide:

$$2Ca(s) + O_2(g) \rightarrow 2CaO(s)$$

The residue dissolves in water to give an alkaline solution.

Magnesium burns with an intense white flame leaving a white residue of magnesium oxide:

$$2Mg(s) + O_2(g) \rightarrow 2MgO(s)$$

The residue is only slightly soluble in water, the resultant solution being weakly alkaline.

Iron wire burns with a shower of sparks and leaves a brown-black residue of iron(II) diiron(III) oxide:

$$3Fe(s) + 2O_2(g) \rightarrow Fe_3O_4(s)$$

The oxide is insoluble in water and so litmus is unaffected.

Carbon burns or glows with a yellow flame to give carbon dioxide; there is no residue:

$$C(s) + O_2(g) \rightarrow CO_2(g)$$

The gas dissolves in water to give a weakly acidic solution.

Phosphorus (either the white or red allotrope) burns with a yellow flame to give dense white clouds of phosphorus(V) oxide:

$$4P(s) + 5O_2(g) \rightarrow 2P_2O_5(s)$$

Some phosphorus(III) oxide is also formed. The residue dissolves in water to give an acidic solution.

Sulphur burns with a bright blue flame and gives white fumes of sulphur dioxide:

$$S(s) + O_2(g) \rightarrow SO_2(g)$$

The gas dissolves in water to give an acidic solution.

Classification of oxides

It is seen from the above results that the metals form oxides which, if soluble in water, give alkaline solutions whilst the oxides of the non-metals give acidic solutions. Oxides may, in fact, be divided into a number of classes but just four types will be discussed here.

1. *Basic oxides*. These are oxides of metals and they dissolve in acids giving a salt and water only, e.g:

$$MgO(s) + 2HCl(aq) \rightarrow MgCl_2(aq) + H_2O(l)$$
i.e. $$O^{2-}(s) + 2H^+(aq) \rightarrow H_2O(l)$$

Some of these oxides, e.g. K_2O, Na_2O, and CaO dissolve in water and the resultant solutions are alkaline.

$$CaO(s) + H_2O(l) \rightarrow Ca(OH)_2(aq)$$

However, many basic oxides such as copper(II) oxide, CuO, are insoluble in water.

2. *Acidic oxides.* These are generally oxides of non-metals. They dissolve in alkali giving a salt and water, e.g:

$$2KOH(aq) + CO_2(g) \rightarrow K_2CO_3(aq) + H_2O(l)$$
$$2NaOH(aq) + SO_2(g) \rightarrow Na_2SO_3(aq) + H_2O(l)$$
Sodium sulphite

If they dissolve in water, the resultant solution is acidic, e.g.:

$$H_2O(l) + CO_2(g) \rightarrow H_2CO_3(aq)$$
Carbonic acid

Silicon(IV) oxide, SiO_2 (sand) is an example of an acidic oxide which is insoluble in water. (Some higher metallic oxides are acidic, e.g. CrO_3 and Mn_2O_7 but a knowledge of these is not required at this level.)

3. *Amphoteric oxides.* These oxides react with acids and bases, i.e. they have both basic and acidic character, and in each case the product is a salt and water. They are the oxides of weak metals such as zinc and aluminium.

$$ZnO(s) + 2HNO_3(aq) \rightarrow Zn(NO_3)_2(aq) + H_2O(l)$$
$$ZnO(s) + 2NaOH(aq) + H_2O(l) \rightarrow Na_2Zn(OH)_4(aq)$$

Sodium zincate (sometimes written as Na_2ZnO_2, i.e. two molecules of water lost)

Amphoteric oxides are insoluble in water.

4. *Neutral oxides.* These are oxides of non-metals. They give neutral solutions when dissolved in water and do not react with acids or bases. Examples are carbon monoxide, dinitrogen oxide, and water, i.e. CO, N_2O, and H_2O respectively.

Test for oxygen

Oxygen relights a glowing splint.

splint

Uses of oxygen

1. Oxygen is used in breathing apparatus such as oxygen tents in hospitals. Divers use oxygen–helium mixtures whilst airmen, at high altitude, and mountaineers use oxygen–air or oxygen–nitrogen mixtures.

2. The use of oxy-hydrogen flames has been mentioned on page 112. Ethyne (acetylene, C_2H_2) burns in oxygen giving a temperature of about 3000 °C and so oxy-acetylene flames also are used in welding and cutting metals.

3. Oxygen is used in the removal of impurities from pig-iron during the production of steel (see page 163).

4. Liquid oxygen is used in various rockets to burn fuels such as hydrogen, kerosene, and hydrazine.

HYDROGEN PEROXIDE, H_2O_2 (H—O—O—H)

Laboratory preparation

A dilute solution of hydrogen peroxide is obtained by adding a cold solution of barium peroxide to ice-cold dilute sulphuric acid until the mixture is only just acidic:

$$BaO_2(aq) + H_2SO_4(aq) \rightarrow BaSO_4(s) + H_2O_2(aq)$$

i.e. $\quad O^{2-}(aq) + 2H^+(aq) \rightarrow H_2O_2(aq)$

If desired, the barium sulphate may be removed by filtration.

Properties and reactions of hydrogen peroxide

Pure hydrogen peroxide is a colourless, syrupy, explosive liquid. The pure compound and its solutions readily decompose into water and oxygen either by adding a catalyst or just by warming:

$$2H_2O_2(l) \rightarrow 2H_2O(l) + O_2(g)$$

The catalytic decomposition is described on page 113.

Dilute solutions of hydrogen peroxide are normally used and the concentration of these is often expressed as its volume strength, this being the volume of oxygen given at s.t.p. when one volume of the solution fully decomposes. Thus, 1 cm³ of '20-volume' hydrogen peroxide would produce 20 cm³ of oxygen if complete decomposition occurred.

Hydrogen peroxide may act as both an oxidising and as a reducing agent.

1. *As an oxidising agent.*

   Oxidising agents are electron acceptors and the relevant equation for hydrogen peroxide is:

   $$H_2O_2 + 2H^+ + 2e^- \rightarrow 2H_2O$$

   (a) Iron(II) sulphate is oxidised to the corresponding iron(III) salt when warmed with hydrogen peroxide solution and a little dilute sulphuric acid:

   $$2FeSO_4(aq) + H_2SO_4(aq) + H_2O_2(aq) \rightarrow Fe_2(SO_4)_3(aq) + 2H_2O(l)$$
   i.e. $\quad 2Fe^{2+}(aq) + 2H^+(aq) + H_2O_2(aq) \rightarrow 2Fe^{3+}(aq) + 2H_2O(l)$

   (b) Hydrogen peroxide oxidises acidified potassium iodide to iodine:

   $$2KI(aq) + H_2SO_4(aq) + H_2O_2(aq) \rightarrow I_2(aq) + K_2SO_4(aq) + 2H_2O(l)$$
   i.e. $\quad 2I^-(aq) + 2H^+(aq) + H_2O_2(aq) \rightarrow I_2(aq) + 2H_2O(l)$

   (c) The use of hydrogen peroxide in oxidising lead(II) sulphide to lead(II) sulphate in old paintings is discussed on page 158.

2. *As a reducing agent.*

   When hydrogen peroxide is added to a powerful oxidising agent, it is itself oxidised, i.e. it acts as an electron donor:

$$H_2O_2 \rightarrow 2H^+ + 2e^- + O_2$$

(a) A purple solution of acidified potassium manganate(VII) is rapidly decolourised on addition of hydrogen peroxide, and oxygen is evolved. The equation is not required at this level.

(b) An orange solution of acidified potassium dichromate(VI) turns green (after an intermediate blue stage) on addition of hydrogen peroxide. Again, oxygen is evolved. The equation is not required.

Uses of hydrogen peroxide

1. Dilute hydrogen peroxide is used as an antiseptic since it destroys coagulated blood and kills germs.

2. Pure hydrogen peroxide is used as a rocket fuel since its catalytic decomposition is explosive.

3. The dilute solution is used to restore old paintings (see page 158).

4. The mild oxidising properties of its dilute solutions are utilised in bleaching hair, wool, silk, and cotton. Stronger oxidising agents such as chlorine may damage these materials.

Questions

1 Name one metal which will readily give hydrogen with cold water. State what you would see during the reaction and give an equation. Draw a fully labelled diagram for the laboratory preparation and collection of dry hydrogen from a named acid. Write an equation for the reaction. Describe two different ways in which hydrogen could be converted into water. In one case say how you would collect a small sample of the water.

[J.M.B.]

2 (a) (i) Name the two reagents normally used in the laboratory preparation of hydrogen from a metal and an acid.
(ii) Write an ionic equation for the reaction.
(iii) The ionic equation indicates that the reaction is an electron transfer between the two reagents. Which of the reagents is oxidised and which reduced?

(b) Hydrogen is also obtained rapidly when steam is passed over heated magnesium. Describe what you would see in this reaction and give a reason why there is such an increase in the rate of reaction under these conditions compared to that with magnesium and cold water.

(c) How would you obtain hydrogen by electrolysis in the laboratory? Name the electrolyte used, and state the name and polarity of each electrode and the material of which it is made. Give the name of the product at each electrode and write equations for the reactions occurring during their discharge.

(d) What is there in common between the reaction at the cathode in (c) and the reaction in (a)?

[J.M.B.]

3 (a) Name the products formed when the following react with an excess of oxygen:
(i) carbon,
(ii) magnesium,
(iii) hydrogen,
(iv) zinc.

(b) Write equations for the reactions, if any, of these products with:
   (i) dilute hydrochloric acid,
   (ii) sodium hydroxide solution.

If no reaction occurs, write 'no reaction'. State the type of oxide formed by each of the four elements.

[J.M.B.]

4 (a) Describe how you would prepare a solution containing hydrogen peroxide. What precautions are taken to slow down the rate of decomposition of hydrogen peroxide solution?

(b) Draw a fully labelled diagram to show how you would prepare and collect oxygen from hydrogen peroxide. What volume of oxygen, measured at s.t.p. could be obtained from 30 cm^3 of 50 volume hydrogen peroxide solution by your method?

(c) Describe and explain what you would see if hydrogen sulphide was bubbled through lead(II) nitrate solution and an excess of hydrogen peroxide solution was then added.

[J.M.B.]

5 In the laboratory, oxygen can be prepared by the catalytic decomposition of a solution of hydrogen peroxide in water.

(a) Write the equation for the reaction and calculate the volume of oxygen that would be obtained at room conditions by the decomposition of a solution containing 0.1 mole of hydrogen peroxide.

(b) When hydrogen peroxide is mixed with the catalyst the evolution of oxygen is rapid at first but then slows down. Briefly explain the cause of this decrease in the rate of reaction.

(c) You are asked to compare the effectiveness of manganese(IV) oxide with that of lead(IV) oxide as a catalyst in this reaction. Sketch the apparatus you would use and outline the experiments you would carry out. Name the method used to obtain oxygen from liquid air and state the physical property which makes the separation possible. Briefly explain the use of oxygen in either steel making or steel cutting and welding.

(1 mole of gas occupies 24.0 litres at room temperature and pressure.)

[A.E.B. June 1975]

10 Thermochemistry, Rate of Change, and Equilibria

THERMOCHEMISTRY

Chemical changes are generally accompanied by heat changes and a study of these heat changes is known as thermochemistry. In reactions such as the combustion of magnesium or coal, the evolution of heat is readily apparent. Similarly, when ammonium nitrite is dissolved in water, there is a noticeable fall in temperature. However, in a reaction such as the formation of silver chloride by mixing solutions of silver nitrate and a chloride, e.g:

$$AgNO_3(aq) + NaCl(aq) \rightarrow AgCl(s) + NaNO_3(aq)$$

the heat change is much less apparent but it does occur.

Reactions which proceed with evolution of heat are said to be *exothermic*, whilst those which involve absorption of heat are said to be *endothermic*. The heat changes in reactions represent increases or decreases in the energy content of the system. Thus, the products from an exothermic process have less energy than the reactants and so they are more stable.

It is not possible to measure absolute values of energy because this involves complex factors such as interionic or intermolecular attractions, etc. However, changes in energy can be readily determined. The energy change in a given reaction is represented by the symbol ΔH (Δ is the capital Greek letter delta) where H is the energy content. According to the current convention, ΔH is given a negative sign in exothermic reactions, i.e. reactions where heat or energy is evolved. Thus,

$$C + O_2 \rightarrow CO_2; \quad \Delta H = -393.7 \text{ kJ}$$

means that when 12 g of carbon react with 32 g of oxygen to give 44 g of carbon dioxide, 393.7 kilojoules are evolved. Conversely, ΔH in endothermic reactions is given a positive sign. Hence,

$$CO + 2H_2 \rightarrow CH_3OH; \quad \Delta H = +90.7 \text{ kJ}$$

indicates that 28 g of carbon monoxide react with 4 g of hydrogen to give 32 g of methanol and 90.7 kJ are absorbed in the process.

The heat change in any reaction will obviously depend upon the state of the reactants and products. For example, in the reaction

$$2H_2 + O_2 \rightarrow 2H_2O$$

the energy change in converting the hydrogen and oxygen, at room temperature and pressure, to steam at 100 °C will be different to the energy change involved when the product is water at room temperature. Energy changes are therefore normally measured with the reactants and products under the standard conditions of 25 °C and 1 atmosphere pressure; the energy change is then represented by the symbol ΔH°.

ΔH is known by a variety of names, depending upon the type of reaction involved. A number of examples are given below.

Heat of combustion

Heat of combustion may be defined as the heat change when one mole of a substance is burnt in excess oxygen. Excess oxygen is necessary to ensure that the combustion is complete.

The heat of combustion of methane is given by

$$CH_4(g) + 2O_2(g) \rightarrow 2H_2O(l) + CO_2(g); \quad \Delta H^\circ = -890.4 \text{ kJ mol}^{-1}$$

It should be noted that the combustion of any substance is an exothermic process and so all heats of combustion have a negative sign. Heats of combustion are important in comparing the quality of different fuels.

An approximate value for the heat of combustion of a liquid fuel may be obtained using the apparatus shown in Figure 10.1. Although the method is not very accurate, it does indicate the principles involved.

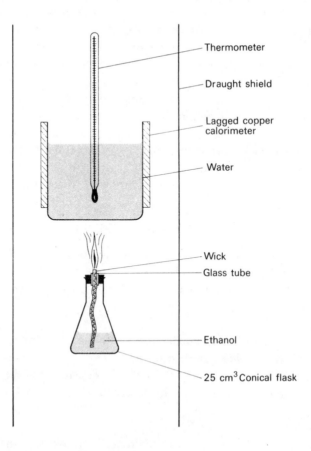

Figure 10.1   Determination of heat of combustion

250 cm³ (250 g) of water are placed in the calorimeter and its temperature is noted. The lamp is about half filled with ethanol and then its mass is determined. The lamp and draught screens are placed in position and then the lamp is lit. The lamp is allowed to burn until the temperature of the water in the calorimeter has risen by about 30 °C. The maximum temperature attained by the water is noted and the lamp is extinguished and immediately weighed to find the mass of ethanol used. The approximate heat of combustion of ethanol is worked out from the results as illustrated in the example below.

Initial mass of lamp + ethanol = 86.75 g
Final mass of lamp + ethanol = 85.32 g
∴ Mass of ethanol used = 1.43 g
Hence moles of ethanol used = $\frac{1.43}{46}$ = 0.0311 mole

Mass of water used = 250 g
Initial temperature of water = 19.7 °C
Final temperature of water = 47.5 °C
∴ Rise in temperature of water = 27.8 °C

Heat gained by the water = mass × temperature rise × 4.2 joules
$$= 250 \times 27.8 \times \frac{4.2}{1000} \text{ kJ}$$
$$= 29.19 \text{ kJ}$$

Since 0.0311 mole of ethanol liberates 29.19 kJ on combustion,

1 mole of ethanol liberates $\frac{29.19}{0.0311}$ = 938 kJ i.e. ΔH_c = −938 kJ mol⁻¹

Note that multiplying the mass in grams by the temperature in degrees Celsius gives the heat change in calories. The factor, 4.2, is used to convert calories to joules.

The heat of combustion obtained above is, in fact, only about two thirds of the true value. There are three main reasons for the error.

(a) Not all the heat produced by the combustion is effectively used in heating the water – hot gases escape.
(b) The water loses some of its heat to the atmosphere.
(c) No account is taken of the heat utilised in raising the temperature of the copper calorimeter.

In accurate work, more sophisticated apparatus is required and the above deficiences have to be eliminated or corrections made for them.

Heat of reaction

The heat of a reaction is the heat change when the moles of reactants in the equation react together. For example,

$$2CO + O_2 \rightarrow 2CO_2; \quad \Delta H^\circ = -566 \text{ kJ}$$

means that 566 kJ of heat are evolved when two moles of carbon monoxide react with one of oxygen to give two moles of carbon dioxide.

Heat of solution

Heat of solution is the heat change when one mole of a substance is dissolved in so much water that further dilution produces no detectable heat change. This may be illustrated by dissolving a few pellets of sodium hydroxide in a little water. The dissolution is accompanied by a rise in temperature. Addition of small portions of water to the concentrated solution results in further temperature rises which become progressively smaller. Eventually, further dilution will produce no increase in temperature.

When an ionic substance is dissolved in water, energy is required to separate the ions in the crystal lattice but, once they are separated, they become hydrated (page 52) and energy is evolved. Whether the reaction overall is exothermic or endothermic will depend on which of the two processes involves the most energy. This may be illustrated by Figures 10.2 and 10.3 which illustrate the energy changes involved in the heat of solution of sodium chloride and sodium hydroxide respectively. The heat of solution of sodium chloride is seen to be endothermic because the energy required to break down the crystal lattice just exceeds the hydration energy. However, the dissolution of sodium hydroxide is exothermic because the hydration of the ions is the dominant factor.

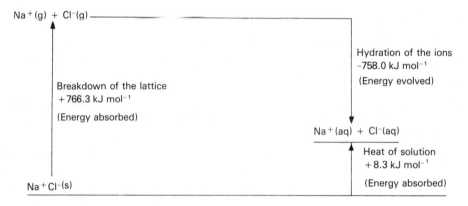

Figure 10.2.   Heat of solution of sodium chloride

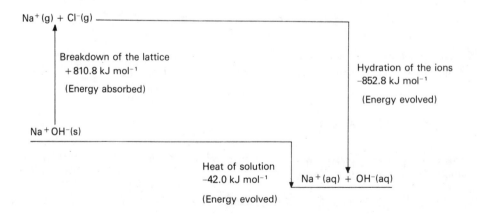

Figure 10.3.   Heat of solution of sodium hydroxide

Heat of vaporisation

When a liquid evaporates, it is the high energy molecules which escape (see page 23) and so the temperature falls. *The heat of vaporisation is the heat required to convert one mole of a liquid to vapour without change in temperature.* Vaporisation is always an endothermic process and the size of the heat of vaporisation indicates the degree of attraction between the molecules of the liquid. Thus, there is little attraction between methane, CH_4, molecules and so the heat of vaporisation is small, i.e. 0.92 kJ mol^{-1}. On the other hand, water molecules attract one another fairly strongly (see page 52) and so energy must be supplied to overcome these attractions before the molecules can escape as vapour. Water has a heat of vaporisation of 33.4 kJ mol^{-1}. This high heat of vaporisation accounts for the comparatively slow rate of evaporation of boiling water.

Heat of fusion

At the melting point, the solid and liquid forms of a substance co-exist indefinitely. *The heat required to convert one mole of solid to liquid without change in temperature is known as the heat of fusion of the substance.* Again the size of ΔH indicates the degree of attraction between the molecules.

Heat of neutralisation

Heat of neutralisation is the heat change when one mole of hydrogen ions from an acid react with one mole of hydroxide ions from a base. All heats of neutralisation are found to be exothermic. It is also found that the heat of neutralisation of strong acids and strong bases is practically constant at about −57 kJ. This may be explained by examining the relevant equations, e.g.

$$Na^+(aq) + HO^-(aq) + H^+(aq) + Cl^-(aq) \rightarrow H_2O(l) + Na^+(aq) + Cl^-(aq)$$
$$Na^+(aq) + HO^-(aq) + H^+(aq) + NO_3^-(aq) \rightarrow H_2O(l) + Na^+(aq) + NO_3^-(aq)$$
$$K^+(aq) + HO^-(aq) + H^+(aq) + \tfrac{1}{2}SO_4^{2-}(aq) \rightarrow H_2O(l) + K^+(aq) + \tfrac{1}{2}SO_4^{2-}(aq)$$

In each case the net change is:

$$H^+(aq) + HO^-(aq) \rightarrow H_2O(l)$$

Weak acids and bases give lower values for their heat of neutralisation because they require some energy to dissociate.

RATE OF CHANGE

Some reactions take place so rapidly that they appear to be almost instantaneous. For example, addition of barium chloride solution to a solution of a sulphate gives an immediate white precipitate of barium sulphate.

$$BaCl_2(aq) + Na_2SO_4(aq) \rightarrow BaSO_4(s) + 2NaCl(aq)$$

In contrast, the rusting of iron in the presence of air and water is a very slow process. Between these extremes, however, there are many reactions which proceed at measurable rates.

The rate of a reaction may be measured in terms of the moles of a reactant used up or the moles of a product formed in a given time. Hence the units are generally mol s^{-1} (i.e. moles per second) or mol l^{-1} s^{-1}.

The rates of reactions may be affected by a number of factors, i.e. temperature, concentration, pressure, catalysts, light, and the particle size in reactions involving solids.

1. Temperature

All reactions proceed more rapidly as the temperature is raised but there is no definite relationship between the rate of a reaction and the temperature. However, it is often found that each 10 °C rise in temperature approximately doubles the rate of the reaction.

The dependence of reaction rate on temperature may be explained as follows. Molecules need to collide with one another before they can react. However, many collisions are ineffective because the molecules: (a) are not in the correct position relative to one another, (b) do not have sufficient energy for reaction to take place. Consider the reaction between hydrogen and iodine to give hydrogen iodide:

$$H_2 + I_2 \rightarrow 2HI$$

For reaction to occur, the hydrogen and iodine molecules must collide as illustrated in Figure 10.4. If the molecules collide end-on as shown in Figure 10.5, they will simply bounce off one another without any reaction taking place. When hydrogen and iodine molecules collide and reaction takes place, new bonds must be formed and old ones broken. This necessitates the formation of an unstable, high energy, intermediate state in which some bonds are half formed and others half broken (Figure 10.6). The molecules must have sufficient energy to overcome this unstable state before hydrogen iodide can be formed. Increasing the temperature of a reaction, increases the number of molecules with sufficient energy to overcome this so-called transition state.

Figure 10.4. Effective collisions between hydrogen and iodine molecules

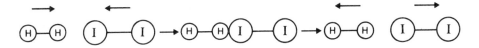

Figure 10.5. Ineffective collisions between hydrogen and iodine molecules

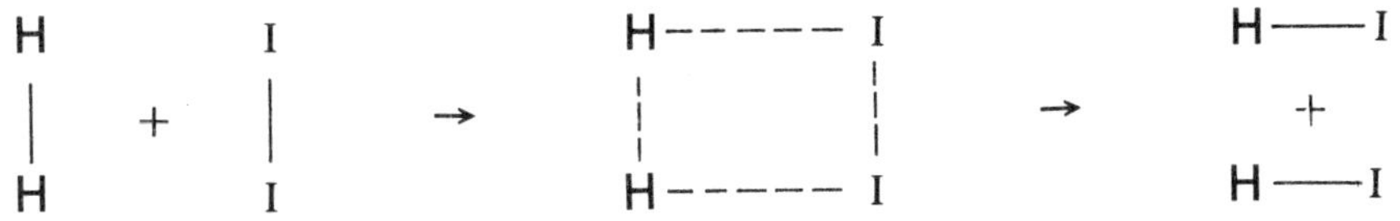

```
H        I         H ----- I         H —— I
|   +    |    →    |       |    →      +
H        I         H ----- I         H —— I
```

Figure 10.6. The formation of an unstable intermediate and its subsequent breakdown

2. Concentration

At any given temperature, the rate of a reaction is directly proportional to the concentration of the reacting substances. This may be readily confirmed by the reaction between sodium thiosulphate solution and nitric acid (see page 262 for full details).

$$Na_2S_2O_3(aq) + 2HNO_3(aq) \rightarrow 2NaNO_3(aq) + S(s) + H_2O(l) + SO_2(g)$$

It is found that, in a series of reactions in which the nitric acid concentration is kept constant, the time taken for sulphur to be precipitated doubles each time the concentration of the sodium thiosulphate solution is halved. Similar results are obtained when the nitric acid concentration is varied whilst the sodium thiosulphate concentration is kept constant. These results are explained by the fact that, at higher concentrations, the reacting molecules are closer together and so they have a greater chance of colliding with one another.

Practical details for investigating the effect of concentration on the rate of the catalytic decomposition of hydrogen peroxide are given on page 263.

3. Pressure

The effect of changing the pressure on reactions just involving liquids, solutions, or solids is negligible. However, the volumes of gases are greatly affected by pressure. Increasing the pressure in a reaction involving gases is equivalent to increasing the concentration. At higher pressures the molecules of gases are closer together and so the chances of collision are enhanced.

4. Catalysts

A catalyst is a substance which affects the rate of a chemical reaction but can be fully recovered unchanged at the end of the reaction. Not all catalysts are positive catalysts, i.e. they do not all increase the rate of a reaction; negative catalysts or inhibitors are possible.

The amount of catalyst required varies widely from reaction to reaction. Thus, in some cases the slightest trace of catalyst is sufficient but some reactions require more catalyst than reactants.

Some catalysts form an intermediate compound during the reaction. This is exemplified by nitrogen oxide which was used as the catalyst in the Lead Chamber process for manufacturing sulphuric acid; the process is now obsolete. The nitrogen oxide owed its catalytic power to the fact that it combines with oxygen to give nitrogen dioxide, which is a powerful oxidising agent. Nitrogen dioxide oxidises water and sulphur dioxide to sulphuric acid and is itself reduced back to nitrogen oxide.

$$2NO(g) + O_2(g) \rightarrow 2NO_2(g)$$
$$H_2O(l) + SO_2(g) + NO_2(g) \rightarrow H_2SO_4(conc) + NO(g)$$

Heterogeneous catalysts are those where reaction occurs at their surface, an example being the decomposition of hydrogen peroxide by manganese(IV) oxide (see page 113). Other examples of this type of catalyst are iron in the Haber process for manufacturing ammonia and vanadium(V) oxide or platinised asbestos in the Contact process for manufacturing sulphur(VI) oxide.

$$N_2(g) + 3H_2(g) \underset{}{\overset{Fe\ catalyst}{\rightleftharpoons}} 2NH_3(g)$$

$$2SO_2(g) + O_2(g) \underset{catalyst}{\overset{V_2O_5\ or\ Pt}{\rightleftharpoons}} 2SO_3(s)$$

The reactants are adsorbed by the catalyst, i.e. attracted to the surface of the catalyst. There is therefore a high concentration of the reactants at the surface of the catalyst and consequently the rate of reaction increases. These catalysts are used in a finely divided form so as to obtain the maximum surface area.

Some living cells can act as catalysts; they are known as *enzymes* – see page 238.

5. Light

Some reactions are greatly accelerated by ultraviolet light. For example, hydrogen and chlorine combine only slowly at room temperature in diffuse light but explosively in ultraviolet light or sunlight.

$$H_2(g) + Cl_2(g) \rightarrow 2HCl(g)$$

Similarly, chlorine does not react with methane under normal conditions but a series of compounds may be formed if the mixture is irradiated with ultraviolet light.

$$CH_4(g) + Cl_2(g) \rightarrow CH_3Cl(g) + HCl(g) \quad \text{(chloromethane)}$$
$$CH_3Cl(g) + Cl_2(g) \rightarrow CH_2Cl_2(l) + HCl(g) \quad \text{(dichloromethane)}$$
$$CH_2Cl_2(l) + Cl_2(g) \rightarrow CHCl_3(l) + HCl(g) \quad \text{(trichloromethane)}$$
$$CHCl_3(l) + Cl_2(g) \rightarrow CCl_4(l) + HCl(g) \quad \text{(tetrachloromethane)}$$

In both of the above examples, the function of the ultraviolet light is to split the chlorine molecules into the much more reactive chlorine atoms.

Photosynthesis is discussed on page 181.

6. Particle size

Reactions involving solids proceed more slowly as the particle size is increased. For example, powdered calcium carbonate reacts very rapidly with dilute hydrochloric acid and, if the reaction is done in a test-tube, the contents will probably bubble out. On the other hand, marble chips (larger lumps of calcium carbonate) react quite steadily with the acid.

$$CaCO_3(s) + 2HCl(aq) \rightarrow CaCl_2(aq) + H_2O(l) + CO_2(g)$$

These observations are explained by the fact that reaction occurs at the surface of the solid and so the larger the surface area, i.e. the smaller the particle size, the greater the contact between the reactants and the greater the scope for reaction. This is illustrated by Figure 10.7 in which it is seen that the large particle has 13 surface atoms available for attack whilst the three small particles, which contain the same total of atoms, have 18 surface atoms.

Figure 10.7. Large particles have fewer exposed atoms than several smaller particles

EQUILIBRIA

Many reactions are reversible, i.e. they can proceed in both directions provided that the conditions are right. For example, hydrogen and iodine combine on heating to give hydrogen iodide, but this compound decomposes on heating to give hydrogen and iodine. The equation is written:

$$H_2 + I_2 \rightleftharpoons 2HI$$

the reversibility being indicated by the $\rightleftharpoons$ sign.

When the hydrogen and iodine are first mixed, the rate of the forward reaction, i.e.

$$H_2 + I_2 \rightarrow 2HI$$

will be high because the concentration of the reactants is at the maximum.

However, as more and more hydrogen and iodine are used up, the rate of formation of hydrogen iodide decreases. On the other hand, the rate of the reverse reaction, i.e.

$$2HI \rightarrow H_2 + I_2$$

gradually increases as the concentration of hydrogen iodide increases. Now, provided that the temperature is kept constant, the rate of formation of hydrogen iodide will eventually become equal to the rate of its decomposition and the system is then said to be at equilibrium. Once equilibrium is attained, the concentration of the reactants and products remains constant. It should be clearly understood that, at equilibrium, the reaction does not stop but rather the product is being formed and decomposed at identical rates.

Le Chatelier's principle

The effect of altering the conditions of a reaction at equilibrium is given by Le Chatelier's principle. This states that, *when the conditions of a system at equilibrium are changed, the system reacts in such a way as to oppose the effects of the change.* The consequences of this are seen in the following examples.

a. The Haber process

Nitrogen and hydrogen take part in a reversible reaction to give ammonia, the process being exothermic,

$$N_2(g) + 3H_2(g) \rightleftharpoons 2NH_3(g); \quad \Delta H \text{ is negative.}$$

If the temperature is raised, then according to Le Chatelier's principle, the system should counteract this by the equilibrium moving to the left. In this way, the temperature will be lowered because the decomposition of ammonia is an endothermic process. Conversely, lowering the temperature should lead to a higher yield of ammonia because this process is exothermic, i.e. the temperature will be increased. However, lowering the temperature lowers the rate of formation of ammonia and so a catalyst is required. Note that catalysts do not affect the position of equilibrium in a reaction. Positive catalysts increase the rate of a reaction by providing a lower energy route for the formation of the products.

If the pressure of the reactants is increased, then Le Chatelier's principle predicts that the equilibrium will move in such a way as to lower the pressure. Thus, the yield of ammonia will be increased because this decreases the volume, and therefore the pressure, of gases present. In this reaction, four volumes of reactants, i.e. one of nitrogen plus three of hydrogen, give two volumes of product (ammonia). Low pressures would result in the equilibrium moving to the left, i.e. ammonia would decompose into nitrogen and hydrogen to increase the volume, and hence the pressure, of the gases present.

In summary, the theoretical requirements for the maximum yield of ammonia are low temperatures, high pressures, and a catalyst. The actual conditions used are 550 °C, 250—1000 atmospheres, and iron as catalyst. It is pertinent to ask therefore, why high temperatures are used when in theory they should be low. The answer is that industrially a balance has to be found between the rate of the reaction and the yield. It is, in fact, more economical to obtain a smaller yield rapidly than a higher yield more slowly. This may be illustrated as follows. Suppose that at 550 °C and 250 atmospheres pressure, in the presence of the catalyst, a 12% yield of ammonia is obtained in 5 minutes, whilst lowering the temperature to 400 °C gives a 20% yield in 10 minutes. Two runs, using the initial conditions, could be done in 10 minutes and so the total yield would be 24%, i.e. a 4% increase on the reaction at the lower temperature over the same period of time.

The unchanged gases are recycled, the ammonia being removed from them either by dissolving it in water or by liquefying it.

b. The Contact process

This involves the exothermic reaction between sulphur dioxide and oxygen to give sulphur(VI) oxide (see also page 209):

$$2SO_2(g) + O_2(g) \rightleftharpoons 2SO_3(g); \quad \Delta H \text{ is negative.}$$

Lowering the temperature results in a greater yield of sulphur(VI) oxide since in this way the system raises the temperature. However, lowering the temperature lowers the rate of the reaction and so a catalyst is required. High pressures are counteracted by more sulphur(VI) oxide being formed because this causes the volume of gas present, and hence the pressure, to be lowered (three volumes of reactants give two volumes of product).

The theoretical requirements for the maximum yield of sulphur(VI) oxide are therefore low temperatures, high pressures, and a catalyst. In practice, the conditions used are 450 °C, atmospheric pressure, and vanadium(V) oxide, V_2O_5, or platinised asbestos as catalyst. Again, the conditions suggested by theory and those used in practice are not in full agreement but nevertheless, the yield is about 99%. High pressures are not used because the small gain in yield possible would not justify the extra cost. High temperatures are used so that the process can be carried out, as the name implies, very rapidly.

c. The Birkeland-Eyde process

This process involved the endothermic reaction between nitrogen and oxygen to give nitrogen oxide:

$$N_2(g) + O_2(g) \rightleftharpoons 2NO(g); \quad \Delta H \text{ is positive.}$$

The process is now obsolete but it is a useful example for study.

The rate and the yield are increased by raising the temperature and so a catalyst is unnecessary. Change in pressure has no effect on the yield since two volumes of reactants give two of product. Increasing the pressure would increase the rate because it would make collisions between the molecules more frequent. However, the rate is not an important factor in this reaction.

The relationship between the temperature and yield is shown by the following figures.

Temperature °C	1500	2000	3000
% Yield	0.1	2	5

The reaction was performed in regions where cheap hydroelectric power was available, an electric spark being passed through the reaction mixture.

Nitrogen oxide is used, on a large scale, in the manufacture of nitric acid and is now made by the catalytic oxidation of ammonia (see page 196).

d. Thermal dissociation of ammonium chloride

The action of heat on ammonium chloride gives ammonia and hydrogen chloride, but these gases recombine on cooling. For this reason, the reaction is classed as *thermal dissociation*, not thermal decomposition.

$$NH_4Cl(s) \rightleftharpoons NH_3(g) + HCl(g); \quad \Delta H \text{ is positive}$$

The dissociation will be increased by raising the temperature. Since a solid is producing two gases, the equilibrium will move to the left if the pressure is increased and to the right if the pressure is decreased.

e. Thermal dissociation of dinitrogen tetraoxide

Dinitrogen tetraoxide is a pale yellow liquid which boils at 22 °C. It slowly dissociates into nitrogen dioxide above its boiling point and at 150 °C the dissociation is complete.

$$N_2O_4(l) \rightleftharpoons 2NO_2(g); \quad \Delta H \text{ is positive}$$

The dissociation is increased by raising the temperature and lowering the pressure.

f. Thermal decomposition of calcium carbonate

Calcium carbonate, like most metal carbonates, decomposes on heating to give the oxide and carbon dioxide. If the reaction is done in a sealed tube, an equilibrium is set up:

$$CaCO_3(s) \rightleftharpoons CaO(s) + CO_2(g); \quad \Delta H \text{ is positive}$$

The equilibrium will move to the right under the influence of high temperatures and low pressures. If the carbon dioxide is allowed to escape, more and more of the carbonate will decompose in an attempt to restore the equilibrium and so the reaction will go to completion.

Questions

1 (a) The heats of dissociation of the halogens are as follows:

$F_2 \rightarrow 2F$ $\Delta H = +313$ kJ mol^{-1}
$Cl_2 \rightarrow 2Cl$ $\Delta H = +231$ kJ mol^{-1}
$Br_2 \rightarrow 2Br$ $\Delta H = +185$ kJ mol^{-1}
$I_2 \rightarrow 2I$ $\Delta H = +144$ kJ mol^{-1}

   (i) State how the stabilities of the halogen molecules vary.
   (ii) At 1700 °C one halogen only is completely dissociated into atoms. Name this element.

(b) The heat change in dilute aqueous solution for the exothermic reaction:

$$NaOH + HCl \rightarrow NaCl + H_2O \quad \text{is} \quad \Delta H = -56.4 \text{ kJ mol}^{-1}.$$

State the heat changes for the following reactions in dilute aqueous solutions and give reasons for your answers:
   (i) $KOH + HNO_3 \rightarrow KNO_3 + H_2O$,
   (ii) $2NaOH + H_2SO_4 \rightarrow Na_2SO_4 + 2H_2O$.

2 (a) Define the term 'catalyst'. Describe a laboratory experiment in which a catalyst is used to speed up the decomposition of a substance.

(b) What is meant by an 'exothermic reaction'? Describe a laboratory experiment which is noticeably exothermic and give the relevant equation.

(c) Briefly describe an industrial process in which a catalyst is used. What effect does the catalyst have on the yield of the product obtained?

3 (a) An aqueous solution of hydrogen peroxide decomposes to water and oxygen. This reaction is rapid at room temperature when a catalyst is added to a moderately concentrated solution.
   (i) Name a suitable catalyst.
   (ii) Write the equation for the reaction.
   (iii) You are asked to investigate the rate at which oxygen is evolved at room temperature when one gram of catalyst is added to 50 cm^3 of a suitable hydrogen peroxide solution. Draw the apparatus you would use and outline the experiment you would do.
   (iv) Sketch a graph showing how you would expect the volume of oxygen to vary with time. (It is not necessary to use graph paper.) Label this graph P.
   (v) On the same axes, sketch another graph to indicate how the results would be changed if the experiment were repeated with warm hydrogen peroxide solution. Label this graph Q.
   (vi) Apart from temperature, what other factors might affect the rate of this reaction?

(b) State two reactions of industrial importance in each of which a named catalyst is used. Give the conditions under which one of these reactions is carried out. (You are not required to describe how the products are isolated.)

[A.E.B. June 1977]

4 (a) The equation for the dissociation of calcium carbonate is

$$CaCO_3 \rightleftharpoons CaO + CO_2; \quad \Delta H = +175.5 \text{ kJ mol}^{-1} \quad \text{(heat taken in)}$$

What will be the effect on the proportion of calcium carbonate in the equilibrium mixture of:
   (i) increasing the temperature,
   (ii) increasing the pressure?
What conditions would be suitable for manufacturing calcium oxide from calcium carbonate on a large scale?

(b) At 400 °C, the three gases, hydrogen, iodine and hydrogen iodide, exist together in equilibrium. The equation for the reaction is

$$H_2 + I_2 \rightleftharpoons 2HI$$

What effect will an increase in pressure have on:
(i) the rate of reaction of hydrogen with iodine,
(ii) the position of equilibrium?
Name the product formed when hydrogen iodide is dissolved in water and describe what would happen if chlorine were passed into the solution.

(c) Write an equation for a reaction which can be catalysed by manganese(IV) oxide. What would be the effect on the rate of this reaction of:
(i) adding more manganese(IV) oxide,
(ii) using the same weight of catalyst but increasing the particle size?

(d) The rate of reaction between hydrogen and chlorine depends on the intensity of light falling on the reagents. For another reaction whose rate also depends on the intensity of the light:
(i) name the reagents,
(ii) either name the products or state the importance or use of the reaction.

[J.M.B.]

5   Sulphuric acid is manufactured by the 'Contact' process which makes use of the equilibrium reaction

$$2SO_2 + O_2 \rightleftharpoons 2SO_3$$

(a) Heat is given out in the formation of sulphur(VI) oxide.
State what effect there would be on the equilibrium concentration of sulphur(VI) oxide if:
(i) the pressure were increased,
(ii) the temperature were raised.

(b) The actual conditions used in the process vary somewhat, but a rough average appears to be:
(i) an excess of air,
(ii) a temperature of about 450 °C,
(iii) a pressure slightly in excess of one atmosphere,
(iv) a catalyst.
Explain why each of the conditions (i) to (iv) is necessary.

(e) Sulphur(VI) oxide reacts with water to give sulphuric acid.

$$SO_3 + H_2O \rightarrow H_2SO_4$$

What is the disadvantage of terminating the process in this way and how are the final stages accomplished in industry?

[J.M.B.]

11 The Chemistry of some Metals

INTRODUCTION

It was seen in Chapter 8, that some metals have a greater tendency than others to lose electrons and go into solution as ions. Thus, the metals with the largest negative electrode potentials are the most reactive and most readily form positive ions; they are said to be the most electropositive elements. Arrangement of the metals in order of decreasing negative electrode potential gives the electrochemical series.

A very similar series, known as the activity series, may be obtained by a study of the action of air, water, and acids on the metals and the ease of reduction of the metal oxides. With regard to the metals encountered in this text, the only difference between the electrochemical and activity series is that the positions of calcium and sodium are reversed. A short version of the activity series is given in Figure 11.1 and the relationship between the position of a metal in the series and its chemical properties is shown. It is seen that, with a few minor exceptions, there are well defined trends as the series is descended. The reactions and the reasons for the trends are discussed in the following pages under the heading of each particular metal.

SODIUM AND POTASSIUM—GROUP I METALS

The Group I metals are characterised by having one electron in their outer shell and they are often known as the *alkali metals*. The first three elements in the group and their electronic structures are lithium, 2.1; sodium, 2.8.1; and potassium, 2.8.8.1. They have a strong tendency to achieve the electronic structure of the noble gases by losing an electron to give a positive ion, i.e. they are strongly electropositive.

$$Na - e^- \rightarrow Na^+$$
$$2.8.1 \qquad 2.8$$

Sodium is more reactive than lithium because its outer electron is further from the nucleus and so less tightly held. Potassium is in turn more reactive than sodium for the same reason. Lithium is relatively unimportant and so only sodium and potassium will be discussed.

SODIUM

Occurrence and extraction

Sodium is far too reactive to occur naturally as the free metal but it occurs in combination with a number of other elements, the most important source

Element	Electro-Positivity	Action of water	Action of acids	Action of air	Action of H_2 on heated oxide	Action of C on heated oxide	Nature of hydroxide	Action of heat on nitrate	Occurs naturally as
K	↓ Steadily decreases	Violent at room temperature	Explosive. Gives off H_2	Tarnishes very rapidly $\rightarrow$ KOH + K_2CO_3	No reduction	Give the carbide at high temperatures	Strong bases stable to heat	Give the nitrite and oxygen	Chloride
Na		Vigorous at room temperature	Violent. Gives off H_2	Tarnishes rapidly $\rightarrow$ NaOH + Na_2CO_3					Chloride
Ca		Steady reaction with cold water	Vigorous reaction giving H_2	Tarnishes rapidly $\rightarrow$ CaO, Ca(OH)$_2$, CaCO$_3$			Strong base, gives oxide on heating		Chloride and carbonate
Mg		Reacts with steam					Weak base, gives oxide on heating		Chloride and carbonate
Al			Give off H_2	Become coated with a protective layer of oxide			Amphoteric. Give oxide on heating		Oxide
Zn		Red hot metal reacts with steam			Reversible reduction		Weak base, gives oxide on heating	Give oxide, nitrogen dioxide and oxygen	Oxide and sulphide
Fe		Reversible		Rusts slowly	Rapid reduction	Reduction to the metal	Amphoteric. Gives oxide on heating		Oxide and sulphide
Pb				Oxidise readily on heating	Rapid reduction		Neutral		Sulphide
H									Oxide
Cu		No reaction	Do not liberate H_2. Attacked only by oxidising acids		Oxide decomposes on heating without the need for H_2	Decomposes by heat alone	Weak base, gives oxide on heating		Sulphide
Ag				Attacked only by traces of sulphur compounds			Unstable at room temperature	Gives the metal, nitrogen dioxide, and oxygen	Sulphide and as free metal

Figure 11.1. The activity series

The Chemistry of some Metals

being sodium chloride. Since sodium has such a strong tendency to exist as Na^+ ions, its compounds are very difficult to reduce and the metal can only be extracted by electrolysis. The manufacture of sodium by electrolysis of the fused chloride is described on page 88.

Properties and reactions of sodium

Sodium is a very soft, light metal which can easily be cut with a knife. It is generally kept under an organic liquid such as liquid paraffin to protect it from the atmosphere. When freshly cut, it has a silvery lustre but it rapidly tarnishes due to reaction with moisture and carbon dioxide in the atmosphere.

Sodium reacts vigorously with cold water giving hydrogen and sodium hydroxide:

$$2Na(s) + 2H_2O(l) \rightarrow 2NaOH(aq) + H_2(g)$$

The heat generated melts the sodium (its m.p. is only 98 °C) which darts about the surface of the water as a molten globule. This ability of sodium to reduce hydrogen ions is predictable because sodium is higher up the electrochemical series than hydrogen. However, only a few metals are sufficiently reactive to reduce the hydrogen ions in cold water.

$$H_2O(l) \rightleftharpoons H^+(aq) + HO^-(aq)$$
$$Na(s) + H^+(aq) \rightarrow Na^+(aq) + H; \quad 2H \rightarrow H_2(g)$$

Sodium reacts with most non-metals on heating.

Uses of sodium

Metallic sodium does not have wide industrial application. It is used as the reducing agent in the extraction of titanium, in the preparation of lead tetraethyl ($(C_2H_5)_4Pb$ – 'anti-knock' in petrol), and in small amounts in vapour discharge lamps for street lighting.

Compounds of sodium

Practically all the compounds of sodium are white or colourless and soluble in water.

Oxides

(a) *Sodium peroxide, Na_2O_2*

Sodium burns, with a golden yellow flame, in excess air or oxygen giving sodium peroxide:

$$2Na(s) + O_2(g) \rightarrow Na_2O_2(s)$$

The valency of sodium here is one, as expected, the structure of the compound being Na^+ $^-O-O^-$ Na^+.

Sodium peroxide is a pale yellow solid which reacts readily with cold water to give a solution containing sodium hydroxide and hydrogen peroxide:

$$Na_2O_2(s) + 2H_2O(l) \rightarrow 2NaOH(aq) + H_2O_2(aq)$$

If the reaction mixture is not cooled, the hydrogen peroxide decomposes into water and oxygen:

$$2H_2O_2(aq) \rightarrow 2H_2O(l) + O_2(g)$$

Sodium peroxide reacts with carbon dioxide, giving sodium carbonate and oxygen,

$$2Na_2O_2(s) + 2CO_2(g) \rightarrow 2Na_2CO_3(s) + O_2(g)$$

and so is used for purifying air in submarines and portable breathing apparatus.

(b) *Disodium oxide, Na_2O*

If sodium is burnt in a limited supply of air or oxygen, or if sodium peroxide is gently heated with sodium, the product is disodium oxide:

$$4Na(s) + O_2(g) \rightarrow 2Na_2O(s)$$
$$Na_2O_2(s) + 2Na(s) \rightarrow 2Na_2O(s)$$

Disodium oxide is a white solid which reacts violently with water to give sodium hydroxide solution.

$$Na_2O(s) + H_2O(l) \rightarrow 2NaOH(aq)$$

Sodium hydroxide, NaOH

Sodium hydroxide solution is formed when sodium reacts with water (see above) but it is manufactured by the electrolysis of brine (page 89).

Sodium hydroxide is a white deliquescent solid, i.e. it absorbs so much moisture from the atmosphere that it forms a solution in it. It is very soluble in water and much heat is evolved in the process. The solution is strongly alkaline, feels soapy, and attacks the skin – this accounts for its old name of caustic soda. Sodium hydroxide is a strong base, i.e. fully dissociated into ions, and reacts with all acids and acidic oxides, e.g:

$$2NaOH(aq) + H_2SO_4(aq) \rightarrow Na_2SO_4(aq) + 2H_2O(l)$$
$$2NaOH(aq) + CO_2(g) \rightarrow Na_2CO_3(aq) + H_2O(l)$$

Sodium hydroxide precipitates the hydroxides of many metals when added to solutions of their salts, e.g.:

$$CuSO_4(aq) + 2NaOH(aq) \rightarrow Cu(OH)_2(s) + Na_2SO_4(aq)$$
i.e. $$Cu^{2+}(aq) + 2HO^-(aq) \rightarrow Cu(OH)_2(s)$$

The hydroxides of most metals decompose on heating to give the metal oxide and water. However, sodium hydroxide is thermally stable.

Sodium carbonate, Na_2CO_3

Sodium carbonate is manufactured by the Solvay process, which consists of a very efficient series of reactions in which the only waste product is calcium chloride. Brine (saturated sodium chloride solution) is saturated with ammonia and then carbon dioxide is bubbled through. The carbon dioxide reacts with the ammonia solution to give ammonium hydrogencarbonate,

$$NH_3(aq) + H_2O(l) + CO_2(g) \rightarrow NH_4HCO_3(aq)$$

which in turn reacts with the sodium chloride:

$$NH_4HCO_3(aq) + NaCl(aq) \rightarrow NaHCO_3(s) + NH_4Cl(aq)$$

Sodium hydrogencarbonate is not very soluble under these conditions and is precipitated. The precipitate is filtered off and warmed:

$$2NaHCO_3(s) \rightarrow Na_2CO_3(s) + H_2O(l) + CO_2(g)$$

The carbon dioxide from this decomposition is recycled along with some obtained from the thermal decomposition of limestone:

$$CaCO_3(s) \rightarrow CaO(s) + CO_2(g)$$

The calcium oxide is warmed with the ammonium chloride to regenerate the ammonia for further use:

$$2NH_4Cl(aq) + CaO(s) \rightarrow CaCl_2(aq) + H_2O(l) + 2NH_3(g)$$

The process is summarised in Figure 11.2, the solid arrows indicating the main reactions whilst the broken arrows show the regeneration of reactants.

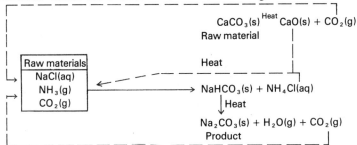

Figure 11.2. Summary of the Solvay process

The anhydrous sodium carbonate produced above is a white powder. Crystallisation from water gives the decahydrate, $Na_2CO_3 \cdot 10H_2O$ which is sold under the name of washing soda. The decahydrate is *efflorescent*, i.e. it loses some of its water of crystallisation on exposure to the atmosphere, and the monohydrate, $Na_2CO_3 \cdot H_2O$ is left as a white powder. Heating the monohydrate gives the anhydrous salt.

In contrast to the carbonates of most metals, anhydrous sodium carbonate does not decompose on heating.

The carbonate ion in aqueous solution undergoes reversible hydrolysis:

$$CO_3^{2-}(aq) + H_2O(l) \rightleftharpoons HCO_3^-(aq) + HO^-(aq)$$
hydrogencarbonate ion

and as a result sodium carbonate solution is alkaline.

Sodium carbonate, like the other carbonates, yields carbon dioxide on treatment with acids, e.g:

$$Na_2CO_3(s) + 2HCl(aq) \rightarrow 2NaCl(aq) + H_2O(l) + CO_2(g)$$

i.e. $$CO_3^{2-} + 2H^+(aq) \rightarrow H_2O(l) + CO_2(g)$$

Sodium carbonate is used to make glass (page 187) and soap, in water softening (page 57), and in the volumetric analysis of acids (page 77). Sodium carbonate solution is also used to prepare the carbonates of other metals, e.g:

$$CaCl_2(aq) + Na_2CO_3(aq) \rightarrow CaCO_3(s) + 2NaCl(aq)$$

i.e. $$Ca^{2+}(aq) + CO_3^{2-}(aq) \rightarrow CaCO_3(s)$$

In several cases, however, it is the basic carbonate (a co-precipitate of the carbonate and hydroxide) which is precipitated since, as explained above, sodium carbonate solution contains hydroxide ions. Thus, addition of sodium carbonate solution to copper(II) sulphate solution gives a pale blue precipitate, $Cu(OH)_2 \cdot 2CuCO_3$.

Sodium hydrogencarbonate, $NaHCO_3$

Sodium hydrogencarbonate may be made in the laboratory by passing excess carbon dioxide through sodium hydroxide solution:

$$NaOH(aq) + CO_2(g) \rightarrow NaHCO_3(aq)$$

or through sodium carbonate solution:

$$Na_2CO_3(aq) + H_2O(l) + CO_2(g) \rightarrow 2NaHCO_3(aq)$$

Industrially, the solid is obtained as an intermediate in the manufacture of sodium carbonate as described above.

Sodium hydrogencarbonate is one of the few hydrogencarbonates that can be obtained in the solid state. It is a white powder which readily decomposes on warming whilst its solution decomposes on boiling.

$$2NaHCO_3(s) \rightarrow Na_2CO_3(s) + H_2O(l) + CO_2(g)$$

Addition of acids to sodium hydrogencarbonate results in the liberation of carbon dioxide, e.g:

$$NaHCO_3(aq) + HNO_3(aq) \rightarrow NaNO_3(aq) + H_2O(l) + CO_2(g)$$

i.e. $$HCO_3^-(aq) + H^+(aq) \rightarrow H_2O(l) + CO_2(g)$$

Sodium hydrogencarbonate is sold as baking soda – the carbon dioxide evolved when it decomposes causes pastry, cakes, etc., to rise. Sodium hydrogencarbonate is also a constituent of health salts.

Sodium chloride, NaCl

This occurs naturally as rock salt and it is present in seawater. It is made in the laboratory by neutralising hydrochloric acid with sodium hydroxide solution by method (b) on page 75.

$$NaOH(aq) + HCl(aq) \rightarrow NaCl(aq) + H_2O(l)$$

Sodium chloride (common salt) is a white solid which is fairly soluble in water. Its reaction with concentrated sulphuric acid is described on page 214. It is essential to the human diet and is used in the manufacture of sodium, sodium hydroxide, and chlorine (see pages 88 to 89).

Sodium nitrate, $NaNO_3$

Sodium nitrate occurs in Chile and is often known as Chilean saltpetre. It may be made in the laboratory by neutralisation of sodium hydroxide with nitric acid:

$$NaOH(aq) + HNO_3(aq) \rightarrow NaNO_3(aq) + H_2O(l)$$

Sodium nitrate is a hygroscopic white solid and is very soluble in water. Like potassium nitrate, it decomposes on heating into the nitrite and oxygen:

$$2NaNO_3(s) \rightarrow 2NaNO_2(s) + O_2(g)$$

whereas most metal nitrates give the metal oxide (see page 198).

Sodium nitrate is used as a fertiliser.

Sodium sulphate, Na_2SO_4

Sodium sulphate may be made by strongly heating sodium chloride with concentrated sulphuric acid:

$$2NaCl(s) + H_2SO_4(conc) \rightarrow Na_2SO_4(s) + 2HCl(g)$$

Alternatively, dilute sulphuric acid may be neutralised with sodium hydroxide solution:

$$H_2SO_4(aq) + 2NaOH(aq) \rightarrow Na_2SO_4(aq) + 2H_2O(l)$$

Sodium sulphate crystallises from aqueous solutions as the decahydrate $Na_2SO_4 \cdot 10H_2O$ but this effloresces to give the anhydrous salt.

The decahydrate is known as Glauber's salt; it has a laxative action and is a constituent of many health salts. The anhydrous salt is used in the manufacture of glass (page 187).

Sodium hydrogensulphate, $NaHSO_4$

This may be made by adding dilute sulphuric acid to half the volume of sodium hydroxide solution required to neutralise it:

$$H_2SO_4(aq) + NaOH(aq) \rightarrow NaHSO_4(aq) + H_2O(l)$$

Sodium hydrogensulphate is an acid salt and gives a strongly acidic solution.

Test for sodium ions

Sodium compounds are detected by the *flame test*. Thus, a clean platinum wire is dipped in concentrated hydrochloric acid, just touched in the sodium compound, and then held in a non-luminous flame. A sodium compound is indicated by an intense golden yellow flame. Hydrochloric acid is used in the flame test since it gives the metal chloride which is generally more volatile than the other salts.

POTASSIUM

Occurrence and extraction

Potassium, like sodium, occurs as the chloride and it is extracted in a similar way. The metal itself has few uses.

Compounds of potassium

These are generally similar to those of sodium but they tend to be less hygroscopic and less soluble in water. Most of the common potassium compounds are white or colourless but two common exceptions are potassium manganate(VII), which is purple, and potassium dichromate(VI), which is orange.

Potassium hydroxide, KOH

This is made in a similar manner as, and has similar reactions to, sodium hydroxide. Aqeuous potassium hydroxide is often used in preference to sodium hydroxide for absorbing carbon dioxide. This is because potassium carbonate is more soluble than sodium carbonate and so there is less chance of it being deposited.

Potassium chloride, KCl

As stated above, potassium chloride occurs naturally. It is an important fertiliser.

Potassium iodide, KI

This may be prepared in the laboratory by neutralising hydriodic acid with potassium hydroxide solution:

$$HI(aq) + KOH(aq) \rightarrow KI(aq) + H_2O(l)$$

Potassium iodide exists as white crystals which are very soluble in water. The aqueous solution dissolves iodine to give potassium triiodide:

$$I_2(s) + KI(aq) \rightleftharpoons KI_3(aq)$$
i.e.
$$I_2(s) + I^-(aq) \rightleftharpoons I_3^-(aq)$$

The reaction is readily reversible and the solution behaves as if it contains iodine – this is important because iodine itself is very sparingly soluble in water. The normal iodine solutions seen in the laboratory are, in fact, solutions of potassium triiodide.

Potassium nitrate, KNO_3

Potassium nitrate may be made in the laboratory by neutralisation of nitric acid with potassium hydroxide:

$$HNO_3(aq) + KOH(aq) \rightarrow KNO_3(aq) + H_2O(l)$$

It is a white solid which, like sodium nitrate, decomposes on strong heating to give the nitrite:

$$2KNO_3(s) \rightarrow 2KNO_2(s) + O_2(g)$$

Potassium nitrate is used as a fertiliser and in fireworks and gunpowder (a mixture of potassium nitrate, carbon, and sulphur). Potassium nitrate, rather than sodium nitrate, is used as the oxidising agent in gunpowder because it is not hygroscopic.

Potassium chlorate(V), $KClO_3$

The reaction between chlorine and hot concentrated potassium hydroxide solution yields potassium chlorate(V) together with potassium chloride:

$$6KOH(aq) + 3Cl_2(g) \rightarrow 5KCl(aq) + KClO_3(aq) + 3H_2O(l)$$

Potassium chlorate(V) is a white solid and a powerful oxidising agent. It is used in matches, fireworks, etc. in preference to sodium chlorate(V) since the latter compound is deliquescent. The thermal decomposition of potassium chlorate(V), with manganese(IV) oxide as catalyst, may be used as a laboratory preparation of oxygen:

$$2KClO_3(s) \xrightarrow{MnO_2} 2KCl(s) + 3O_2(g)$$

Test for potassium ions

The flame test with potassium salts gives a lilac flame which is visible even through blue glass.

MAGNESIUM AND CALCIUM-GROUP II METALS

The first three elements in Group II are beryllium, magnesium, and calcium and they each have two electrons in their outer shell: Be, 2.2; Mg, 2.8.2; and Ca, 2.8.8.2. The elements in this group are often known as the *alkaline earths*. The chemistry of beryllium is beyond the scope of this text but the general tendency in the group is to achieve the electronic structure of the noble gases by losing two electrons, e.g:

$$Mg - 2e^- \rightarrow Mg^{2+}$$
$$2.8.2 \qquad 2.8$$

It is more difficult for magnesium to achieve the noble gas electronic structure than sodium. This is because the second electron has to be removed from an ion which is already positively charged and which holds on to its remaining electrons more tightly:

$$Mg \xrightarrow{\text{relatively easy}} Mg^+ + e^- \xrightarrow{\text{more difficult}} Mg^{2+} + e^-$$

Magnesium is therefore less reactive than sodium and this is illustrated by, for example, their reactions with water (see Figure 11.1 and below). Calcium is more reactive than magnesium because its two outer electrons are further from the nucleus and so less tightly held. However, the reactions of magnesium and calcium are basically similar and the valency of both elements is always two.

MAGNESIUM

Occurrence and extraction

Magnesium is too reactive to occur as the free metal but it is found as the carbonate, sulphate, chloride, etc. It is present in seawater, in low concentrations, as the bromide, chloride, and sulphate. Large quantities of magnesium are extracted from seawater by converting the salt to the hydroxide and then heating this to give the oxide:

$$Mg(OH)_2(s) \rightarrow MgO(s) + H_2O(l)$$

The magnesium oxide is heated with coke in a current of chlorine:

$$MgO(s) + C(s) + Cl_2(g) \rightarrow MgCl_2(s) + CO(g)$$

and the resultant chloride is fused and electrolysed:

$$\text{Cathode} \quad Mg^{2+} + 2e^- \rightarrow Mg$$
$$\text{Anode} \quad Cl^- - e^- \rightarrow Cl; \quad 2Cl \rightarrow Cl_2$$

The extraction of magnesium involves the use of very large quantities of sea water and so the plant is obviously situated near the coast to eliminate transport costs.

Properties and reactions of magnesium

Magnesium is a silvery coloured metal with a very low density, i.e. only about 1.7 times that of water. It is a highly electropositive divalent metal and so it has a strong affinity for oxygen and other electronegative elements.

Magnesium tarnishes in air due to the formation of an oxide film but this prevents further attack. However, when heated above its melting point (650 °C) in air, magnesium ignites with an intense white flame to give the oxide and a little of the nitride:

$$2Mg(s) + O_2(g) \rightarrow 2MgO(s)$$
$$3Mg(s) + N_2(g) \rightarrow Mg_3N_2(s)$$

Since magnesium is a very electropositive metal near the top of the electrochemical and activity series, it may have been expected to react readily with cold water. Its failure to do so is a result of the protective oxide film. Reaction does occur slowly with boiling water and it burns when heated strongly in steam (Figure 11.3):

$$Mg(s) + H_2O(g) \rightarrow MgO(s) + H_2(g)$$

The hydroxide is not formed as it would decompose at this temperature.

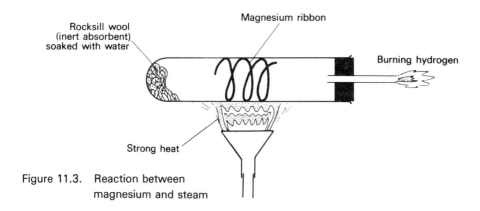

Figure 11.3. Reaction between magnesium and steam

Magnesium is well above hydrogen in the electrochemical and activity series and so predictably it reacts rapidly with dilute acids, e.g:

$$Mg(s) + H_2SO_4(aq) \rightarrow MgSO_4(aq) + H_2(g)$$
i.e.
$$Mg(s) + 2H^+(aq) \rightarrow Mg^{2+}(aq) + H_2(g)$$

Hydrogen is also evolved when magnesium is added to very dilute (less than 4%) nitric acid. With more concentrated nitric acid, oxides of nitrogen are produced instead of hydrogen. The other metals will not liberate hydrogen from nitric acid under any conditions.

Magnesium displaces copper, and the other metals below it in the electrochemical and activity series, from solutions of their salts (see page 105). There is no reaction between magnesium and alkalis.

Uses of magnesium

Strong light alloys of magnesium with zinc, aluminium, etc. are used in cars and aircraft. Magnesium is also used in flares, photographic flash bulbs, fireworks, and incendiary bombs.

Compounds of magnesium

All the common magnesium compounds are white or colourless. The oxide, hydroxide, and carbonate have low solubilities in water.

Magnesium oxide, MgO

This is made by any of the general methods for preparing metal oxides, i.e. by heating the element in oxygen or by the thermal decomposition of the hydroxide, carbonate, or nitrate.

Magnesium oxide is a white powder, slightly soluble in water, giving a mildly alkaline solution. It reacts with acids giving the corresponding salts. It has a very high melting point (2900 °C) and is extremely stable and so is used as a refactory lining in furnaces. Magnesium oxide is also used in various preparations to counter gastric acidity.

Magnesium hydroxide, $Mg(OH)_2$

This is formed as a white precipitate when sodium hydroxide or ammonia solution is added to a solution of a magnesium salt, e.g:

$$MgSO_4(aq) + 2NaOH(aq) \rightarrow Mg(OH)_2(s) + Na_2SO_4(aq)$$
i.e. $$Mg^{2+}(aq) + 2HO^-(aq) \rightarrow Mg(OH)_2(s)$$

Magnesium hydroxide is only slightly soluble in water and, like the oxide, is used to neutralise gastric acidity.

Magnesium carbonate, $MgCO_3$

This occurs naturally. Addition of sodium carbonate solution to a solution of a magnesium salt results in the precipitation of a basic carbonate, i.e. carbonate and hydroxide are precipitated together (see page 140). The simple carbonate is formed when solutions of sodium hydrogencarbonate and a magnesium salt are mixed and warmed, e.g:

$$MgSO_4(aq) + 2NaHCO_3(aq) \rightarrow Na_2SO_4(aq) + Mg(HCO_3)_2(aq)$$
$$\downarrow warm$$
$$MgCO_3(s) + H_2O(l) + CO_2(g)$$

Magnesium carbonate goes into solution as the hydrogencarbonate when it comes into contact with water containing dissolved carbon dioxide:

$$MgCO_3(s) + H_2O(l) + CO_2(g) \rightarrow Mg(HCO_3)_2(aq)$$

This causes temporary hardness in water (page 56).

Magnesium carbonate is a mild abrasive and is used in toothpastes. It also finds use as a filler for paper and paints.

Magnesium sulphate, $MgSO_4$

This occurs naturally and is present in seawater. It can be made in the laboratory from dilute sulphuric acid solution and the metal, oxide, hydroxide, or carbonate, e.g:

$$MgCO_3(s) + H_2SO_4(aq) \rightarrow MgSO_4(aq) + H_2O(l) + CO_2(g)$$

It crystallises as the heptahydrate, $MgSO_4 \cdot 7H_2O$ which is sold as Epsom salt – a laxative. Magnesium sulphate is one of the causes of permanent hardness (page 56).

Test for magnesium ions

Addition of sodium hydroxide or ammonia solution to an aqueous solution of magnesium ions results in a white precipitate of magnesium hydroxide; the precipitate is not soluble in excess of either reagent.

CALCIUM

Occurrence and extraction

Calcium is far too reactive to occur naturally in the free state but it is abundant as the carbonate, sulphate, fluoride, and silicate. It is extracted by electrolysis of the fused chloride, a little calcium fluoride being added to lower the melting point. The cathode is steel but the anode has to be graphite since chlorine attacks steel.

$$\text{Cathode} \quad Ca^{2+} + 2e^- \rightarrow Ca$$
$$\text{Anode} \quad Cl^- - e^- \rightarrow Cl; \quad 2Cl \rightarrow Cl_2$$

Properties and reactions of calcium

Calcium is a fairly hard grey metal. It tarnishes rapidly on exposure to air, forming a layer of oxide and hydroxide. It is a highly electropositive metal and so readily combines with electronegative elements such as oxygen, sulphur, and the halogens on heating.

Calcium reacts steadily with cold water and rapidly with hot:

$$Ca(s) + 2H_2O(l) \rightarrow Ca(OH)_2(aq) + H_2(g)$$

The mixture may turn milky because calcium hydroxide is not very soluble in water. Rapid evolution of hydrogen also occurs when calcium is added to dilute acids (except nitric acid), e.g.:

$$Ca(s) + 2HCl(aq) \rightarrow CaCl_2(aq) + H_2(g)$$
i.e. $$Ca(s) + 2H^+(aq) \rightarrow Ca^{2+}(aq) + H_2(g)$$

These reactions with water and acids are consistent with the position of calcium in the electrochemical and activity series.

Uses of calcium

The metal does not find wide application. Some calcium is used to manufacture calcium hydride, CaH_2 – this reacts with water to give hydrogen and so is used as a convenient portable source of this gas.

Compounds of calcium

All the common calcium compounds are white. They are also generally soluble but the sulphate and carbonate are exceptions.

Calcium oxide, CaO

This is made industrially by heating limestone at about 800 °C in a stream of air, the function of the air being to remove the carbon dioxide and prevent the reverse reaction.

$$CaCO_3(s) \rightleftharpoons CaO(s) + CO_2(g)$$

It may be made in the laboratory by any of the general methods (see page 72).

Calcium oxide is a white powder often known as *quicklime*. It undergoes a strongly exothermic reaction with water to give the hydroxide, this compound often being known as *slaked lime*. The reaction is accompanied by a hissing sound, steam is formed, and the volume increases visibly.

$$CaO(s) + H_2O(l) \rightarrow Ca(OH)_2(s)$$

Calcium oxide reacts vigorously with dilute acids, at room temperature, to give the corresponding salts, e.g:

$$CaO(s) + 2HNO_3(aq) \rightarrow Ca(NO_3)_2(aq) + H_2O(l)$$

It reacts with acidic oxides on heating, the reaction with silicon(IV) oxide being of great industrial importance (see page 163):

$$CaO(s) + SiO_2(s) \rightarrow CaSiO_3(s)$$

Calcium oxide is used in the manufacture of glass and as a fertiliser. Its stability and very high melting point (2600 °C) also make it a useful refractory lining in furnaces.

Calcium hydroxide, Ca(OH)$_2$

This is made by adding water to calcium oxide:

$$CaO(s) + H_2O(l) \rightarrow Ca(OH)_2(s)$$

It is a white solid only slightly soluble in water and, unlike most substances, its solubility decreases on raising the temperature. The solution is alkaline and is called *limewater*; it is used to test for carbon dioxide (see page 45). An aqueous suspension of calcium hydroxide is known as *milk of lime* in view of its appearance.

Calcium hydroxide reacts with acids and acidic oxides giving the calcium salt and water. Passing carbon dioxide through limewater gives a milky precipitate of calcium carbonate:

$$Ca(OH)_2(aq) + CO_2(g) \rightarrow CaCO_3(s) + H_2O(l)$$

However, on continued passage of carbon dioxide, the precipitate redissolves to give a colourless solution of calcium hydrogencarbonate:

$$CaCO_3(s) + H_2O(l) + CO_2(g) \rightarrow Ca(HCO_3)_2(aq)$$

This process results in temporary hardness in water (see page 56); boiling the solution reverses the reaction.

The passage of chlorine through moist calcium hydroxide, with the temperature kept below 35 °C, produces bleaching powder. The exact composition of this product is open to doubt but it is thought to contain, amongst other things, calcium chlorate(I), $Ca(OCl)_2$.

Mixtures of calcium hydroxide and ammonium salts yield ammonia on heating, e.g:

$$Ca(OH)_2(s) + 2NH_4Cl(s) \rightarrow CaCl_2(s) + 2H_2O(l) + 2NH_3(g)$$

i.e.
$$HO^- + NH_4^+ \rightarrow H_2O + NH_3$$

Calcium hydroxide is used in builders' mortar and plaster, water softening (page 57), to counter acidity in soil, and in the manufacture of bleaching powder (see above).

Calcium carbonate, $CaCO_3$

This occurs naturally as chalk, limestone, marble, etc. It is insoluble in pure water but it dissolves slowly in the presence of dissolved carbon dioxide, giving calcium hydrogencarbonate, thus accounting for temporary hard water in limestone areas.

$$CaCO_3(s) + H_2O(l) + CO_2(g) \rightarrow Ca(HCO_3)_2(aq)$$

Calcium carbonate is used to make quicklime and in the manufacture of cement. It is a constituent of some toothpastes.

Calcium chloride, $CaCl_2$

This is obtained as a by-product from the Solvay process for manufacturing sodium carbonate (page 139). The hexahydrate, $CaCl_2 \cdot 6H_2O$ may be obtained in the laboratory from the reaction between calcium carbonate and hydrochloric acid:

$$CaCO_3(s) + 2HCl(aq) \rightarrow CaCl_2(aq) + H_2O(l) + CO_2(g)$$

whilst the anhydrous salt is obtained by direct combination of the elements.

Calcium chloride is used as a drying agent in the laboratory but there is no large scale use.

Calcium sulphate, $CaSO_4$

This occurs naturally both as the anhydrous salt and as the dihydrate. It is only slightly soluble in water but it is the major cause of permanent hardness in water.

When heated just above 100 °C, the dihydrate loses three quarters of its water of crystallisation to give the hemihydrate, $CaSO_4 \cdot \frac{1}{2}H_2O$, i.e. $(CaSO_4)_2 \cdot H_2O$. The hemihydrate, known as Plaster of Paris, sets to a hard mass of the dihydrate on addition of water.

$$2CaSO_4 \cdot 2H_2O \rightleftharpoons (CaSO_4)_2 \cdot H_2O + 3H_2O$$

Calcium sulphate is used to manufacture Plaster of Paris for plaster boards in building, making plaster casts, and for blackboard chalk.

Calcium phosphate, $Ca_3(PO_4)_2$

This compound occurs naturally and is a constituent of bones. Large quantities are used to make 'superphosphate', an important fertiliser. 'Superphosphate' is made by treating calcium sulphate with concentrated sulphuric acid:

$$Ca_3(PO_4)_2(s) + 2H_2SO_4(conc) \rightarrow \underbrace{Ca(H_2PO_4)_2(s) + 2CaSO_4(s)}_{\text{'Superphosphate'}}$$

Calcium dicarbide, CaC_2

This is made industrially by heating calcium oxide and powdered coke in an electric furnace at 2000 °C:

$$CaO(s) + 3C(s) \rightarrow CaC_2(s) + CO(g)$$

Calcium dicarbide reacts with cold water to give ethyne; this reaction is important both in industry and in the laboratory:

$$CaC_2(s) + 2H_2O(l) \rightarrow Ca(OH)_2(s) + CH\equiv CH(g)$$

Test for calcium ions

The flame test on calcium compounds results in a brick-red flame.

ALUMINIUM - A GROUP III METAL

The Group III elements all have three electrons in their outer shell. The first two members and their electronic structures are boron, 2.3 and aluminium, 2.8.3 but only aluminium will be discussed here.

Occurrence and extraction

Aluminium occurs as the silicate in rocks and clays but the main source is the oxide. Aluminium is extracted by electrolysis of the fused oxide, some sodium hexafluoroaluminate (*cryolite*), Na_3AlF_6, being added to lower the melting point from over 2000 °C to just under 1000 °C. Both electrodes are made of graphite.

Cathode $\quad Al^{3+} + 3e^- \rightarrow Al$
Anode $\quad\quad O^{2-} - 2e^- \rightarrow O; \quad 2O \rightarrow O_2$

This method of extraction is necessary because the oxide is unaffected by hydrogen even at high temperatures, heating with carbon gives the carbide, Al_4C_3, and the chloride is covalent. The process is normally carried out where cheap hydroelectric power is available.

Properties and reactions of aluminium

Aluminium is fairly high up the electrochemical and activity series and so it may have been expected to react with both air and water. However, there is no apparent reaction in either case, the reason being that the aluminium rapidly becomes covered with a thin layer of oxide which inhibits further attack. This oxide layer may be made thicker and stronger by the process of anodising (page 92). The oxide layer also accounts for the lack of reaction between aluminium and cold dilute hydrochloric and sulphuric acids. However, aluminium reacts vigorously with warm dilute hydrochloric acid once the oxide layer has dissolved:

$$2Al(s) + 6HCl(aq) \rightarrow 2AlCl_3(aq) + 3H_2(g)$$
i.e.
$$2Al(s) + 6H^+(aq) \rightarrow 2Al^{3+}(aq) + 3H_2(g)$$

The reaction with hot dilute sulphuric acid is similar.

There is little reaction with nitric acid, regardless of the concentration, since as an oxidising acid it aids the formation of the protective layer of oxide.

The reaction between aluminium and hot concentrated sulphuric acid gives aluminium sulphate, water, and sulphur dioxide:

$$2Al(s) + 6H_2SO_4(conc) \rightarrow Al_2(SO_4)_3(s) + 6H_2O(l) + 3SO_2(g)$$

Note that hot concentrated sulphuric acid behaves as an oxidising agent (page 101) and is itself reduced to sulphur dioxide during the process: hydrogen is never evolved under these conditions.

Aluminium, like most of the other elements near the centre of the Periodic Table, reacts with alkalis as well as acids. The reaction is rather complex but it may be written as:

$$2Al(s) + 2NaOH(aq) + 6H_2O(l) \rightarrow 2NaAl(OH)_4(aq) + 3H_2(g)$$
$$\text{Sodium aluminate}$$

Aluminium combines with oxygen, sulphur, chlorine, etc., on heating.

Uses of aluminium

Aluminium is resistant to corrosion and it and its compounds are non-toxic, and so it is an ideal metal for cooking utensils. Aluminium foil is used for milk bottle tops. Alloys of aluminium and magnesium or copper are very strong and have low densities, and so they are used extensively in aircraft and ships. It is used to a smaller extent in the *thermit process* for welding iron. In the thermit process, a mixture of iron(III) oxide and aluminium powder is ignited with a magnesium fuse. The iron(III) oxide is reduced by the aluminium and so much heat is evolved in the process that the iron melts.

$$Fe_2O_3(s) + 2Al(s) \rightarrow Al_2O_3(s) + 2Fe(l)$$
i.e.
$$Fe^{3+}(s) + Al(s) \rightarrow Al^{3+}(s) + Fe(l)$$

Compounds of aluminium

The Al^{3+} ion is very small and highly charged and so it has a great attraction for the electrons of anions. The outer electrons of the anions are therefore pulled towards the aluminium and so covalent character will result to a greater or lesser degree, i.e. there is some sharing of the electrons. This process is illustrated in Figure 11.4 which shows the effect of the attraction of the Al^{3+} ion for Cl^- ions; aluminium chloride is in fact a typical covalent compound (see below). Many aluminium compounds are predominantly covalent but the fluoride, oxide, and sulphate are ionic.

Aluminium compounds are generally white and many are insoluble.

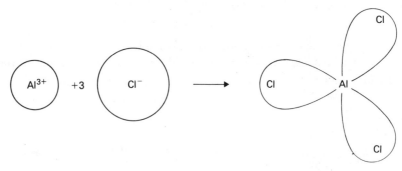

Figure 11.4. The covalency of aluminium chloride

Aluminium oxide, Al_2O_3

This occurs naturally in several forms. Ruby, emerald, sapphire, and topaz are all aluminium oxide along with traces of impurity which provide the colour. Aluminium oxide is a white crystalline solid which can be made by the action of heat on the hydroxide:

$$2Al(OH)_3(s) \rightarrow Al_2O_3(s) + 3H_2O(g)$$

The oxide is amphoteric when it is freshly prepared but it 'ages' to become very unreactive. The amphoteric behaviour is to be expected since metals have basic oxides, non-metals have acidic oxides, and aluminium is in the centre of the Periodic Table where the change from metallic to non-metallic behaviour takes place.

Aluminium oxide is used as the source of aluminium, as an abrasive (e.g. in emery paper), and as an adsorbent in chromatography (see page 6). As a result of its stability and very high melting point (2040 °C), it is used as a refractory lining in furnaces.

Aluminium hydroxide, $Al(OH)_3$

This is best made by the addition of excess ammonia solution to a solution of an aluminium salt, e.g:

$$Al_2(SO_4)_3(aq) + 6NH_3(aq) + 6H_2O(l) \rightarrow 2Al(OH)_3(s) + 3(NH_4)_2SO_4(aq)$$

The aluminium hydroxide is obtained as a white gelatinous precipitate. It is amphoteric, e.g:

$$Al(OH)_3(s) + 3HNO_3(aq) \rightarrow Al(NO)_3(aq) + 3H_2O(l)$$
$$Al(OH)_3(s) + NaOH(aq) \rightarrow NaAl(OH)_4(aq)$$

Addition of aqueous sodium hydroxide to a solution of an aluminium salt will therefore initially give a white precipitate of aluminium hydroxide but this will redissolve, as more sodium hydroxide is added, and sodium aluminate is formed.

Aluminium hydroxide is used in mordant dyeing, i.e. it combines with both the cloth and the dye, thus making the dye 'fast' on the material.

Aluminium chloride, AlCl₃

The hexahydrate, $AlCl_3 \cdot 6H_2O$, is made by dissolving aluminium in hot dilute hydrochloric acid, concentrating the solution, and then allowing it to crystallise.

Anhydrous aluminium chloride is prepared by heating aluminium in a current of chlorine (Figure 11.5):

$$2Al(s) + 3Cl_2(g) \rightarrow 2AlCl_3(s)$$

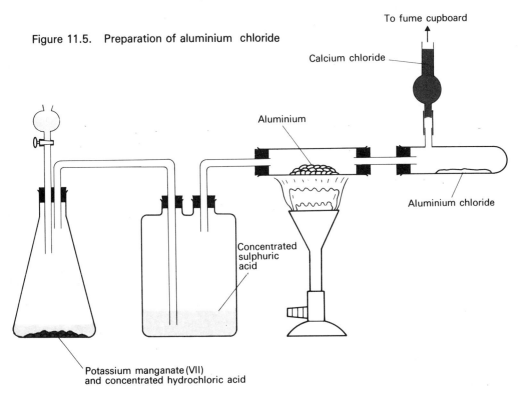

Figure 11.5. Preparation of aluminium chloride

It is extremely deliquescent, sublimes just below 200 °C, and dissolves in organic solvents. It fumes in moist air owing to reaction with it, i.e. it is hydrolysed:

$$AlCl_3(s) + 6H_2O(l) \rightarrow [Al(H_2O)_6]^{3+}(aq) + 3Cl^-(aq)$$

This is not unexpected because, as a general rule, metals form chlorides which dissolve in water without hydrolysis, non-metals form chlorides which are readily hydrolysed, and aluminium is in the intermediate region of the Periodic Table.

From the above properties, and the fact that it will not pass an electric current, it is apparent that aluminium chloride is covalent. If the molecular formula was $AlCl_3$, the aluminium would have only six electrons in its outer shell. However, relative molecular mass measurements show that both the solid and its vapour exist as a dimer, i.e. a double molecule, Al_2Cl_6. Each aluminium therefore has a share of eight electrons in its outer shell as a result of two co-ordinate bonds being formed:

$$\begin{array}{c} Cl \diagdown \quad \diagup Cl \diagdown \quad \diagup Cl \\ Al \quad \quad Al \\ \diagup \quad \diagdown \quad \diagup \quad \diagdown \\ Cl \quad \quad Cl \quad \quad Cl \end{array}$$

Anhydrous aluminium chloride is used as a catalyst in a number of reactions in organic chemistry.

Aluminium sulphate, $Al_2(SO_4)_3$

This may be prepared by dissolving aluminium hydroxide in dilute sulphuric acid. The salt crystallises as colourless crystals, $Al_2(SO_4)_3 \cdot 18H_2O$.

$$2Al(OH)_3(s) + 3H_2SO_4(aq) \rightarrow Al_2(SO_4)_3(aq) + 6H_2O(l)$$

Aluminium sulphate is used for sizing paper, i.e. giving it body and strength, in tanning leather, and for preparing the aluminium hydroxide for mordant dyeing. It is also used for sewage treatment since it coagulates the very small particles suspended in water effluents.

Test for aluminium ions

Addition of aqueous ammonia to a solution of an aluminium salt gives a white precipitate of the hydroxide. The precipitate is insoluble in excess ammonia solution, but it does dissolve in sodium hydroxide solution.

LEAD – A GROUP IV METAL

The Group IV elements all have four electrons in their outer shell. The first two elements in the group are carbon and silicon and they are typical non-metals. However, metallic character increases down all groups as the outer electrons of the atoms become further from the nucleus and so more easily lost. Lead, the last element in the group, is a weak metal.

Occurrence and extraction

The main source of lead is the sulphide (*galena*). The lead is extracted by heating the ore in air to give the oxide which is then reduced by heating with carbon:

$$2PbS(s) + 3O_2(g) \rightarrow 2PbO(s) + 2SO_2(g)$$
$$PbO(s) + C(s) \rightarrow Pb(s) + CO(g)$$

The sulphur dioxide is used to make sulphuric acid.

Properties and reactions of lead

Lead is a shiny soft grey metal which tarnishes slowly in air, the resultant film of hydroxide and carbonate preventing further attack. It is very dense, low melting (328 °C), and has a low tensile strength.

Lead is not very reactive and it is unaffected by water or steam. It is just above hydrogen in the electrochemical and activity series and so should liberate hydrogen from acids. However, there is negligible reaction with dilute hydrochloric and sulphuric acids because lead chloride and sulphate are insoluble and so a protective layer is formed. Lead dissolves slowly in cold concentrated hydrochloric acid but rapidly on heating. (Lead(II) chloride is considerably more soluble as the temperature is raised and, in fact, it forms a soluble complex, H_2PbCl_4, in concentrated hydrochloric acid. A knowledge of this complex is not required at this level of work.)

$$Pb(s) + 2HCl(conc) \rightarrow PbCl_2 + H_2(g)$$

Lead undergoes the usual reaction of metals with hot concentrated sulphuric acid, i.e. it is oxidised to the sulphate whilst the acid is reduced to sulphur dioxide:

$$Pb(s) + 2H_2SO_4(conc) \rightarrow PbSO_4(s) + 2H_2O(l) + SO_2(g)$$

The metal also dissolves in warm dilute nitric acid:

$$3Pb(s) + 8HNO_3(aq) \rightarrow 3Pb(NO_3)_2(aq) + 4H_2O(l) + 2NO(g)$$

and in concentrated nitric acid:

$$Pb(s) + 4HNO_3(conc) \rightarrow Pb(NO_3)_2(aq) + 2H_2O(l) + 2NO_2(g)$$

Uses of lead

The use of lead and its compounds is tending to go out of favour in view of the cost and toxicity. Before the advent of plastics, it was widely used for water and gas pipes because there are no corrosion problems, it is malleable, and it readily bends. Cracks are easily mended and joints readily made because it melts just above 300 °C.

Lead is used as a shield against radioactive material and X-rays. The metal is also used in car batteries and accumulators. A number of lead alloys are important; for example, solder contains lead and tin in the ratio of two to one, whilst type metal for use in printing is an alloy of lead, antimony, and tin.

Compounds of lead

Lead can have valencies of two or four. The divalent state is most stable and important and these compounds are predominantly ionic. Compounds containing tetravalent lead are covalent since too much energy would be required to remove four electrons from the atom to give a Pb^{4+} ion.

Many lead compounds are white or colourless but the oxides, iodide, and sulphide are exceptions (see below). Most lead salts are insoluble in water, the only common soluble ones being the nitrate and ethanoate (acetate).

Oxides

(a) *Lead(II) oxide, PbO*

Heating lead in air at 800 °C gives lead(II) oxide as an orange powder:

$$2Pb(s) + O_2(g) \rightarrow 2PbO(s)$$

The same product is obtained by heating the nitrate, carbonate, or hydroxide, but in this case a yellow powder is formed, the colour difference being due to the smaller particle size.

$$2Pb(NO_3)_2(s) \rightarrow 2PbO(s) + 4NO_2(g) + O_2(g)$$
$$PbCO_3(s) \rightarrow PbO(s) + CO_2(g)$$
$$Pb(OH)_2(s) \rightarrow PbO(s) + H_2O(g)$$

Lead(II) oxide is amphoteric. It is used in the manufacture of lead glass.

(b) *Dilead(II) lead(IV) oxide (red lead), Pb_3O_4*

This oxide is prepared as an orange-red powder by heating lead(II) oxide in air at 450 °C, but care has to be taken since the reaction is reversed at 550 °C:

$$6PbO(s) + O_2(g) \underset{550\ °C}{\overset{450\ °C}{\rightleftharpoons}} 2Pb_3O_4(s)$$

Dilead(II) lead(IV) oxide behaves as a mixture of lead(II) oxide and lead(IV) oxide, i.e. $2PbO + PbO_2$. When mixed with linseed oil, it is sold as a paint – red lead – for preventing the corrosion of iron.

(c) *Lead(IV) oxide, PbO_2*

Addition of dilute nitric acid to dilead(II) lead(IV) oxide leaves a purple-brown residue of lead(IV) oxide:

$$Pb_3O_4(s) + 4HNO_3(aq) \rightarrow PbO_2(s) + 2Pb(NO_3)_2(aq) + 2H_2O(l)$$

Lead(IV) oxide is amphoteric and it is an oxidising agent, e.g. it oxidises warm concentrated hydrochloric acid to chlorine:

$$PbO_2(s) + 4HCl(conc) \rightarrow PbCl_2(s) + 2H_2O(l) + Cl_2(g)$$

Lead(II) hydroxide, $Pb(OH)_2$

This is prepared by the addition of sodium hydroxide or ammonia solution to a solution of a lead(II) salt, e.g.:

$$Pb(NO_3)_2(aq) + 2NaOH(aq) \rightarrow Pb(OH)_2(s) + 2NaNO_3(aq)$$
i.e. $$Pb^{2+}(aq) + 2HO^-(aq) \rightarrow Pb(OH)_2(s)$$

The white precipitate is amphoteric and so is soluble in excess sodium hydroxide giving a colourless solution of sodium plumbate(II):

$$Pb(OH)_2(s) + 2NaOH(aq) \rightarrow Na_2Pb(OH)_4(aq)$$

Lead(II) hydroxide is not soluble in ammonia solution. It has no important uses.

Lead(II) carbonate, PbCO$_3$

This is precipitated when sodium hydrogencarbonate solution is added to a solution of a lead(II) salt, e.g.:

$$Pb(NO_3)_2(aq) + 2NaHCO_3(aq) \rightarrow PbCO_3(s) + 2NaNO_3(aq) + H_2O(l) + CO_2(g)$$
i.e. $\quad Pb^{2+}(aq) + 2HCO_3^-(aq) \rightarrow PbCO_3(s) + H_2O(l) + CO_2(g)$

Sodium carbonate precipitates the basic carbonate, $Pb(OH)_2 \cdot 2PbCO_3$.

Lead(II) carbonate is a white solid which has found use in some white paint. However, the cost, toxicity, and the slow darkening on exposure to the atmosphere (see lead(II) sulphide below) all contribute to its declining use.

Lead(II) chloride, PbCl$_2$

This is obtained as a white precipitate when a solution of a chloride is added to a solution of a lead(II) salt, e.g:

$$Pb(NO_3)_2(aq) + 2HCl(aq) \rightarrow PbCl_2(s) + 2HNO_3(aq)$$
i.e. $\quad Pb^{2+}(aq) + 2Cl^-(aq) \rightarrow PbCl_2(s)$

Lead(II) chloride is slightly soluble in cold water but fairly soluble in hot water and so it may be purified by recrystallisation.

Lead(II) iodide, PbI$_2$

A golden yellow precipitate of lead(II) iodide is formed when aqueous solutions of potassium iodide and a lead(II) salt are mixed, e.g:

$$Pb(NO_3)_2(aq) + 2KI(aq) \rightarrow PbI_2(s) + 2KNO_3(aq)$$
i.e. $\quad Pb^{2+}(aq) + 2I^-(aq) \rightarrow PbI_2(s)$

It is fairly soluble in hot water and may be recrystallised from it.

Lead(II) nitrate, Pb(NO$_3$)$_2$

This may be obtained by the action of nitric acid on lead or lead(II) oxide, hydroxide, or carbonate, e.g:

$$PbCO_3(s) + 2HNO_3(aq) \rightarrow Pb(NO_3)_2(aq) + H_2O(l) + CO_2(g)$$

Lead(II) nitrate is fairly soluble in water but it may be obtained as colourless crystals when its solution is concentrated and cooled. It is used in the laboratory preparation of nitrogen dioxide (see page 199).

Lead(II) sulphate, PbSO$_4$

This is obtained as a white precipitate when dilute sulphuric acid is added to lead(II) nitrate solution:

$$Pb(NO_3)_2(aq) + H_2SO_4(aq) \rightarrow PbSO_4(s) + 2HNO_3(aq)$$
i.e. $\quad Pb^{2+}(aq) + SO_4^{2-}(aq) \rightarrow PbSO_4(s)$

Lead(II) sulphide, PbS

This is best prepared by passing hydrogen sulphide through a solution of a lead(II) salt, e.g.:

$$Pb(NO_3)_2(aq) + H_2S(g) \rightarrow PbS(s) + 2HNO_3(aq)$$

i.e. $Pb^{2+}(aq) + S^{2-}(aq) \rightarrow PbS(s)$

It is a black solid which, as mentioned above, is slowly formed when paints containing lead(II) carbonate react with traces of hydrogen sulphide in the atmosphere. Old paintings discoloured in this way may be restored by treating them with hydrogen peroxide solution – this oxidises the sulphide to sulphate which is white:

$$PbS(s) + 4H_2O_2(aq) \rightarrow PbSO_4(s) + 4H_2O(l)$$

Test for lead(II) ions

Solutions containing lead(II) ions give a white precipitate on addition of sodium hydroxide solution, the precipitate being soluble in excess of the alkali but not in ammonia solution. A black precipitate is formed when hydrogen sulphide is passed through an aqueous solution of lead(II) ions. (This reaction is also used as a test for hydrogen sulphide—see page 206).

TRANSITION METALS

Transition metals are those in which an inner electron shell is being filled up after an outer one has been started. It was mentioned earlier (page 28) that the third shell initially behaves as if it were full when it contains eight electrons but that it can eventually hold eighteen. One transition series occurs when the third shell is completed:

Element	Electronic structure	
Potassium	2.8.8.1	
Calcium	2.8.8.2	
Scandium	2.8.9.2	
⋮	⋮	Transition elements
Zinc	2.8.18.2	
Gallium	2.8.18.3	

Other transition series occur lower down the Periodic Table.

Transition metals have a number of characteristic properties, some of which are listed below.

(a) Variable valencies.
(b) Coloured ions in solution.
(c) Form complex ions.
(d) Possess catalytic activity.

Four transitional metals will be discussed: zinc, iron, copper, and silver. However, as explained below, zinc is not a typical transition metal.

ZINC

Occurrence and extraction

Zinc occurs as the carbonate and the sulphide (zinc blende). The oxide is obtained from these compounds by the action of heat in the former case and by roasting in air in the latter:

$$ZnCO_3(s) \rightarrow ZnO(s) + CO_2(g)$$
$$2ZnS(s) + 3O_2(g) \rightarrow 2ZnO(s) + 2SO_2(g)$$

The oxide is then reduced by heating it with coke:

$$ZnO(s) + C(s) \rightarrow Zn(s) + CO(g)$$

The sulphur dioxide, obtained by roasting the zinc sulphide, is converted into sulphuric acid via the Contact process. Both processes are performed on the same site so as to eliminate transport costs.

Properties and reactions of zinc

Zinc is a bluish white metal which is brittle at room temperature but malleable and ductile at 100–150 °C. It is not a typical transition metal because it is the last one in the first transition series, i.e. its electronic structure is 2.8.18.2, and so it has the third shell full. Unlike the majority of transition metals, it has a fairly low melting point (420 °C) and a single valency, i.e. two.

Zinc tarnishes on exposure to air due to the formation of a thin protective layer of the oxide and basic carbonate. Heating zinc in air gives the oxide.

Despite its position in the electrochemical and activity series, zinc does not react with cold water, the reason again being the protective layer of oxide. The hot metal does react with steam:

$$Zn(s) + H_2O(g) \rightarrow ZnO(s) + H_2(g)$$

Zinc is above hydrogen in the electrochemical and activity series and so it will displace hydrogen ions from acids. The pure metal reacts very sluggishly with cold dilute hydrochloric and sulphuric acids but the commercial product reacts rapidly:

$$Zn(s) + 2HCl(aq) \rightarrow ZnCl_2(aq) + H_2(g)$$
i.e.
$$Zn(s) + 2H^+(aq) \rightarrow Zn^{2+}(aq) + H_2(g)$$

Hot concentrated sulphuric acid oxidises the zinc to Zn^{2+} and is itself reduced to sulphur dioxide:

$$Zn(s) + 2H_2SO_4(conc) \rightarrow ZnSO_4(s) + 2H_2O(l) + SO_2(g)$$

Nitric acid reacts with zinc to give the nitrate but the other products depend on the concentration and the temperature of the acid used.

Zinc is similar to aluminium in that it dissolves in hot sodium hydroxide solution with the liberation of hydrogen; a solution of sodium zincate remains.

$$Zn(s) + 2NaOH(aq) + 2H_2O(l) \rightarrow Na_2Zn(OH)_4(aq) + H_2(g)$$

Uses of zinc

The main use of zinc is in the galvanisation of iron (see page 105). It is also used for making alloys, e.g. brass.

Compounds of zinc

As mentioned above, zinc is not a typical transition metal. It is restricted to a valency of two and very few of its compounds are coloured.

Several of the common zinc compounds have very limited solubility in water but the sulphate, chloride, and nitrate are exceptions.

Zinc oxide, ZnO

This may be prepared by burning zinc in air or by heating the carbonate, nitrate, or hydroxide. It is a white solid but it turns yellow when hot. Zinc oxide, like the hydroxide, is insoluble in water but is amphoteric, e.g:

$$ZnO(s) + H_2SO_4(aq) \rightarrow ZnSO_4(aq) + H_2O(l)$$
$$ZnO(s) + 2NaOH(aq) + H_2O(l) \rightarrow Na_2Zn(OH)_4(aq)$$

Zinc oxide is used as a pigment in paints and has the advantage over lead paints in that its colour is unaffected by traces of hydrogen sulphide in the atmosphere because zinc sulphide is white. It also finds use as a constituent in various cosmetic powders, creams, and ointments.

Zinc hydroxide, Zn(OH)$_2$

This may be obtained as a white gelatinous precipitate by adding sodium hydroxide solution to a solution of a zinc salt, e.g.:

$$ZnSO_4(aq) + 2NaOH(aq) \rightarrow Zn(OH)_2(s) + Na_2SO_4(aq)$$

i.e.
$$Zn^{2+}(aq) + 2HO^-(aq) \rightarrow Zn(OH)_2(s)$$

Excess sodium hydroxide must not be added because zinc hydroxide is amphoteric:

$$Zn(OH)_2(s) + 2HCl(aq) \rightarrow ZnCl_2(aq) + 2H_2O(l)$$
$$Zn(OH)_2(s) + 2NaOH(aq) \rightarrow Na_2Zn(OH)_4(aq)$$

Zinc hydroxide also dissolves in ammonia solution owing to the formation of the tetraammine zinc ion – a complex ion:

$$Zn(OH)_2(s) + 4NH_3(aq) \rightarrow [Zn(NH_3)_4]^{2+}(aq) + 2HO^-(aq)$$

Zinc carbonate, ZnCO$_3$

This occurs naturally. It is a white solid, insoluble in water. Zinc carbonate is a constituent of calomine lotion which is used for the treatment of inflammation of the skin.

Zinc chloride, $ZnCl_2$

The anhydrous salt may be prepared by heating zinc in chlorine or hydrogen chloride. Dissolving zinc oxide, hydroxide, or carbonate in dilute hydrochloric acid, and then allowing the solution to crystallise, gives the monohydrate, $ZnCl_2 \cdot H_2O$.

Zinc chloride is a very deliquescent white solid which gives an acidic solution due to hydrolysis, a simplified equation being

$$ZnCl_2 + 2H_2O \rightleftharpoons Zn(OH)_2 + 2HCl$$

On mixing with moist zinc oxide, the chloride sets to a hard mass of zinc chloride hydroxide, $Zn(OH)Cl$ – this finds application as a filling for teeth. Zinc chloride is also used as a flux in soldering.

Zinc nitrate, $Zn(NO_3)_2$

This may be obtained as the hexahydrate, $Zn(NO_3)_2 \cdot 6H_2O$, from the reaction between zinc or its oxide, hydroxide, or carbonate with dilute nitric acid. It is a deliquescent white solid which has no important commercial use.

Zinc sulphate, $ZnSO_4$

Colourless crystals of the heptahydrate, $ZnSO_4 \cdot 7H_2O$, may be obtained from reaction between zinc or its oxide, hydroxide, or carbonate and dilute sulphuric acid. The heptahydrate loses its water of crystallisation on gentle heating and decomposes into the oxide on stronger heating:

$$ZnSO_4 \cdot 7H_2O \xrightarrow{250\ °C} ZnSO_4 + 7H_2O \xrightarrow{750\ °C} ZnO + SO_3$$

Zinc sulphate is used in the manufacture of zinc sulphide, a white pigment used in some paints.

Zinc sulphide, ZnS

This may be prepared in the laboratory by passing hydrogen sulphide through a solution of a zinc salt, e.g:

$$ZnSO_4(aq) + H_2S(g) \rightarrow ZnS(s) + H_2SO_4(aq)$$

Zinc sulphide is a white solid which has found some use in white paints, but there are now better substances available.

Test for zinc ions

Solutions containing zinc ions give a white precipitate of zinc hydroxide on addition of either sodium hydroxide or ammonia solution. The precipitate redissolves in the presence of excess of either reagent.

IRON

Occurrence and extraction

Iron is the second most abundant metal (page 3). Its most important ores consist of iron(II) carbonate, $FeCO_3$; iron(III) oxide (*haematite*), Fe_2O_3; and iron(II) diiron(III) oxide (*magnetite*), Fe_3O_4. Iron is extracted by reduction of its oxides in a blast furnace and so, if the carbonate is the ore used, it needs an initial roasting in air to convert it to the oxide:

$$4FeCO_3(s) + O_2(g) \rightarrow 2Fe_2O_3(s) + 4CO_2(g)$$

This process may be regarded as involving two reactions, i.e:

$$FeCO_3(s) \rightarrow FeO(s) + CO_2(g)$$
$$4FeO(s) + O_2(g) \rightarrow 2Fe_2O_3(s)$$

In the blast furnace, preheated air is forced through a mixture of iron ore, coke, and limestone. The process is continuous and a complex series of reactions takes place as outlined below and summarised in Figure 11.6.

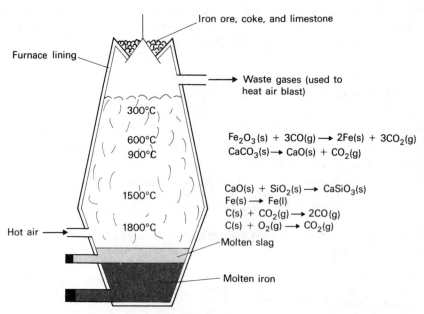

Figure 11.6. Reactions occurring in a blast furnace

A blast of hot air causes the coke near the bottom of the furnace to burn in a very exothermic reaction, to give carbon dioxide:

$$C(s) + O_2(g) \rightarrow CO_2(g)$$

As the carbon dioxide passes up the furnace, it is reduced to carbon monoxide:

$$C(s) + CO_2(g) \rightarrow 2CO(g)$$

The carbon monoxide in turn reduces the iron ore in the cooler (about 600 °C) upper regions of the furnace:

$$Fe_2O_3(s) + 3CO(g) \rightarrow 2Fe(s) + 3CO_2(g)$$
$$Fe_3O_4(s) + 4CO(g) \rightarrow 3Fe(s) + 4CO_2(g)$$

The iron sinks down the furnace, melts in the hotter zone, and is run off at the base.

The ores contain appreciable amounts of silicon(IV) oxide and this combines with calcium oxide formed by the thermal decomposition of the limestone:

$$CaCO_3(s) \xrightarrow{850\,°C} CaO(s) + CO_2(g)$$
$$CaO(s) + SiO_2(s) \longrightarrow CaSiO_3(s)$$

The resultant calcium silicate slag melts in the hot part of the furnace and floats on the molten iron, thus preventing oxidation of the latter by the air blast. The slag is run off as necessary.

The gases leaving the top of the furnace contain about 25% of carbon monoxide and this is burnt to preheat the air used in the blast.

The molten iron from the furnace is allowed to cool in sand moulds. The so-called 'pig-iron' produced is hard but brittle because the uncombined impurities cause defects in the crystal lattice. It contains about 4% of carbon, both combined and dissolved, together with smaller amounts of manganese, phosphorus, silicon, and sulphur.

Cast iron is pig-iron which has been melted with scrap steel and cooled in a mould of some definite shape. It is not very pure or strong but it is relatively cheap. Before the advent of plastics, cast iron found considerable use as drain pipes. However, its main use now is in such things as radiators, cookers, etc.

Removal of most of the impurities from pig-iron gives wrought iron. The purification is brought about by heating the pig-iron, under a current of air, in a furnace lined with iron(III) oxide. The carbon is used up in the reduction of iron(III) oxide:

$$Fe_2O_3(s) + 3C(s) \rightarrow 2Fe(s) + 3CO(g)$$

whilst the other major impurities are converted into gaseous oxides, e.g:

$$4P(s) + 5O_2(g) \rightarrow 2P_2O_5(g)$$

Wrought iron is tough, malleable, and ductile; it is used for making nails, wire, and ornamental objects such as gates, etc.

Steel is an alloy of iron with carbon and other elements, the amounts being strictly controlled to give the required properties. The carbon content is normally up to about 1.5%.

The first step in steel production involves the removal of the impurities from pig-iron. Thus, molten pig-iron is placed in a furnace lined with silicon(IV) oxide if the impurities form basic oxides as is the case with magnesium or manganese. Alternatively, if the predominant impurities form acidic oxides, as do phosphorus and silicon, the lining is of calcium oxide/magnesium oxide. Oxygen is forced through the molten metal and the impurities are converted to their oxides which either volatilise or react with the furnace lining to give a slag. The required mass of carbon and other elements is then added to the pure iron.

There are a number of different types of steel and a few are mentioned below. Stainless steel contains nickel and 12—18% chromium; it is very resistant to corrosion and is used for cutlery, car bumpers, and chemical plant, etc. Steel containing 5% tungsten is used for high speed drills since it remains hard even when hot. Invar steel contains about 36% nickel and is used for making watches because it has a very small coefficient of expansion.

Steel can be hardened by heating it to about 850 °C and then plunging it into cold water or oil. This process is known as quenching and it makes the steel very hard but brittle – it alters the crystal lattice of the steel. If the steel is then heated to between 200 °C and 300 °C and allowed to cool slowly, it retains much of its hardness but loses its brittleness and it is said to be 'tempered' or 'annealed'.

It will have been noted from above that large quantities of coke are utilised in the extraction of iron. Coke is made by the destructive distillation of coal between about 1200—1600 °C. Fortunately, iron ore and coal often occur in the same locality and this considerably reduces the cost of transportation. Steel production requires the use of oxygen (see above) and so, as a general rule, coal distillation and air liquefaction plant are situated on the same site as the blast furnaces.

Properties and reactions of iron

Iron has the electronic structure 2.8.14.2 and is a typical transition metal. Thus it can have a valency of two or three (note that it does not attain the noble gas electronic structure in these states), it forms coloured salts (see below), and it exhibits catalytic activity (see the Haber process, page 130).

Its position in the electrochemical and activity series indicates that iron is a fairly reactive metal. It rusts fairly rapidly in moist air, forming hydrated iron(III) oxide (see page 48). The rusting may be prevented by covering the iron with oil or grease, by painting, by galvanising or tin plating (page 105), or by alloying with other metals (see stainless steel above).

When heated in air above 150 °C, iron becomes coated with iron(II) diiron(III) oxide, Fe_3O_4. Red hot iron reacts with steam giving hydrogen:

$$3Fe(s) + 4H_2O(g) \rightleftharpoons Fe_3O_4(s) + 4H_2(g)$$

This used to be a major industrial source of hydrogen before the cracking of petroleum (page 238) became so important.

Iron dissolves in dilute hydrochloric and sulphuric acids giving hydrogen and the corresponding iron(II) salt, e.g.:

$$Fe(s) + H_2SO_4(aq) \rightarrow FeSO_4(aq) + H_2(g)$$

With dilute nitric acid, iron(II) nitrate and oxides of nitrogen are formed but concentrated nitric acid renders iron 'passive' (cf. aluminium) due to the formation of a thin protective layer of iron(II) diiron(III) oxide.

Iron is unaffected by alkalis. Its reactions with chlorine and sulphur are discussed below.

Compounds of iron

As a general rule, iron(II) salts tend to be green and iron(III) salts yellow. With the exception of the oxides and hydroxides, the common iron compounds are soluble. Iron(II) compounds are readily oxidised to iron(III) compounds.

Oxides

(a) *Iron(II) oxide, FeO*

This may be made by the thermal decomposition of some iron(II) compounds. However, it is unimportant and is not normally encountered in the laboratory since it inflames in air to give iron(III) oxide.

(b) *Iron(III) oxide, Fe_2O_3*

This may be made by the action of heat on iron(III) hydroxide or nitrate, or iron(II) sulphate:

$$2Fe(OH)_3(s) \rightarrow Fe_2O_3(s) + 3H_2O(g)$$
$$4Fe(NO_3)_3(s) \rightarrow 2Fe_2O_3(s) + 12NO_2(g) + 3O_2(g)$$
$$2FeSO_4(s) \rightarrow Fe_2O_3(s) + SO_3(g) + SO_2(g)$$

It is used as an abrasive polishing powder (jeweller's rouge) and as a red pigment.

(c) *Iron(II) diiron(III) oxide, Fe_3O_4*

This oxide is made by heating iron in air or steam, e.g:

$$3Fe(s) + 4H_2O(g) \rightleftharpoons Fe_3O_4(s) + 4H_2(g)$$

It is a black solid, inert to acids but readily reduced to iron by heating with carbon or carbon monoxide.

Hydroxides

(a) *Iron(II) hydroxide, $Fe(OH)_2$*

This is obtained as a dirty green precipitate by addition of alkali to a solution of an iron(II) salt, e.g:

$$FeSO_4(aq) + 2NaOH(aq) \rightarrow Fe(OH)_2(s) + Na_2SO_4(aq)$$
i.e. $\quad Fe^{2+}(aq) + 2HO^-(aq) \rightarrow Fe(OH)_2(s)$

The precipitate turns brown on exposure to air owing to oxidation to iron(III) hydroxide. It is unaffected by excess alkali or ammonia solution.

(b) *Iron(III) hydroxide, $Fe(OH)_3$*

This is obtained as a red-brown gelatinous precipitate by the addition of sodium hydroxide or ammonia solution to an iron(III) salt solution:

$$FeCl_3(aq) + 3NaOH(aq) \rightarrow Fe(OH)_3(s) + 3NaCl(aq)$$
i.e. $\quad Fe^{3+}(aq) + 3HO^-(aq) \rightarrow Fe(OH)_3(s)$

Again the precipitate is entirely basic and so is unaffected by excess alkali.

Chlorides

(a) *Iron(II) chloride, $FeCl_2$*

Anhydrous iron(II) chloride is obtained as a deliquescent white solid by heating iron filings in a stream of dry hydrogen chloride:

$$Fe(s) + 2HCl(g) \rightarrow FeCl_2(s) + H_2(g)$$

The apparatus is basically similar to that shown in Figure 11.5. Crystals of the tetrahydrate, $FeCl_2 \cdot 4H_2O$ are obtained from the reaction between iron and dilute hydrochloric acid but precautions have to be taken to prevent their oxidation by air to iron(III) chloride.

(b) *Iron(III) chloride, $FeCl_3$*

Anhydrous iron(III) chloride is formed when chlorine is passed over heated iron in an apparatus similar to that shown in Figure 11.5.

$$2Fe(s) + 3Cl_2(g) \rightarrow 2FeCl_3(s)$$

It is a deep red-black solid which is similar to aluminium chloride in that it is covalent, forms a dimer, Fe_2Cl_6, sublimes on heating, and gives acidic solutions due to hydrolysis.

Note that, since chlorine is an oxidising agent (page 100), it gives the higher chloride when it reacts with metals which can have more than one valency.

Sulphates

(a) *Iron(II) sulphate, $FeSO_4$*

Green crystals of iron(II) sulphate heptahydrate, $FeSO_4 \cdot 7H_2O$, may be obtained from the reaction between iron and dilute sulphuric acid:

$$Fe(s) + H_2SO_4(aq) \rightarrow FeSO_4(aq) + H_2(g)$$
i.e.
$$Fe(s) + 2H^+(aq) \rightarrow Fe^{2+}(aq) + H_2(g)$$

Ideally, the reaction should be done under an atmosphere of nitrogen to prevent the slow oxidation by air of iron(II) to iron(III) sulphate.

The pale green heptahydrate loses its water of crystallisation on gentle heating whilst stronger heating results in thermal decomposition to give red-brown iron(III) oxide:

$$FeSO_4 \cdot 7H_2O(s) \rightarrow FeSO_4(s) + 7H_2O(g)$$
$$2FeSO_4(s) \rightarrow Fe_2O_3(s) + SO_3(g) + SO_2(g)$$

Iron(II) sulphate is used in the manufacture of ink and in the tanning and dyeing industries. It forms a brown addition compound with nitrogen oxide and this is utilised in the *brown ring test* for nitrates. Thus, the suspected nitrate is added to a mixture of iron(II) sulphate solution and dilute sulphuric acid. Concentrated sulphuric acid is then carefully run in so that it forms a dense lower layer; a brown ring at the junction of the layers (Figure 11.7) indicates the presence of a nitrate. The brown ring arises as follows. Sulphuric acid reacts with the nitrate to give nitric acid:

$$2NaNO_3(aq) + H_2SO_4(aq) \rightarrow Na_2SO_4(aq) + 2HNO_3(aq)$$

The nitric acid oxidises some of the iron(II) sulphate to iron(III) sulphate and is itself reduced to nitrogen oxide:

$6FeSO_4(aq) + 3H_2SO_4(aq) + 2HNO_3(aq) \rightarrow 3Fe_2(SO_4)_3(aq) + 4H_2O(l) + 2NO(g)$

The nitrogen oxide then adds on to unchanged iron(II) sulphate:

$FeSO_4(aq) + NO(g) \rightarrow FeSO_4 \cdot NO(aq)$

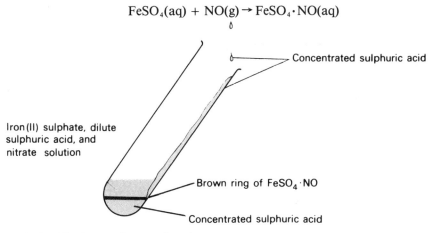

Figure 11.7. The brown ring test for nitrates

(b) *Iron(III) sulphate, $Fe_2(SO_4)_3$*

This is obtained as an off-white solid by oxidising iron(II) sulphate with hot concentrated sulphuric acid:

$2FeSO_4(s) + 2H_2SO_4(conc) \rightarrow Fe_2(SO_4)_3(s) + 2H_2O(l) + SO_2(g)$

Iron(II) sulphide, FeS

This black solid may be prepared by heating iron powder and sulphur:

$Fe(s) + S(s) \rightarrow FeS(s)$

or better, by passing hydrogen sulphide through a solution of an iron(II) salt:

$FeSO_4(aq) + H_2S(g) \rightarrow FeS(s) + H_2SO_4(aq)$
i.e. $Fe^{2+}(aq) + S^{2-}(aq) \rightarrow FeS(s)$

Iron(II) sulphide is used in the preparation of hydrogen sulphide (page 204).

Tests for iron(II) and iron(III) ions

Iron(II) and iron(III) ions give dirty green and red-brown precipitates respectively of the corresponding hydroxide on addition of a solution of alkali. Alternatively, the ions may be distinguished by adding ammonium thiocyanate solution; there is no apparent change with Fe^{2+} ions but Fe^{3+} ions give a blood red solution:

$Fe^{3+}(aq) + 6NH_4CNS(aq) \rightarrow [Fe(CNS)_6]^{3-}(aq) + 6NH_4^+(aq)$

COPPER

Occurrence and extraction

Copper occurs in a number of ores, the most important of which is $CuFeS_2$. After a number of processes involving roasting in air, etc. a mixture of copper(I) sulphide and copper(I) oxide is obtained and these react on heating to give crude copper which is about 97% pure. The crude copper is purified by electrolysis as described on page 90.

Properties and reactions of copper

The electronic structure of copper is 2.8.18.1; the reason for this arrangement, rather than 2.8.17.2, is beyond the level of this text. Copper is a typical transition metal and as such exhibits variable valency (1 or 2), forms coloured salts, and readily forms complex ions. It is fairly soft, malleable, and ductile but, in contrast to most metals, it is a reddish gold in colour. It slowly tarnishes on exposure to air and rapidly reacts with air on strong heating to give a mixture of copper(I) and copper(II) oxide.

Copper is below hydrogen in the electrochemical and activity series and so it is unattacked by water or by acids unless they are oxidising agents. There is, therefore, no reaction between copper, and dilute or concentrated hydrochloric acid or dilute sulphuric acid. With hot concentrated sulphuric acid (an oxidising agent) the reaction is:

$$Cu(s) + 2H_2SO_4(conc) \rightarrow CuSO_4(s) + 2H_2O(l) + SO_2(g)$$

With nitric acid, the products depend on the concentration of acid used.

Dilute acid $\quad 3Cu(s) + 8HNO_3(aq) \rightarrow 3Cu(NO_3)_2(aq) + 4H_2O(l) + 2NO(g)$
Concentrated acid $\quad Cu(s) + 4HNO_3(conc) \rightarrow Cu(NO_3)_2(aq) + 2H_2O(l) + 2NO_2(g)$

Uses of copper

Copper is an excellent conductor of electricity and is used extensively in electrical wiring. Its inertness and workability account for its use as water tanks and pipes. Copper atoms are similar in size to those of nickel and zinc and so it forms strong alloys with these metals. Brass consists of roughly two-thirds copper and one-third zinc. British 'silver' coins contain 75% of copper and 25% of nickel whilst 'bronze' coins contain 95% of copper, the balance being tin and zinc. Bronze for bearings, etc. contains 80—90% of copper, the rest being tin.

Compounds of copper

Copper can have valencies of one or two, copper(I) being the most stable state at high temperatures. The copper(II) compounds are generally the more important and are often blue or green.

Oxides

(a) *Copper(I) oxide, Cu_2O*

This is a red solid and it may be prepared by heating copper(II) oxide above 1000 °C:

$$4CuO(s) \rightarrow 2Cu_2O(s) + O_2(g)$$

It can also be made the reduction of some copper(II) salts but the reactions are more involved.

(b) *Copper(II) oxide, CuO*

This may be prepared by thermal decomposition of the nitrate, hydroxide, or carbonate, e.g.:

$$CuCO_3(s) \rightarrow CuO(s) + CO_2(g)$$

It is a black solid which is readily reduced to copper by heating it in a stream of hydrogen.

Copper(II) hydroxide, $Cu(OH)_2$

This may be prepared as a pale blue precipitate by the addition of sodium hydroxide solution to a solution of a copper(II) salt, e.g:

$$CuSO_4(aq) + 2NaOH(aq) \rightarrow Cu(OH)_2(s) + Na_2SO_4(aq)$$
i.e. $$Cu^{2+}(aq) + 2HO^-(aq) \rightarrow Cu(OH)_2(s)$$

Dropwise addition of dilute aqueous ammonia to a solution of a copper(II) salt initially gives a pale blue precipitate of the hydroxide but this redissolves in excess of the reagent to give a deep blue solution containing the tetraammine copper(II) ion (a complex ion):

$$Cu(OH)_2(s) + 4NH_3(aq) \rightarrow [Cu(NH_3)_4]^{2+}(aq) + 2HO^-(aq)$$

Note that, since four neutral ammonia molecules combine with the Cu^{2+} ion, the complex ion also has two positive charges.

The deep blue complex is used in the manufacture of viscose rayon.

Copper(II) carbonate

Addition of sodium carbonate solution to a solution of a copper(II) salt gives a pale blue precipitate of the basic carbonate, $Cu(OH)_2 \cdot 2CuCO_3$. As explained previously (page 139) sodium carbonate solution is alkaline due to hydrolysis and addition of it to a copper(II) salt results in the simultaneous precipitation of the carbonate and hydroxide.

Copper chlorides

(a) *Copper(I) chloride, CuCl*

This may be prepared by refluxing copper(II) oxide with copper and concentrated hydrochloric acid:

$$CuO(s) + Cu(s) + 2HCl(conc) \rightarrow 2CuCl + H_2O(l)$$

The copper(I) chloride is present as a complex ion in this acid solution but it is precipitated as a white solid on the addition of excess water.

Copper(I) chloride dissolves in concentrated hydrochloric acid and in ammonia solution; these solutions are used in gas analysis to absorb carbon monoxide (see page 183).

(b) *Copper(II) chloride, $CuCl_2$*

The anhydrous compound is obtained as a brown solid by heating copper in a stream of chlorine:

$$Cu(s) + Cl_2(g) \rightarrow CuCl_2(s)$$

Alternatively, blue-green crystals of the dihydrate may be obtained from the reaction between copper(II) oxide and hydrochloric acid:

$$CuO(s) + 2HCl(aq) \rightarrow CuCl_2(aq) + H_2O(l)$$

Copper(II) nitrate, $Cu(NO_3)_2$

This is obtained by adding copper or copper(II) oxide, hydroxide, or carbonate to dilute nitric acid, e.g:

$$CuCO_3(s) + 2HNO_3(aq) \rightarrow Cu(NO_3)_2(aq) + H_2O(l) + CO_2(g)$$

Crystallisation gives deep blue crystals of the trihydrate, $Cu(NO_3)_2 \cdot 3H_2O$.

Copper(II) sulphate, $CuSO_4$

Boiling copper(II) oxide with dilute sulphuric acid gives a blue solution of copper(II) sulphate and crystallisation gives the blue pentahydrate, $CuSO_4 \cdot 5H_2O$. On heating to 250 °C, the water of crystallisation is lost and the white anhydrous salt is obtained. Anhydrous copper(II) sulphate is used to detect water (page 48). Prolonged strong heating of copper(II) sulphate results in its slow decomposition:

$$CuSO_4(s) \rightarrow CuO(s) + SO_3(g)$$

Copper(II) sulphate is used as a fungicide and also as the electrolyte in copper plating.

Copper(II) sulphide, CuS

This is best prepared by passing hydrogen sulphide through a solution of a copper(II) salt:

$$CuSO_4(aq) + H_2S(g) \rightarrow CuS(s) + H_2SO_4(aq)$$
i.e.
$$Cu^{2+}(aq) + S^{2-}(aq) \rightarrow CuS(s)$$

The black solid may also be obtained by heating a mixture of the two elements. Once reaction has started, the mixture glows due to the heat of reaction.

Test for copper(II) ions

Dropwise addition of dilute ammonia solution to a solution of a copper(II) salt initially gives a pale blue precipitate and then a deep blue solution (see copper(II) hydroxide).

SILVER

Occurrence and extraction

Some silver occurs as the free metal but the chief source is the sulphide. However, the extraction of the metal from the sulphide is beyond the scope of this text.

Properties and reactions of silver

Silver belongs to a different transition metal series than zinc, iron, and copper. It is very ductile and the best conductor of heat and electricity known. It is low in the electrochemical and activity series and so is relatively unreactive. Silver does not combine directly with oxygen even on heating. However, it does tarnish in the presence of sulphur compounds. Being below hydrogen in the electrochemical and activity series, it is attacked only by oxidising acids, e.g. nitric acid and hot concentrated sulphuric acid.

Uses of silver

The metal is used as a constituent of coins, for electroplating, and for making jewellery and mirrors. It is also the catalyst in a number of important industrial processes.

Compounds of silver

Silver is restricted to a valency of one. The common compounds, with the exception of the nitrate, are insoluble in water.

Silver oxide, Ag_2O

This is obtained as a brown precipitate when sodium hydroxide solution is added to silver nitrate solution:

$$2AgNO_3(aq) + 2NaOH(aq) \rightarrow Ag_2O(s) + 2NaNO_3(aq) + H_2O(l)$$

The expected product here would have been silver hydroxide, but this compound is unstable and spontaneously decomposes into silver oxide and water.

Silver oxide readily decomposes into silver and oxygen on heating and so these are the products obtained on thermal decomposition of silver nitrate.

Silver chloride, AgCl

The addition of silver nitrate solution to a solution of any chloride gives a white precipitate of silver chloride, e.g:

$$AgNO_3(aq) + HCl(aq) \rightarrow AgCl(s) + HNO_3(aq)$$

i.e. $$Ag^+(aq) + Cl^-(aq) \rightarrow AgCl(s)$$

The reaction is used as a test for chlorides (see page 216).

Silver nitrate, $AgNO_3$

This is obtained as colourless crystals by dissolving silver in nitric acid, concentrating the solution, and then cooling it.

Silver nitrate is one of the few silver salts which are soluble in water. The solution is used in the detection and in the volumetric estimation of halide ions, i.e. chlorides, bromides, and iodides (pages 216, 217, and 218).

Test for silver ions

Solutions of silver salts give a white precipitate on addition of dilute hydrochloric acid. The precipitate is insoluble in nitric acid but it dissolves in ammonia solution.

Questions

1   (a) Give, with equations, an account of the manufacture of sodium carbonate by the Solvay process. Include in your account:
   (i) the names of the raw materials,
   (ii) the reactions by which the sodium carbonate is produced from the raw materials.
   (b) State two factors which make the Solvay process efficient and name the by-product formed.
   (c) How does sodium carbonate react with:
   (i) dilute nitric acid,
   (ii) a solution of calcium chloride?
   For each reaction state what would be seen and either name the products or write an equation.
   [J.M.B.]

2   (a) How would you prepare dry crystals of sodium carbonate starting from sodium hydroxide solution?
   (b) How, and under what conditions, does sodium carbonate react with:
   (i) carbon dioxide,
   (ii) calcium hydroxide,
   (iii) nitric acid,
   (iv) lead(II) nitrate?
   [J.M.B.]

3   (a) Calcium oxide (quicklime) is manufactured by heating limestone in a stream of air.
   (i) Write the equation for the reaction and explain why it is necessary to heat the limestone in a stream of air.
   (ii) State how calcium oxide is converted to calcium hydroxide and indicate two points of interest concerning the reaction that takes place.
   (b) Explain one laboratory use and one large scale application of calcium hydroxide.
   (c) Give the reactions that take place when:
   (i) a mixture of coke and calcium oxide is heated in an electric furnace,
   (ii) water is slowly added to the solid product obtained in (i).
   [A.E.B. Nov 1976]

4   Crystals of magnesium sulphate ($MgSO_4 \cdot 7H_2O$) can be prepared by the reaction between magnesium carbonate and dilute sulphuric acid.

(a) Explain why the sulphuric acid must be dilute.
(b) Describe how you would carry out this preparation.
(c) Give the equation.
(d) Calculate:
(i) the volume of 0.5 M sulphuric acid required to react with 0.1 mole of magnesium carbonate,
(ii) the maximum mass of magnesium sulphate crystals that would be obtained.
(e) Why would the mass of crystals obtained be less than your calculated value?
(f) Name another magnesium compound that could be used in place of magnesium carbonate.
(g) Explain why it is not possible to prepare lead(II) sulphate by a similar method and state (without further description) the reactions you could use to prepare this compound from lead carbonate.

[A.E.B. Nov 1974]

5 Dilead(II) lead(IV) oxide (red lead oxide), Pb_3O_4 will react with 2 M nitric acid to give a precipitate of lead(IV) oxide, PbO_2, and a solution of lead(II) nitrate, $Pb(NO_3)_2$.
(a) Write the equation for this reaction.
(b) Starting from dilead(II) lead(IV) oxide, describe how you would obtain pure, dry samples of:
(i) lead(IV) oxide,
(ii) crystalline lead(II) nitrate.
(c) Calculate the maximum masses of lead(IV) oxide and lead(II) nitrate which could be obtained from 13.7 g of dilead(II) lead(IV) oxide.
(d) Starting with 13.7 g of dilead(II) lead(IV) oxide, a student found that he had obtained a good specimen of lead(II) nitrate crystals by the method described in (b) (ii) but his yield was only about 50% of that calculated in (c). Suggest one reason for this low yield.

[J.M.B.]

6 Describe how you would prepare a pure, dry sample of lead(II) chloride by precipitation, giving an equation for the reaction used.
A small amount of lead(II) chloride was placed in a boiling tube which was then about half filled with water. The tube and contents were heated until a colourless solution A was obtained. Hydrogen sulphide was bubbled through the hot solution and a silvery black precipitate B was formed. After filtering and washing, the precipitate was treated with hydrogen peroxide solution which converted the black precipitate into a white solid C.
(a) Identify A, B and C and write an equation for the formation of B from A.
(b) What type of reaction is the conversion of B into C? Write an equation for this reaction.

[J.M.B.]

7 The raw materials used in the blast furnace for the production of pig iron are iron ore, coke, limestone and a supply of hot air.
(a) Explain carefully, with equations, the reactions which take place in the blast furnace.
(b) State why pig iron is brittle and explain very briefly how iron is converted to steel.
(c) Describe the reactions of iron with:
(i) copper(II) sulphate,
(ii) pure oxygen.
(d) For each reaction in (c), give the conditions, state what you would see and either name the products or give an equation.

[J.M.B.]

8   Pig-iron is obtained by feeding a mixture of iron ore, coke and limestone into a blast furnace. Give an account of the reactions that take place in the furnace. Calculate the mass of iron obtainable from 1000 kilogrammes of pure iron(III) oxide. What are the essential differences between pig-iron and steel? Give two instances in which steel is a more useful material than pig-iron.
[A.E.B. June 1974]

9   (a) Describe how you would prepare pure dry iron(II) sulphate crystals starting from iron filings.
   (b) State and explain the changes you would observe when:
      (i) chlorine is passed into a solution of iron(II) chloride,
      (ii) hydrogen sulphide is passed into a solution of iron(III) chloride.
   (c) How would you distinguish chemically between solutions containing Fe^{2+} and Fe^{3+} ions?

10 (a) Crude iron is produced in the blast furnace from a mixture of iron ore, coke and limestone.
      (i)   Give the common name and formula of one iron ore used.
      (ii) What happens chemically to the coke?
      (iii) Give the equation for one reaction by which iron ore is converted to iron.
      (iv) Why is limestone used? What happens to it?
      (v)   What are the essential differences between the composition of crude iron and that of steel?
   (b) State, without further description, one reaction in each case by which you could:
      (i) obtain a solution containing iron(II) ions (Fe^{2+}) from metallic iron,
      (ii) convert this solution into one containing iron(III) ions (Fe^{3+}).
      Briefly describe a chemical test to confirm that the change in (ii) has taken place.
[A.E.B. June 1976]

11 (a) Both calcium and sodium react with cold water. In their reaction with water,
      (i) describe two different tests you could make to show that the reactions are similar, giving the results in each case,
      (ii) state two observations you would make which lead you to the conclusion that sodium is the more active metal of the two.
   (b) What conditions are necessary to enable magnesium to react rapidly with water? Write the equation for the reaction.
   (c) Describe what is seen when sodium hydroxide solution is added, until present in excess, to:
      (i) copper(II) sulphate solution,
      (ii) zinc sulphate solution. Name the products in each case.
   (d) Describe briefly one single reaction which illustrates that zinc is higher in the activity series than copper.
[J.M.B.]

12 (a) Briefly outline the apparatus and conditions that you would use to prepare and collect hydrogen by the action of water on:
      (i) calcium,
      (ii) iron.
      Name the other products of the reactions.
   (b) State which of the following metals would give hydrogen when added to dilute hydrochloric acid, and give equations for the reactions:
      (i)   calcium,
      (ii) copper,
      (iii) iron,
      (iv) aluminium.
      Describe how aluminium and iron(III) oxide react together. On the basis of the

evidence from these reactions, arrange the four metals in order of increasing chemical reactivity.

[J.M.B.]

13 (a) Describe in detail how, starting from copper(II) oxide, you would make a pure sample of copper(II) sulphate crystals. Write an equation for the reaction.

(b) Calculate the maximum mass of copper(II) sulphate crystals, $CuSO_4 \cdot 5H_2O$, which could be obtained from 16 g copper(II) oxide.
Give two reasons why the maximum mass is never obtained.

(c) When copper(II) sulphate crystals are heated gently, a white solid is formed. When the white solid is heated strongly for a long time, a black solid remains. Name the white and black solids and write an equation for each reaction.

[J.M.B.]

14 (a) (i) Explain briefly how aluminium, lead and zinc are normally obtained from their oxides and give an equation for one of the reactions.
(ii) Place these three metals in descending order of activity and justify your answer by reference to the ease with which their oxides can be reduced.

(b) State which of the metals calcium, copper, and magnesium will react:
(i) readily with cold water,
(ii) very slowly with cold water, but vigorously when heated with steam,
(iii) with neither cold water nor steam.
Place these metals in descending order of activity and justify your answer by reference to your statements.

(c) Describe what you would see if a burning strip of magnesium ribbon were plunged into a gas jar of carbon dioxide.

[J.M.B.]

15 (a) How and under what conditions do the metals copper, iron, and magnesium react with:
(i) water (steam if necessary),
(ii) oxygen,
(iii) dilute sulphuric acid?
Name all the products or give an equation for each reaction. Indicate clearly if there is no reaction.

(b) Aluminium and zinc are both above iron in the electrochemical series. Explain why:
(i) aluminium does not appear to react as vigorously as iron with dilute sulphuric acid,
(ii) zinc is used to protect iron from reacting with the air.

[J.M.B.]

16 Describe, including relevant equations, how the following conversions can be brought about either on an industrial or a laboratory scale:
(i) iron(III) oxide to iron,
(ii) sodium chloride to sodium hydroxide,
(iii) magnesium chloride to magnesium oxide.

17 Explain the following observations, naming the substances involved, and giving relevant equations where possible.

(a) Addition of dilute nitric acid to an orange-red solid gave a colourless solution and a purple-brown residue. Addition of dilute hydrochloric acid to the colourless solution results in a white precipitate being formed.

(b) On strong heating, a white solid gave off a colourless gas which relit a glowing splint. On cooling, a white residue remained and a flame test on this gave a golden yellow flame.

(c) A white solid dissolved in water to give a blue solution. Dropwise addition of aqueous ammonia to this solution initially gave a pale blue precipitate but this redissolved in excess of the reagent to give a deep blue solution.

(d) On strong heating, a white solid gave off a gas which turned limewater milky. The hot residue was yellow but it turned white on cooling.

18 (a) The following list of elements is arranged in order of an 'activity series':

Na, Ca, Mg, Zn, Fe, H, Cu, Hg.

From these elements name:
(i) a metal which reacts with cold water,
(ii) a metal which burns in steam but does not react with cold water,
(iii) any other elements which react when heated in steam,
(iv) an element which has an oxide which decomposes on heating,
(v) those metals which do not displace hydrogen from dilute hydrochloric acid.

(b) State and explain the results of placing pieces of zinc in aqueous solutions of:
(i) copper(II) sulphate,
(ii) magnesium sulphate.
Explain why zinc is often used as a protective coating for iron.

(c) Outline one method by which dry crystals of hydrated magnesium sulphate can be obtained from another magnesium compound.

[A.E.B. Nov 1976]

19 The reaction between zinc and aqueous copper(II) sulphate solution can be represented by the equation

$$Zn + Cu^{2+} \rightarrow Zn^{2+} + Cu \qquad \Delta H = -217 \text{ kJ}$$

(a) Explain the reaction in terms of oxidation/reduction.

(b) What energy change would you expect if a slight excess of zinc were added to 100 cm³ of 0.2 M copper(II) sulphate solution?

(c) State two other observations that would indicate that a chemical reaction had taken place.
State and explain what happens when:
(i) dilute sulphuric acid is added separately to copper and zinc,
(ii) aqueous sodium hydroxide is added separately to aqueous solutions of copper(II) sulphate and zinc sulphate respectively until it is present in excess.

[A.E.B. June 1975]

12 The Chemistry of some Non-Metals

INTRODUCTION

It was seen in the previous chapter that metals have a distinct tendency to lose electrons and form positive ions. Non-metals, however, often attain the electronic structure of the noble gases by sharing electrons in the formation of covalent bonds. Some non-metals can also attain the noble gas electronic structure by accepting electrons to produce negative ions. The tendency to form negative ions increases from left to right across the Periodic Table and is greatest with the element at the top of Group VII, i.e. fluorine. A number of non-metals are discussed below.

CARBON AND SILICON-GROUP IV ELEMENTS

The elements of Group IV all have four electrons in their outer shell and so are tetravalent. The first two members are carbon, 2.4 and silicon, 2.8.4 and they are typical non-metals. Neither of these elements can form ions because the energy requirement is too great (page 30).

Carbon is unique in that its atoms can form long stable chains with one another. A consequence of this is that there are far more compounds of carbon than of all the other elements put together. These carbon compounds, with the exception of carbon monoxide, the dioxide, the disulphide, and the metal carbonates and carbides, are generally classed as *organic compounds*. Some simple organic chemistry is discussed in the next chapter. The chemistry of elemental carbon and the few of its compounds mentioned above, together with all the other elements and their compounds, is classed as *inorganic chemistry*.

CARBON

Occurrence, manufacture, and uses

Carbon occurs in the free state as diamond and graphite. It also occurs in combination with other elements in all living things, and in natural gas, oil, coal, limestone, etc.

Elements which can exist in two or more forms without changing state are said to exhibit allotropy. Diamond and graphite are allotropes of carbon. Allotropes of a particular element differ in their physical properties and may also have some different chemical properties. The differences in the physical properties of diamond and graphite are summarised in Table 12.1.

Table 12.1 Physical properties of diamond and graphite

Allotrope	Graphite	Diamond
Density	2.3 g cm^{-3}	3.5 g cm^{-3}
Colour	Grey-black and opaque	Colourless and transparent
Hardness	Very soft	Extremely hard
Melting point	Sublimes at about 3700 °C	3550 °C

Diamonds are found in old volcanic regions such as South Africa, South America, and Russia. Small ones may be made industrially by heating graphite at very high temperature and pressure. Each carbon atom in diamond is joined by covalent bonds to four others, the four bonds pointing towards the corners of a regular tetrahedron (Figure 12.1). Diamonds are therefore macro-molecules and they contain millions of carbon atoms in a three dimensional network. They are extremely hard and have a very high melting point because a large amount of energy is required to break the vast number of covalent bonds. Diamonds are used as glass cutters and drill points and for jewellery.

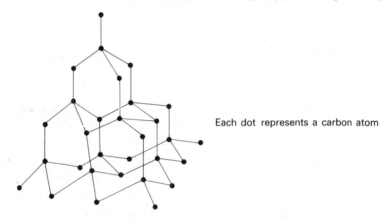

Each dot represents a carbon atom

Figure 12.1. The structure of diamond

Graphite occurs in many areas but the demand exceeds the supply from natural sources. It is manufactured by heating powdered coke at high temperatures. Each carbon atom in graphite is linked by covalent bonds to three others so that they form layers of regular hexagons (Figure 12.2). The distance

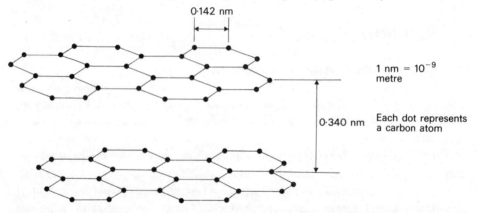

1 nm = 10^{-9} metre

Each dot represents a carbon atom

Figure 12.2. The structure of graphite

between the layers is more than twice the distance between adjacent carbon atoms and so the layers are only weakly attracted to one another by van de Waals forces. The unshared electron in the outer shell of each carbon atom is free to move anywhere along the layer. This delocalisation of the unshared electrons results in graphite being a good conductor of heat and electricity. Graphite electrodes are in fact used in a number of industrial electrolyses. The layers of graphite readily slide over one another and this is why it feels greasy. It is used as a lubricant, particularly at high temperatures when oils would decompose. Graphite mixed with clay is used as pencil leads.

It can be shown that diamond and graphite are different forms of carbon by heating a known mass of each in a stream of oxygen and determining the mass of carbon dioxide produced by absorbing it in weighed bulbs of potassium hydroxide solution (Figure 12.3).

$$C(s) + O_2(g) \rightarrow CO_2(g)$$
$$2KOH(aq) + CO_2(g) \rightarrow K_2CO_3(aq) + H_2O(l)$$

Equal masses of each allotrope are found to produce equal masses of carbon dioxide.

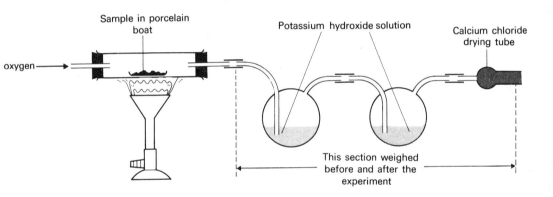

Figure 12.3. To show that diamond and graphite are different forms of carbon

Carbon was thought to exist in an amorphous or non-crystalline state in coke, charcoal, and carbon black. However, it appears that these forms of carbon consist of randomly orientated minute crystals of graphite.

Coke is formed by heating coal, in the absence of air, until all the volatile compounds (coal gas) have been driven off. Coke is used as a fuel and as the reducing agent in the extraction of iron, lead, and zinc. It is also used to make producer gas and water gas (see below).

Charcoal may be obtained in two different ways. Wood charcoal is made by heating wood in the absence of air. It can absorb fairly large quantities of gases and is used in gas masks. Animal charcoal is obtained by heating bones in the absence of air; it consists mainly of calcium phosphate, $Ca_3(PO_4)_2$, and contains only about 10% of carbon. Animal charcoal adsorbs colouring matter in solution and is used industrially to convert brown sugar into white.

Carbon black is obtained by burning cheap or unwanted oil in a limited amount of air so that its hydrogen is oxidised but the carbon is not. It is used as a filler in tyres and to make printer's ink and black shoe polish.

Reactions of carbon

1. Carbon burns in excess air or oxygen giving carbon dioxide:
$$C(s) + O_2(g) \rightarrow CO_2(g)$$

2. Red hot carbon reacts with sulphur vapour to give carbon disulphide:
$$C(s) + 2S(g) \rightarrow CS_2(l)$$

3. On strong heating, carbon will reduce the oxides of zinc and the metals below it in the electrochemical and activity series. This reaction is utilised industrially, e.g:
$$PbO(s) + C(s) \rightarrow Pb(s) + CO(g)$$

4. Red hot carbon reduces carbon dioxide to carbon monoxide:
$$C(s) + CO_2(g) \rightarrow 2CO(g)$$

   The reaction is used to manufacture *producer gas*. Air is blown up through a furnace containing red hot coke. The coke burns in a very exothermic reaction to give carbon dioxide which is reduced, as it passes up the furnace, to carbon monoxide. The gas obtained at the top of the furnace contains about one-third by volume of carbon monoxide and two-thirds of nitrogen. Producer gas is a cheap fuel (carbon monoxide will burn to the dioxide) which is normally used, as it is made, to heat furnaces, boilers, etc.

5. Red hot coke reduces steam:
$$C(s) + H_2O(g) \rightarrow CO(g) + H_2(g)$$

   The mixture of gases obtained from this endothermic reaction is known as *water gas* and it consists of about 45% by volume of carbon monoxide and 50% of hydrogen. The reaction has industrial importance. Producer gas and water gas are generally made alternately in the same plant, the former process raising the temperature of the coke whilst the latter lowers it. Water gas is used in the manufacture of methanol (page 183) and of hydrogen by the Bosch process:

$$CO(g) + H_2O(g) \xrightarrow{Fe_2O_3/500\ °C} CO_2(g) + H_2(g)$$
(In water gas)

   The carbon dioxide is removed by passing the gases through water under pressure.

6. Carbon is slowly oxidised to carbon dioxide by hot concentrated sulphuric or nitric acids.

The carbon or carbon dioxide cycle

The carbon dioxide content of the atmosphere remains practically constant at about 0.03% as a result of the carbon or carbon dioxide cycle. Carbon dioxide is added to the atmosphere by two main processes, i.e. by respiration (see page 46) and by combustion of various materials such as coal, gas, oil, and petrol, etc. Both of these processes simultaneously remove oxygen from the atmosphere. The two major ways by which carbon dioxide is removed from the atmosphere are by photosynthesis and by dissolution in water, e.g. rain and the sea. *Photosynthesis* is the process by which plants in sunlight convert carbon dioxide and water into starchy materials; chlorophyll, the green colouring matter in plants, acts as a catalyst. Photosynthesis restores oxygen to the atmosphere:

$$\text{Carbon dioxide} + \text{water} \xrightarrow{\text{sunlight/chlorophyll}} \text{starch} + \text{oxygen}$$

The essentials of the carbon cycle are summarised in Figure 12.4.

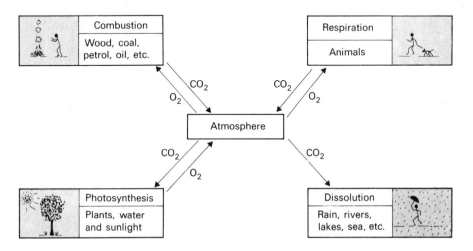

Figure 12.4. Simple version of the carbon cycle

Carbon monoxide, CO

Laboratory preparation

1. Carbon monoxide is prepared in the laboratory by the action of cold concentrated sulphuric acid on methanoic (formic) acid (Figure 12.5).

$$\text{HCOOH(l)} \xrightarrow{\text{H}_2\text{SO}_4} \text{H}_2\text{O(l)} + \text{CO(g)}$$

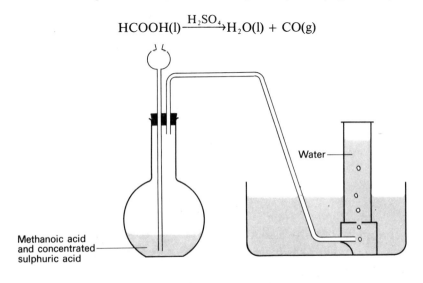

Figure 12.5. Laboratory preparation of carbon monoxide

The sulphuric acid acts as a dehydrating agent, i.e. it removes the elements of water from the methanoic acid. The gas has to be collected over water because its density is similar to that of air.

The method of collection may be deduced as follows.

$$\text{Relative molecular mass of CO} = 28$$
$$\therefore \text{Relative density (see page 63)} = \frac{28}{2} = 14$$

Air contains $\frac{4}{5} N_2$ with relative molecular mass = 28
and $\frac{1}{5} O_2$ with relative molecular mass = 32

$$\therefore \text{Relative density} = \frac{28.8}{2} = 14.4$$

Since the relative densities of carbon monoxide and air are very close, collection by displacement of air is not feasible. The mode of collection of other gases may be deduced in a similar manner.

2. An alternative laboratory preparation of carbon monoxide is to heat ethanedioic (oxalic) acid with concentrated sulphuric acid (Figure 12.6).

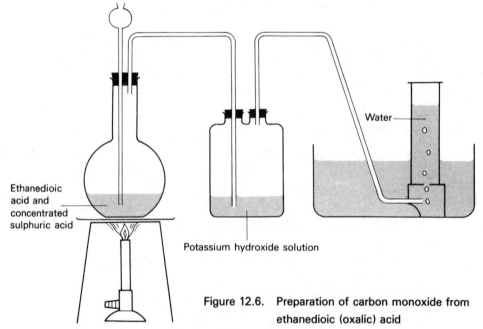

Figure 12.6. Preparation of carbon monoxide from ethanedioic (oxalic) acid

A mixture of carbon monoxide and the dioxide is formed and the latter is removed by passing the gases through potassium hydroxide solution.

$$\begin{array}{c} \text{COOH} \\ | \\ \text{COOH(s)} \end{array} \xrightarrow{H_2SO_4} H_2O(l) + CO(g) + CO_2(g)$$

Industrial preparation

Carbon monoxide is made industrially by passing air through red hot coke – see producer gas (page 180).

Physical properties

Carbon monoxide is a colourless gas which is slightly soluble in water but does not affect litmus. It is very poisonous since it combines irreversibly with the haemoglobin (red corpuscles) of the blood. This prevents the haemoglobin from carrying out its normal function of picking oxygen up in the lungs and carrying it round the veins for use in the body. Carbon monoxide is particularly dangerous in view of the fact that it is odourless.

Chemical properties
1. Carbon monoxide burns with a blue flame to give carbon dioxide:

$$2CO(g) + O_2(g) \rightarrow 2CO_2(g)$$

In contrast to the dioxide, it has no effect on limewater.

2. Carbon monoxide reduces the hot oxides of some metals, e.g:

$$Fe_2O_3(s) + 3CO(g) \rightarrow 2Fe(s) + 3CO_2(g)$$

3. Carbon monoxide is absorbed by solutions of copper(I) chloride in concentrated hydrochloric acid or ammonia solution:

$$CuCl + CO + 2H_2O \rightarrow CuCl \cdot CO \cdot 2H_2O$$

Uses of carbon monoxide
The gas is used as a fuel (producer gas), as a reducing agent in the extraction of some metals, e.g. iron and lead, and in the manufacture of methanol:

$$CO(g) + 2H_2(g) \xrightarrow{\text{catalyst/heat}} CH_3OH(l)$$

Carbon dioxide, CO_2

Laboratory preparation
Carbon dioxide is prepared in the laboratory by the addition of dilute hydrochloric acid to marble chips (calcium carbonate):

$$CaCO_3(s) + 2HCl(aq) \rightarrow CaCl_2(aq) + H_2O(l) + CO_2(g)$$
i.e. $$CO_3^{2-}(s) + 2H^+(aq) \rightarrow H_2O(l) + CO_2(g)$$

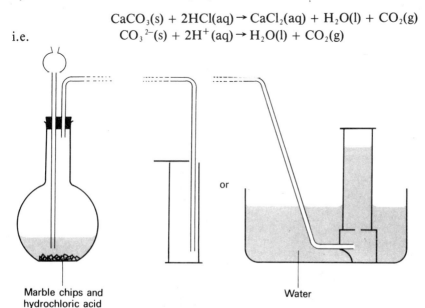

Figure 12.7. Laboratory preparation of carbon dioxide

The reaction may be performed in a Kipps apparatus (Figure 9.5) or in the apparatus shown in Figure 12.7. The gas may be collected by upward displacement of air or over water.

Industrial preparation

Carbon dioxide may be obtained by heating limestone in lime kilns:

$$CaCO_3(s) \rightarrow CaO(s) + CO_2(g)$$

It is also obtained as a by-product from the Bosch process (page 180) and from fermentations (page 239).

Physical properties

Carbon dioxide is colourless and odourless and is about one and a half times as dense as air. It can be poured from one vessel to another and diffisuion is rather slow. It is sparingly soluble in water at atmospheric pressure.

Carbon dioxide liquefies under moderate pressure (about 50 atmospheres) at room temperature. However, the liquid is unstable at pressures of less than 5 atmospheres and consequently carbon dioxide solidifies directly when cooled at atmospheric pressure. Solid carbon dioxide, often known as *dry ice*, sublimes at −78 °C.

Chemical properties

1. Carbon dioxide dissolves in water to a small extent to give a weakly acidic solution of carbonic acid:

$$H_2O(l) + CO_2(g) \rightleftharpoons H_2CO_3(aq)$$

2. The gas does not burn and, as a general rule, it does not support combustion (Figure 12.8). Magnesium, however, burns at a very high temperature and this is sufficient to decompose the carbon dioxide and allow the magnesium to burn feebly:

$$2Mg(s) + CO_2(g) \rightarrow 2MgO(s) + C(s)$$

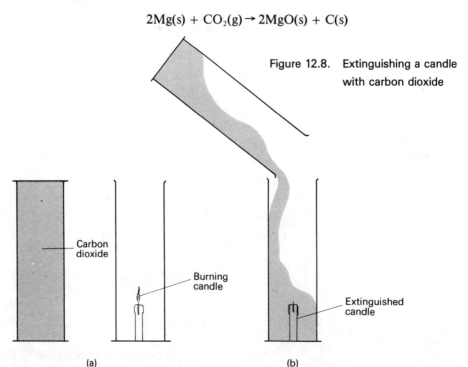

Figure 12.8. Extinguishing a candle with carbon dioxide

3. Carbon dioxide turns limewater milky due to the formation of calcium carbonate:

$$Ca(OH)_2(aq) + CO_2(g) \rightarrow CaCO_3(s) + H_2O(l)$$

This is used as a test for carbon dioxide. The precipitate redissolves on the passage of excess carbon dioxide to give a colourless solution of calcium hydrogencarbonate:

$$CaCO_3(s) + H_2O(l) + CO_2(g) \rightarrow Ca(HCO_3)_2(aq)$$

The reaction between carbon dioxide and the other alkalis also produces the carbonate and then the hydrogencarbonate.

Uses of carbon dioxide

The density and inertness of carbon dioxide make it a useful fire fighting agent. The gas forms a blanket over the burning material and so excludes the air and extinguishes the flames. Fire extinguishers may be obtained which contain carbon dioxide under pressure or reagents for making carbon dioxide rapidly, e.g. concentrated sulphuric acid and sodium hydrogencarbonate solution.

Solid carbon dioxide or *dry ice* is used for refrigeration purposes. It has the advantage over ordinary ice in that it is much colder (-78 °C) and it sublimes, thus eliminating the problem of a residual liquid.

The gas in soda-water and mineral waters is carbon dioxide dissolved under pressure.

Carbonates and hydrogencarbonates

When carbon dioxide is passed into water a weakly acidic solution of carbonic acid is formed:

$$H_2O(l) + CO_2(g) \rightleftharpoons H_2CO_3(aq)$$

This is a dibasic acid and so it will form two series of salts depending on whether one or both of the hydrogens are replaced by a metal.

$$H_2CO_3 \swarrow \searrow$$

MHCO$_3$ M_2CO_3 M = a monovalent metal
Hydrogencarbonates   Carbonates

Hydrogencarbonates may be made by passing carbon dioxide through solutions or suspensions of carbonates in water, e.g:

$$Na_2CO_3(aq) + H_2O(l) + CO_2(g) \rightarrow 2NaHCO_3(aq)$$

All hydrogencarbonates are soluble in water. They liberate carbon dioxide on treatment with acids, e.g:

$$NaHCO_3(s) + HCl(aq) \rightarrow NaCl(aq) + H_2O(l) + CO_2(g)$$

i.e. $$HCO_3^-(s) + H^+(aq) \rightarrow H_2O(l) + CO_2(g)$$

and decompose on heating, e.g:

$$2NaHCO_3(s) \rightarrow Na_2CO_3(s) + H_2O(l) + CO_2(g)$$

Similar decomposition occurs if solutions of hydrogencarbonates are boiled; temporary hardness in water is removed in this way. Only sodium, potassium, and ammonium hydrogencarbonates can be obtained in the solid state.

Carbonates have to be prepared in various ways as described in the previous chapter. The only soluble ones are sodium, potassium, and ammonium carbonates. They all liberate carbon dioxide on addition of dilute acids, e.g.:

i.e.
$$ZnCO_3(s) + 2HCl(aq) \rightarrow ZnCl_2(aq) + H_2O(l) + CO_2(g)$$
$$CO_3^{2-}(s) + 2H^+(aq) \rightarrow H_2O(l) + CO_2(g)$$

and with the exception of sodium and potassium carbonates, they decompose on heating, e.g:

$$CaCO_3(s) \rightarrow CaO(s) + CO_2(g)$$

Solutions of carbonates and hydrogencarbonates can be distinguished by adding magnesium sulphate solution to each. The carbonate will give an immediate white precipitate of magnesium carbonate, e.g:

$$Na_2CO_3(aq) + MgSO_4(aq) \rightarrow MgCO_3(s) + Na_2SO_4(aq)$$

whilst the hydrogencarbonate will not give a white precipitate until the mixture is boiled, e.g:

$$2NaHCO_3(aq) + MgSO_4(aq) \rightarrow Mg(HCO_3)_2(aq) + Na_2SO_4(aq)$$
$$Mg(HCO_3)_2(aq) \xrightarrow{boil} MgCO_3(s) + H_2O(l) + CO_2(g)$$

Tetrachloromethane, CCl₄

Preparation

Methane can react with chlorine under the influence of ultraviolet light to give a variety of products (see page 229). However, if excess chlorine is used the major product is tetrachloromethane:

$$CH_4(g) + 4Cl_2(g) \rightarrow CCl_4(l) + 4HCl(g)$$

Properties

Tetrachloromethane is a dense, non-flammable liquid which is immiscible with water. It is unreactive and, unlike the chlorides of most non-metals, it does not react with water even on heating.

Uses

Tetrachloromethane is used in some fire extinguishers but it has the disadvantage that its vapour is toxic and toxic products may be formed. It is a very good solvent for grease and so it is used in dry cleaning.

SILICON

Silicon is the second most abundant element. It is difficult to obtain in the pure state and so it is fortunate that it does not have a great deal of use other than in transistors.

Some mention has already been made (Chapter 3) of silicon oxide, chloride, and hydride but the oxide merits further study.

Silicon(IV) oxide (silica), SiO_2

Occurrence
Silicon(IV) oxide occurs naturally in several forms, for example quartz, sand, flint, amethyst, agate, and opal. The colours of red sand, amethyst, agate, and opal are due to traces of impurity.

Properties and uses
The difference in physical properties between carbon dioxide and silicon(IV) oxide may appear rather surprising. Thus, carbon dioxide is a colourless gas whilst silicon(IV) oxide is a solid with a melting point in excess of 1600 °C. The explanation is that carbon dioxide is monomolecular, i.e. it exists as single molecules, whilst silicon(IV) oxide is macromolecular. The structure of silicon(IV) oxide is rather similar to that of diamond, the silicon atoms being linked via oxygen atoms (Figure 12.9).

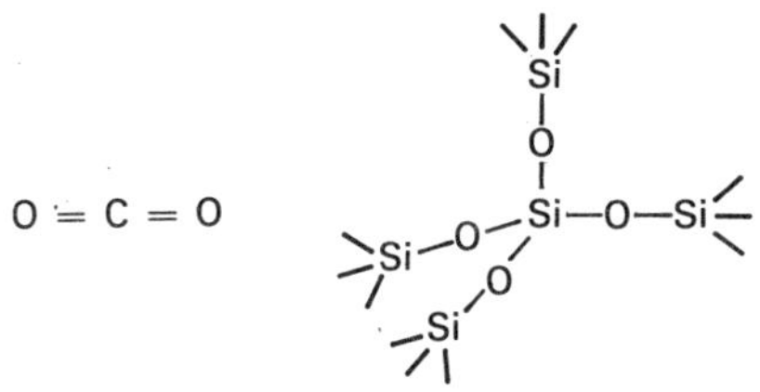

Figure 12.9. The structures of carbon dioxide and silicon(IV) oxide

Silicon(IV) oxide is acidic and reacts slowly with hot concentrated solutions of alkalis to give silicates, e.g:

$$SiO_2(s) + 2NaOH(aq) \rightarrow Na_2SiO_3(aq) + H_2O(l)$$

Concentrated solutions of sodium silicate are sold under the name of *water glass* and this can be used for preserving eggs.

Silicon(IV) oxide displaces more volatile acidic oxides from their salts when heated with them, e.g:

$$Na_2CO_3(s) + SiO_2(s) \rightarrow Na_2SiO_3(s) + CO_2(g)$$
$$CaSO_4(s) + SiO_2(s) \rightarrow CaSiO_3(s) + SO_3(g)$$

Reactions such as these are used in the manufacture of glass. Soda glass, for use as window panes, is made by heating sodium and calcium salts with white sand; it is a mixture of silicon(IV) oxide and sodium and calcium silicates. Lead glass, for making cut glass articles, consists of silicon(IV) oxide with potassium and lead silicates, whilst borosilicate glass, for use as ovenware, consists of silicon(IV) oxide, boron oxide, and sodium and aluminium silicates.

Quartz can be melted at high temperature to give a glass which is resistant to cracking when subjected to sudden changes in temperature.

Silicon(IV) oxide (sand) is also important as a constituent of concrete and mortar.

NITROGEN AND PHOSPHORUS-GROUP V ELEMENTS

INTRODUCTION

The first two elements in Group V and their electronic structures are nitrogen, 2.5 and phosphorus, 2.8.5. They generally attain the electronic structure of the noble gases by sharing three of their electrons in the formation of covalent bonds. However, they can both gain three electrons to give nitride or phosphide ions, i.e. N^{3-} or P^{3-} respectively. Phosphorus, unlike nitrogen, can have a valency of five as well as the normal group valency of three.

NITROGEN

Occurrence

Nitrogen accounts for about 78% by volume of the atmosphere. It occurs in combination with other elements in all living matter in the form of proteins but the only nitrogen compounds obtained in bulk are sodium and potassium nitrates.

Laboratory preparations

1. Nitrogen may be obtained by warming a solution of sodium nitrite and ammonium chloride (Figure 12.10).

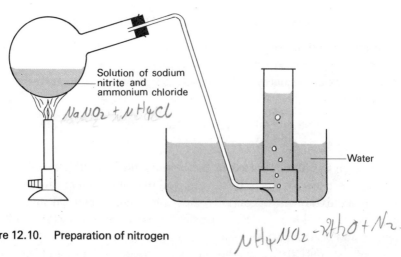

Figure 12.10. Preparation of nitrogen

$$NaNO_2(aq) + NH_4Cl(aq) \rightarrow NaCl(aq) + NH_4NO_2(aq)$$
$$NH_4NO_2(aq) \rightarrow 2H_2O(l) + N_2(g)$$

The ammonium nitrite has to be made as required since the solid compound is potentially explosive.

2. Nitrogen containing about 1% of the noble gases can be obtained by passing air, which has had its carbon dioxide removed, over hot copper (Figure 12.11). The copper and oxygen combine but the nitrogen is unaffected.

$$2Cu(s) + O_2(g) \rightarrow 2CuO(s)$$

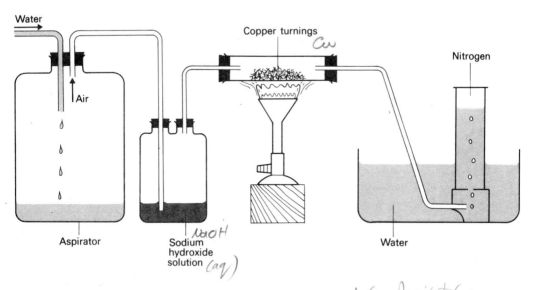

Figure 12.11 The preparation of nitrogen from air

Manufacture

Nitrogen is obtained industrially by the fractional distillation of liquid air (see page 113).

Properties of nitrogen

Nitrogen is a colourless, odourless gas which is only slightly soluble in water. It does not burn or support combustion. Nitrogen exists as diatomic molecules, N_2 (:N≡N:) which are very stable and so it is unreactive. It only combines with other elements on strong heating, e.g:

$$N_2(g) + O_2(g) \xrightleftharpoons{3000\ °C} 2NO(g)$$

$$N_2(g) + 3H_2(g) \xrightleftharpoons[\text{Iron catalyst}]{550\ °C,\ 250\ atm} 2NH_3(g)$$

Burning magnesium glows in nitrogen for a few seconds as the nitride is formed:

$$3Mg(s) + N_2(g) \rightarrow Mg_3N_2(s)$$

i.e. $\quad 3Mg(s) + N_2(g) \rightarrow 3Mg^{2+}(s) + 2N^{3-}(s)$

This reaction is used to test for nitrogen (see page 45).

Uses of nitrogen

The major use of nitrogen is in the manufacture of ammonia (see page 130). It is also used to provide an inert atmosphere in various processes.

The nitrogen cycle

Leguminous plants, e.g. peas, beans, and clover, are able to utilise nitrogen from the atmosphere to form proteins. However, these plants are the exceptions and most make proteins by absorbing nitrates from the soil. Generally, therefore, cultivation results in nitrogen being taken from the soil and, unless this is replaced, the soil becomes less fertile.

The circulation of nitrogen between the soil and living things is known as the nitrogen cycle and a simple version is illustrated in Figure 12.12.

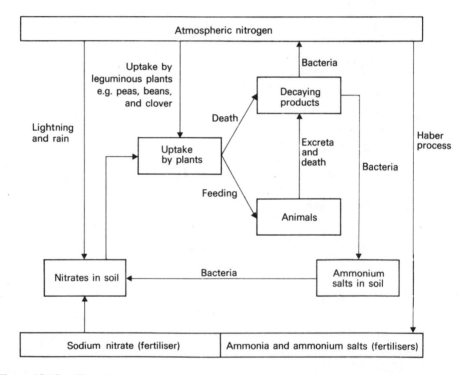

Figure 12.12. The nitrogen cycle

The conversion of atmospheric nitrogen to nitrates in the soil takes place as follows. Lightning causes nitrogen and oxygen to form nitrogen oxide which then combines with oxygen in the air to give nitrogen dioxide:

$$N_2(g) + O_2(g) \rightarrow 2NO(g)$$
$$2NO(g) + O_2(g) \rightarrow 2NO_2(g)$$

The nitrogen dioxide combines with rain and oxygen to give nitric acid which can react with various substances in the soil to give nitrates.

$$2H_2O(l) + 4NO_2(g) + O_2(g) \rightarrow 4HNO_3(aq)$$
$$CaCO_3(s) + 2HNO_3(aq) \rightarrow Ca(NO_3)_2(aq) + H_2O(l) + CO_2(g)$$

Ammonia, NH_3

Laboratory preparation

Ammonia may be prepared by heating any ammonium salt with any alkali, but generally ammonium chloride and calcium hydroxide are used:

$$2NH_4Cl(s) + Ca(OH)_2(s) \rightarrow CaCl_2(s) + 2H_2O(l) + 2NH_3(g)$$

i.e. $\quad NH_4^+(s) + HO^-(s) \rightarrow H_2O(l) + NH_3(g)$

The reaction flask is tilted as shown in Figure 12.13 because, if condensed water dropped back on to the hot glass, the flask could crack.

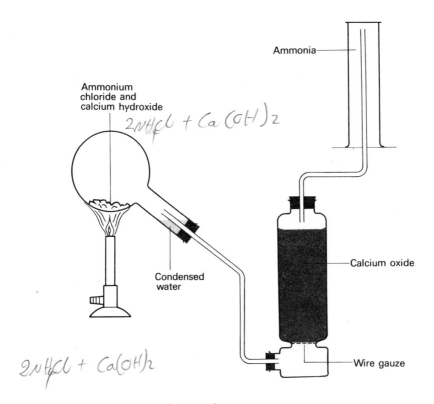

Figure 12.13. Preparation of ammonia

The ammonia cannot be dried with phosphorus(V) oxide or concentrated sulphuric acid since it is an alkaline gas and would react with these acidic drying agents. Calcium chloride cannot be used either because it forms an addition compound with ammonia, i.e. $CaCl_2 \cdot 8NH_3$. The drying agent is therefore calcium oxide, a basic oxide.

The relative molecular mass of ammonia is 17 and so its relative density is 8.5. Hence it is much lighter than air (relative density 14.4 (see page 182)) and so it is collected by downward displacement.

Manufacture

Ammonia is manufactured from nitrogen and hydrogen by the Haber process as described on page 130.

Physical properties

Ammonia is a colourless gas which is less dense than air and has a very pungent smell. It is readily liquefied either by cooling or by pressure. It is extremely soluble in water as illustrated by the fact that, at 15 °C, 800 volumes of ammonia dissolve in 1 volume of water. The high solubility may be demonstrated by the fountain experiment (Figure 12.14). The round bottomed flask is filled with ammonia and connected up to the Buchner flask as shown. The spring clip is released and a little water is blown up the tube so that it enters the flask. The water dissolves so much ammonia that a partial vacuum is set up and more water is rapidly sucked into the flask as a fountain. The red litmus turns blue as it enters the flask because an alkaline solution is formed.

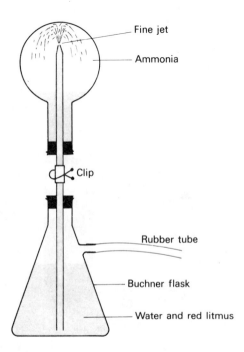

Figure 12.14. The fountain experiment with ammonia

Chemical properties

1. Ammonia does not burn in air or support combustion. However, it burns in oxygen to give nitrogen and water:

$$4NH_3(g) + 3O_2(g) \rightarrow 6H_2O(l) + 2N_2(g)$$

2. When ammonia is passed into water, most of it simply dissolves, but a little reacts to give ammonium and hydroxide ions. An equilibrium is set up, the position of the equilibrium being very much on the left hand side, i.e. the concentration of ions is low.

$$H_2O(l) + NH_3(aq) \rightleftharpoons NH_4^+(aq) + HO^-(aq)$$

This solution is often called ammonium hydroxide and given the formula NH_4OH, but it is much better described as ammonia solution or aqueous ammonia.

3. Ammonia is oxidised to nitrogen oxide by air in the presence of a platinum catalyst:

$$4NH_3(g) + 5O_2(g) \rightarrow 6H_2O(l) + 4NO(g)$$

The colourless nitrogen oxide subsequently reacts with oxygen in the air to give brown fumes of nitrogen dioxide:

$$2NO(g) + O_2(g) \rightarrow 2NO_2(g)$$

These reactions are utilised in the manufacture of nitric acid (see page 196).

The oxidation of ammonia may be illustrated using the apparatus shown in Figure 12.15. Oxygen is bubbled through the dilute ammonia solution (1 volume 0.880 ammonia solution:1 volume of water) and a red hot platinum or copper coil is held just above. The coil continues to glow owing to the exothermic reaction, i.e. the oxidation of ammonia, occurring at its surface. Brown fumes of nitrogen dioxide are also seen.

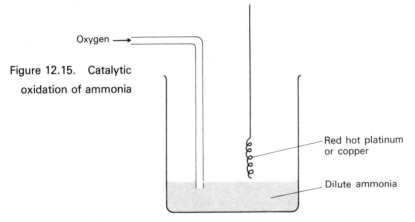

Figure 12.15. Catalytic oxidation of ammonia

4. Ammonia is oxidised by hot copper(II) oxide to nitrogen and water (Figure 12.16). The black copper(II) oxide turns reddish gold as it is reduced to copper.

$$3CuO(s) + 2NH_3(g) \rightarrow 3Cu(s) + 3H_2O(l) + N_2(g)$$

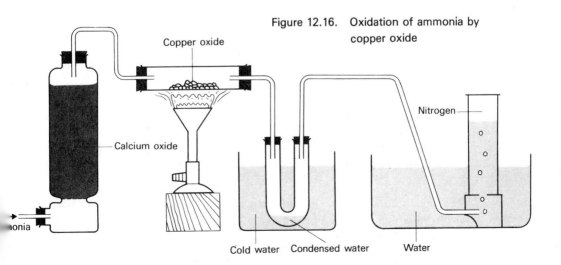

Figure 12.16. Oxidation of ammonia by copper oxide

5. Ammonia forms dense white fumes of ammonium chloride with hydrogen chloride.

$$NH_3(g) + HCl(g) \rightarrow NH_4Cl(s)$$

This is used as a test for ammonia (in conjunction with its property of turning moist red litmus paper blue), the normal procedure being to dip a glass rod in concentrated hydrochloric acid and then to hold it in the ammonia.

Ammonia solution

Preparation

This cannot be prepared simply by placing the delivery tube from the reaction flask straight into water because the solubility of ammonia is so great that the water would suck back. Instead, an inverted funnel is fitted to the delivery tube (Figure 12.17) so that it just dips under the surface of the water. As soon as sucking back starts, the level of the water drops below the rim of the funnel and so air is sucked in instead.

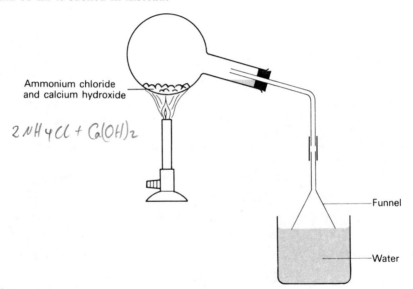

Figure 12.17. Preparation of ammonia solution

The usual concentrated ammonia solution encountered in the laboratory is often known as eight eighty ammonia because its density is 0.880 g cm^{-3}.

Properties

1. Ammonia solution readily splits up on warming into ammonia and water.
2. Ammonia solution neutralises acids to give salts; the solutions may be titrated using an indicator such as methyl orange.

$$2NH_3(aq) + H_2SO_4(aq) \rightarrow (NH_4)_2SO_4(aq)$$
$$NH_3(aq) + HNO_3(aq) \rightarrow NH_4NO_3(aq)$$

Reaction occurs because the nitrogen of the ammonia has a lone pair of electrons which attracts a proton from the acid, i.e:

$$\ddot{N}H_3(aq) + H^+(aq) \rightarrow NH_4^+(aq)$$

3. Ammonia solution precipitates the hydroxides of many metals from solutions of their salts, e.g:

$$Mg(NO_3)_2(aq) + 2NH_3(aq) + 2H_2O(l) \rightarrow Mg(OH)_2(s) + 2NH_4NO_3(aq)$$
White

$$FeSO_4(aq) + 2NH_3(aq) + 2H_2O(l) \rightarrow Fe(OH)_2(s) + (NH_4)_2SO_4(aq)$$
Dirty green

$$Fe_2(SO_4)_3(aq) + 6NH_3(aq) + 6H_2O(l) \rightarrow 2Fe(OH)_3(s) + 3(NH_4)_2SO_4(aq)$$
Red-brown

$$Al_2(SO_4)_3(aq) + 6NH_3(aq) + 6H_2O(l) \rightarrow 2Al(OH)_3(s) + 3(NH_4)_2SO_4(aq)$$
White

In a few cases the hydroxide redissolves owing to the formation of a complex ion, notable examples being $[Cu(NH_3)_4]^{2+}$ and $[Zn(NH_3)_4]^{2+}$ (see pages 169 and 160 respectively).

Ammonium salts

All ammonium salts are ionic and soluble in water. They may be obtained from their solutions by recrystallisation. The ammonium ion, NH_4^+, in many ways resembles a monovalent metal ion.

All ammonium salts yield ammonia on heating with an alkali, e.g:

$$NH_4Cl(aq) + NaOH(aq) \rightarrow NaCl(aq) + H_2O(l) + NH_3(g)$$
$$(NH_4)_2SO_4(s) + Ca(OH)_2(s) \rightarrow CaSO_4(s) + 2H_2O(l) + 2NH_3(g)$$

Ammonium salts decompose on heating, e.g:

$$NH_4NO_3(s) \rightarrow 2H_2O(l) + N_2O(g) \quad \text{(dinitrogen oxide)}$$
$$NH_4NO_2(s) \rightarrow 2H_2O(l) + N_2(g)$$
$$(NH_4)_2SO_4(s) \rightarrow NH_4HSO_4(s) + NH_3(g)$$

Some ammonium salts sublime. Ammonium carbonate sublimes at room temperature and is used in smelling salts whilst ammonium chloride sublimes on heating:

$$NH_4Cl(s) \rightleftharpoons NH_3(g) + HCl(g)$$

This is an example of thermal dissociation as opposed to thermal decomposition. The sublimation can be demonstrated simply by heating ammonium chloride in a test tube; a ring of the solid will appear on the cooler parts of the tube (Figure 12.18).

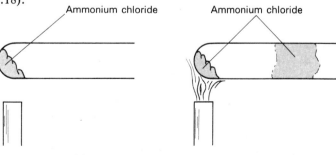

Figure 12.18. The sublimation of ammonium chloride

Uses of ammonia and ammonium salts

Ammonia is used to manufacture nitirc acid (see below) and for making many other products such as sodium carbonate (page 139), etc. Ammonium salts, particularly the sulphate, are used extensively as fertilisers. Ammonium nitrate is used in the manufacture of some explosives.

Nitric acid, HNO_3

Laboratory preparation

Nitric acid may be prepared by heating a nitrate with concentrated sulphuric acid. Generally, sodium or potassium nitrates are used, e.g:

$$NaNO_3(s) + H_2SO_4(conc) \rightarrow NaHSO_4(s) + HNO_3(l)$$

The product is yellow due to dissolved nitrogen dioxide resulting from some thermal decomposition of the acid. Nitric acid is very corrosive and so an all glass apparatus has to be used (Figure 12.19).

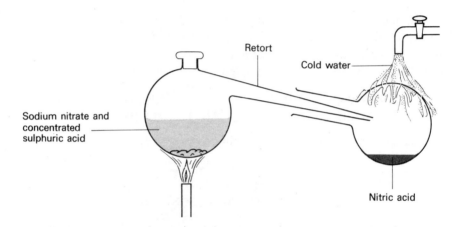

Figure 12.19. The preparation of nitric acid

Manufacture

Nitric acid is manufactured by the Ostwald process which is based on the catalytic oxidation of ammonia. A mixture of ammonia and about ten times its volume of air is passed over a platinum-rhodium catalyst at about 700—800 °C under slight pressure. The ammonia is oxidised to nitrogen oxide:

$$4NH_3(g) + 5O_2(g) \rightarrow 6H_2O(g) + 4NO(g)$$

The gases are cooled and then diluted with air. The nitrogen oxide reacts with oxygen from the air to give nitrogen dioxide which combines with water and oxygen to yield nitric acid.

$$2NO(g) + O_2(g) \rightarrow 2NO_2(g)$$
$$2H_2O(l) + 4NO_2(g) + O_2(g) \rightarrow 4HNO_3(aq)$$

Concentrated nitric acid (68% by mass of nitric acid and 32% water) is obtained by distilling the solution. Pure nitric acid is made by distilling the concentrated acid from concentrated sulphuric acid.

Properties

Nitric acid is a strong, very corrosive acid. The pure acid is a colourless fuming liquid whilst the concentrated acid is often yellow due to dissolved nitrogen dioxide; the latter substance arises as a result of slight decomposition of the acid.

Nitric acid decomposes on heating to give water, nitrogen dioxide, and oxygen:

$$4HNO_3(g) \rightarrow 4NO_2(g) + O_2(g) + 2H_2O(g)$$

Nitric acid undergoes most of the reactions associated with strong acids. For example, it turns blue litmus red, neutralises acids to give salts and water, and it liberates carbon dioxide from carbonates and hydrogencarbonates.

Nitric acid is a powerful oxidising agent and it reacts with most metals giving a variety of products. These reactions are often complex and the equations given represent the major process. Normally, acids react with metals to give a salt and hydrogen. This, in fact, does occur when very dilute nitric acid (less than about 4%) is added to magnesium:

$$Mg(s) + 2HNO_3(aq) \rightarrow Mg(NO_3)_2(aq) + H_2(g)$$

i.e.
$$Mg(s) + 2H^+(aq) \rightarrow Mg^{2+}(aq) + H_2(g)$$

However, at higher concentrations, or with other metals, oxides of nitrogen, etc. are formed instead of hydrogen, the products depending upon the temperature and concentration of the acid. For example, cold dilute (1:1) nitric acid reacts with copper at room temperature to give copper(II) nitrate, water, and nitrogen oxide:

$$3Cu(s) + 8HNO_3(aq) \rightarrow 3Cu(NO_3)_2(aq) + 4H_2O(l) + 2NO(g)$$

The nitrogen oxide may be collected over water (Figure 12.20); it is a colourless gas which reacts with oxygen in the air to give brown fumes of nitrogen dioxide. Addition of concentrated nitric acid to copper, at room temperature, gives nitrogen dioxide:

$$Cu(s) + 4HNO_3(conc) \rightarrow Cu(NO_3)_2(aq) + 2H_2O(l) + 2NO_2(g)$$

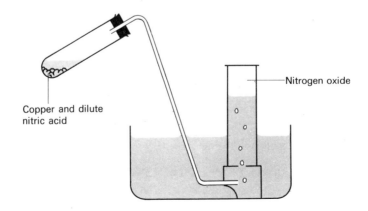

Figure 12.20. Preparation of nitrogen oxide

When nitric acid is added to iron or aluminium, reaction ceases after a few seconds due to the formation of a thin protective film of oxide being formed on the metal. The metals are said to be rendered passive.

Nitric acid oxidises a number of non-metals. For example, hot concentrated nitric acid oxidises sulphur to sulphuric acid:

$$S(s) + 6HNO_3(conc) \rightarrow H_2SO_4(aq) + 2H_2O(l) + 6NO_2(g)$$

The acid is often used to oxidise iron(II) to iron(III) salts. A pale green solution of iron(II) sulphate turns yellow-brown on boiling with a few drops of concentrated nitric acid owing to the formation of iron(III) sulphate:

$$6FeSO_4(aq) + 3H_2SO_4(aq) + 2HNO_3(aq) \rightarrow 3Fe_2(SO_4)_3(aq) + 4H_2O(l) + 2NO(g)$$

i.e. $\qquad Fe^{2+} - e^- \rightarrow Fe^{3+}$

Uses of nitric acid

Nitric acid is extensively used to make nitrates, e.g. calcium, sodium, and ammonium nitrates, for use as fertilisers. It is also used in the manufacture of explosives such as TNT and nitroglycerine, and in the dyestuff and pharmaceutical industries.

Nitrates

Nitrates can be made by the action of nitric acid on metals (there are a few exceptions), oxides, hydroxides, or carbonates, e.g:

$$ZnO(s) + 2HNO_3(aq) \rightarrow Zn(NO_3)_2(aq) + H_2O(l)$$
$$NaOH(aq) + HNO_3(aq) \rightarrow NaNO_3(aq) + H_2O(l)$$
$$CuCO_3(s) + 2HNO_3(aq) \rightarrow Cu(NO_3)_2(aq) + H_2O(l) + CO_2(g)$$

All the nitrates are soluble in water and many have water of crystallisation.

The nitrates can be divided into four classes according to the products they give on thermal decomposition.

(a) The nitrates of the elements at the top of the electrochemical and activity series, i.e. sodium and potassium, yield the nitrite and oxygen on heating. Both of these nitrates melt before they decompose, e.g:

$$2KNO_3(l) \rightarrow 2KNO_2(l) + O_2(g)$$

(b) Most nitrates are decomposed by heat into the metal oxide, nitrogen dioxide, and oxygen, e.g:

$$2Cu(NO_3)_2(s) \rightarrow 2CuO(s) + 4NO_2(g) + O_2(g)$$

(c) The oxides of the metals at the bottom of the electrochemical and activity series are unstable and so the action of heat on the nitrates of these metals yields the metal, nitrogen dioxide, and oxygen. The nitrates of silver and mercury behave in this manner:

$$2AgNO_3(s) \rightarrow 2Ag(s) + 2NO_2(g) + O_2(g)$$
$$Hg(NO_3)_2(s) \rightarrow Hg(l) + 2NO_2(g) + O_2(g)$$

(d) Ammonium nitrate decomposes on heating to give dinitrogen oxide and water. The reaction is inclined to be explosive:

$$NH_4NO_3(s) \rightarrow N_2O(g) + 2H_2O(g)$$

Many nitrates yield nitrogen dioxide on heating but this can react with their water of crystallisation giving some nitric acid and nitrogen oxide:

$$H_2O(l) + 3NO_2(g) \rightarrow 2HNO_3(aq) + NO(g)$$

(The nitrogen oxide reacts with oxygen and water to give more nitric acid, which accounts for the net equation given on page 196, i.e.

$$2H_2O(l) + 4NO_2(g) + O_2(g) \rightarrow 4HNO_3(aq))$$

Nitrogen dioxide is therefore prepared in the laboratory by heating lead(II) nitrate because this has no water of crystallisation and so contamination with nitrogen oxide does not occur. The apparatus used is illustrated in Figure 12.21.

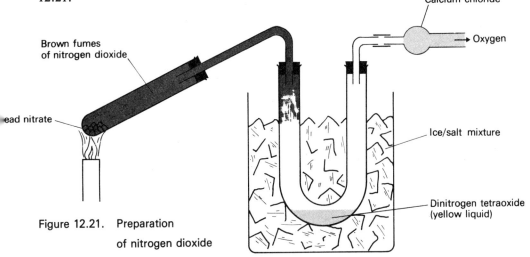

Figure 12.21. Preparation of nitrogen dioxide

$$2Pb(NO_3)_2(s) \rightarrow 2PbO(s) + 4NO_2(g) + O_2(g)$$

The brown fumes of nitrogen dioxide condense to a pale yellow liquid whilst the oxygen is unaffected and escapes to the atmosphere. Nitrogen dioxide exists in equilibrium with dinitrogen tetraoxide, N_2O_4:

$$N_2O_4 \underset{\text{cool}}{\overset{\text{heat}}{\rightleftharpoons}} 2NO_2$$

The dimer, or double molecule, becomes more predominant as the temperature is lowered so that the pale yellow liquid obtained above is mainly dinitrogen tetraoxide.

Test for nitrates
Nitrates are detected by the brown ring test as described on page 166.

PHOSPHORUS

Phosphorus occurs only in combination with other elements, the main source being calcium phosphate, $Ca_3(PO_4)_2$. The extraction and chemistry of the element is beyond the scope of this text but a few of its compounds will be discussed.

Phosphorus(V) oxide (phosphorus pentoxide), P_2O_5

Phosphorus(V) oxide is prepared by burning phosphorus in excess air or oxygen:

$$4P(s) + 5O_2(g) \rightarrow 2P_2O_5(s)$$

It is an extremely hygroscopic white solid. It reacts vigorously with water, in two stages, to yield phosphoric acid as the final product:

$$P_2O_5(s) + H_2O(l) \rightarrow 2HPO_3(aq)$$
$$HPO_3(aq) + H_2O(l) \rightarrow H_3PO_4(aq)$$

Phosphorus(V) oxide is used as a drying and dehydrating agent.

Phosphoric acid, H_3PO_4, and phosphates

Phosphoric acid may be made by the reaction between phosphorus(V) oxide and water as described above. It is generally obtained as a concentrated solution in the form of a colourless, syrupy liquid.

Phosphoric acid is a weak tribasic acid. The three hydrogen atoms can be replaced, one at a time, to give a phosphate as the final product.

$$H_3PO_4(aq) + 3NaOH(aq) \rightarrow Na_3PO_4(aq) + 3H_2O(l)$$

The acid is used to rust-proof steel for car bodies, etc. as it forms a protective layer of iron phosphate. A further very important use is the manufacture of phosphates, e.g. ammonium phosphate, $(NH_4)_3PO_4$, for use as fertilisers.

Phosphorus pentachloride, PCl_5

Warming the white allotrope of phosphorus with dry chlorine (Figure 12.22) gives phosphorus trichloride:

$$2P(s) + 3Cl_2(g) \rightarrow 2PCl_3(l)$$

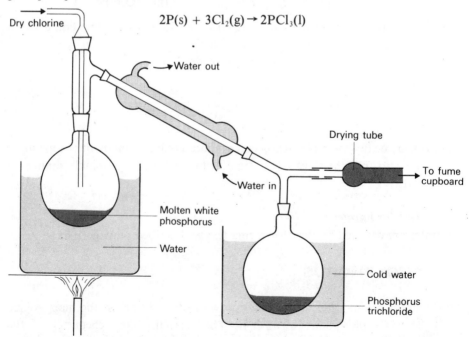

Figure 12.22. Preparation of phosphorus trichloride

Treatment of cold phosphorus trichloride with chlorine (Figure 12.23) gives the pentachloride:

$$PCl_3(l) + Cl_2(g) \rightarrow PCl_5(s)$$

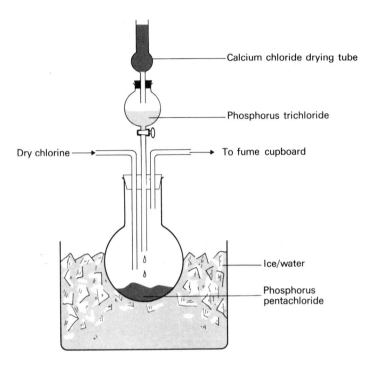

Figure 12.23. Preparation of phosphorus pentachloride

Phosphorus pentachloride is a pale yellow solid which fumes in moist air and reacts violently with water, giving first phosphorus trichloride oxide and then phosphoric acid:

$$PCl_5(s) + H_2O(l) \rightarrow POCl_3(l) + 2HCl(g)$$
$$POCl_3(l) + 3H_2O(l) \rightarrow H_3PO_4(aq) + 3HCl(g)$$

Phosphorus pentachloride reacts with all compounds containing a hydroxyl group; white fumes of hydrogen chloride are evolved and the hydroxyl group is replaced by a chlorine atom. This reaction is used as a test for hydroxyl groups (white fumes are evolved) and for chlorinating hydroxy compounds.

OXYGEN AND SULPHUR – GROUP VI ELEMENTS

The first two elements in Group VI are oxygen and sulphur, their electronic structures being 2.6 and 2.8.6 respectively. Both elements exhibit a valency of two and can form covalent bonds, as for example in their hydrides, H_2O and H_2S, or divalent negative ions, i.e. O^{2-} and S^{2-}. In addition to its valency of two, sulphur can have a valency of four as in sulphur dioxide, SO_2 and a valency of six as in sulphur(VI) oxide, SO_3.

The chemistry of oxygen was discussed in Chapter 9 and so only sulphur will be considered here.

SULPHUR

Occurrence

Sulphur deposits occur in past or present volcanic areas, for example, in Texas, Italy, Sicily, and Japan. It occurs in combination with other elements as the sulphide and sulphate, e.g. iron(II) disulphide (*iron pyrites*), FeS_2; lead(II) sulphide (*galena*), PbS; zinc sulphide (*zinc blende*), ZnS; and calcium sulphate (*gypsum*), $CaSO_4 \cdot 2H_2O$.

Extraction

Sulphur is obtained from the sulphur deposits by the Frasch process. A hole is bored down to the deposit and three concentric pipes are inserted (Figure 12.24). Superheated water at 170 °C under pressure is passed down the outer pipe in order to melt the sulphur. Hot compressed air is then passed down the centre pipe so that molten sulphur and water is forced up the middle pipe. The water and sulphur mixture is allowed to cool in large vats and the water is run off when the sulphur has solidified. The sulphur is 99% pure; it is remelted, allowed to cool in moulds and then sold as *roll sulphur*.

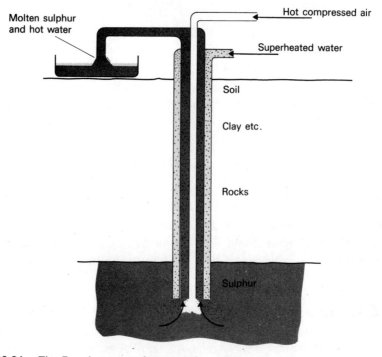

Figure 12.24. The Frasch process for extracting sulphur

Allotropes of sulphur

Sulphur can exist in four forms: rhombic, monoclinic, plastic, and amorphous sulphur.

Rhombic (octahedral) sulphur is prepared by shaking powdered sulphur with carbon disulphide (note that this solvent is highly flammable as well as having a foul smell), filtering to remove amorphous sulphur, and then allowing the solution to evaporate slowly. Small yellow diamond shaped crystals form as the solvent evaporates. Best results are obtained when the carbon disulphide evaporates very slowly. The usual procedure, therefore, is to cover the vessel containing the solution with a perforated filter paper and leave it in a fume cupboard until all the solvent has disappeared. This allotrope is stable up to 96.5 °C.

Monoclinic (prismatic) sulphur is prepared by melting sulphur and then allowing it to cool. As soon as a crust forms on the surface, two holes are pierced in it and the molten sulphur underneath is poured out (Figure 12.25). When the crust is removed, pale yellow needle-like crystals of monoclinic sulphur are seen adhering to the walls of the container.

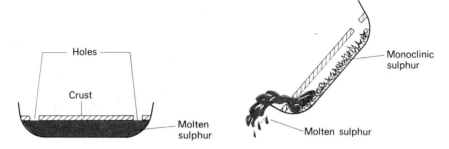

Figure 12.25. Preparation of monoclinic sulphur

This allotrope is only stable between 96.5 °C and its melting point. When the temperature falls below 96.5 °C, which is known as the *transition temperature*, the crystals slowly revert to rhombic sulphur.

$$\text{rhombic sulphur} \underset{}{\overset{96.5\ °C}{\rightleftharpoons}} \text{monoclinic sulphur}$$

The crystals retain their overall needle-like shape but become a mass of very small rhombic crystals.

Plastic sulphur is obtained when boiling sulphur is poured in a thin stream into cold water. A brownish plastic mass is formed but it slowly becomes brittle and reverts to rhombic sulphur.

Amorphous sulphur is sulphur with no definite shape. It is made in various chemical reactions, for example, by the addition of dilute nitric acid to sodium thiosulphate solution (see page 127).

Effect of heat on sulphur

Some interesting changes occur when sulphur is heated. The pale yellow solid melts at 113 °C giving a mobile, straw coloured liquid. This liquid consists of S_8 molecules, the eight atoms in each molecule forming a ring. At about

200 °C, the liquid becomes very dark and viscous as the rings open and the ends join to give long chains which become entwined. On raising the temperature further, the colour lightens somewhat, and the liquid becomes less viscous, due to the long chains of sulphur atoms breaking down to give shorter ones. The liquid boils at 444 °C. Sulphur vapour condenses directly, on cool surfaces, to a yellow solid which is known as *flowers of sulphur.*

Properties of sulphur

Sulphur is a typical non-metal: it has a low melting point, it is a poor conductor of heat, and a non-conductor of electricity. Rhombic and monoclinic sulphur are readily soluble in carbon disulphide and many organic solvents but insoluble in water. It forms acidic oxides and negative ions.

Reactions of sulphur

1. Sulphur burns in air or oxygen with a pale blue flame forming sulphur dioxide:

$$S(s) + O_2(g) \rightarrow SO_2(g)$$

2. Sulphur combines with many metals on heating, e.g:

$$Fe(s) + S(s) \rightarrow FeS(s)$$

3. If hydrogen is passed through boiling sulphur a low yield of hydrogen sulphide is obtained:

$$S(l) + H_2(g) \rightarrow H_2S(g)$$

4. Hot concentrated sulphuric acid slowly oxidises sulphur to sulphur dioxide:

$$S(s) + 2H_2SO_4(conc) \rightarrow 2H_2O(l) + 3SO_2(g)$$

Sulphur is oxidised to sulphuric acid by hot concentrated nitric acid:

$$S(s) + 6HNO_3(conc) \rightarrow H_2SO_4(aq) + 2H_2O(l) + 6NO_2(g)$$

Uses of sulphur

The major use of sulphur is in the manufacture of sulphuric acid. However, it also finds considerable use in the vulcanisation of rubber. This process involves heating soft pliable rubber with sulphur to make a harder and tougher rubber suitable for tyres. Sulphur is used as a fungicide, as a constituent of gunpowder, and for making a number of other products, particularly drugs.

Hydrogen sulphide, H_2S

Preparation

Hydrogen sulphide is prepared by the action of dilute hydrochloric acid on iron(II) sulphide:

$$FeS(s) + 2HCl(aq) \rightarrow FeCl_2(aq) + H_2S(g)$$

The reaction can be performed using a flask, thistle funnel, and delivery tube but a Kipp's apparatus (page 110) is generally used so as to minimise escape of the gas. The hydrogen sulphide may be collected over warm water or by upward displacement of air. If the dry gas is required, calcium chloride is used as the drying agent because hydrogen sulphide reduces concentrated sulphuric acid – see below.

Since iron(II) sulphide is made by heating iron with sulphur, there is always a little free iron present. Addition of dilute acid to iron(II) sulphide will therefore give hydrogen sulphide containing a small amount of hydrogen as impurity. This contamination is unimportant for most purposes.

Physical properties

Hydrogen sulphide is a colourless, highly poisonous gas with a smell of bad eggs. It is fairly soluble in cold water but less so in hot. It is denser than air: its relative density is 17 compared to 14.4 for air.

Reactions of hydrogen sulphide

1. The gas burns in a limited supply of air to give water and sulphur:

$$2H_2S(g) + O_2(g) \rightarrow 2S(s) + 2H_2O(g)$$

With excess air or oxygen, however, further oxidation occurs and the products are water and sulphur dioxide:

$$2H_2S(g) + 3O_2(g) \rightarrow 2SO_2(g) + 2H_2O(g)$$

2. Hydrogen sulphide dissolves in water to give a weakly acidic solution which will turn blue litmus red. It is a weak dibasic acid and so will form acid or normal salts depending on whether one or both of the hydrogens are replaced by a metal.

$$H_2S(aq) \rightarrow H^+(aq) + HS^-(aq)$$
$$HS^-(aq) \rightarrow H^+(aq) + S^{2-}(aq)$$

Many metal sulphides are insoluble in water and are generally prepared by passing hydrogen sulphide through solutions of their salts, e.g:

$$CuSO_4(aq) + H_2S(g) \rightarrow CuS(s) + H_2SO_4(aq)$$
Blue   Dark brown

$$ZnSO_4(aq) + H_2S(g) \rightarrow ZnS(s) + H_2SO_4(aq)$$
Colourless   White

$$Pb(NO_3)_2(aq) + H_2S(g) \rightarrow PbS(s) + 2HNO_3(aq)$$
Colourless   Black

In all of these examples, the basic reaction is:

$$M^{2+}(aq) + S^{2-}(aq) \rightarrow MS(s) \quad \text{where } M \text{ is a divalent metal.}$$

The only soluble sulphides are those of sodium, potassium, ammonium, and calcium. If excess hydrogen sulphide is passed through aqueous solutions of the alkalis, the hydrogensulphide is produced, e.g:

$$KOH(aq) + H_2S(g) \rightarrow KHS(aq) + H_2O(l)$$

The normal sulphide is obtained by neutralising the hydrogensulphide with an equivalent amount of alkali:

$$KHS(aq) + KOH(aq) \rightarrow K_2S(aq) + H_2O(l)$$

3. Hydrogen sulphide is a powerful reducing agent:

$$H_2O(l) + H_2S(g) \rightleftharpoons 2H_3O^+(aq) + S^{2-}(aq)$$
$$S^{2-}(aq) \rightarrow S(s) + 2e^-$$

(a) It reduces iron(III) salts to iron(II) salts, e.g:

$$2FeCl_3(aq) + H_2S(g) \rightarrow S(s) + 2FeCl_2(aq) + 2HCl(aq)$$
Yellow Yellow Pale green

i.e. $2Fe^{3+}(aq) + S^{2-}(aq) \rightarrow S(s) + 2Fe^{2+}(aq)$

(b) Concentrated sulphuric acid is reduced to sulphur:

$$H_2SO_4(conc) + 3H_2S(g) \rightarrow 4S(s) + 4H_2O(l)$$

(c) Chlorine is reduced to hydrogen chloride:

$$Cl_2(g) + H_2S(g) \rightarrow S(s) + 2HCl(g)$$

(d) When hydrogen sulphide is passed into an acidified solution of potassium manganate(VII), the purple solution turns colourless and a very pale yellow deposit of sulphur is formed.

(e) Passage of hydrogen sulphide into an acidified solution of potassium dichromate(VI) results in the orange solution turning green and sulphur being deposited.

Test for hydrogen sulphide

Hydrogen sulphide turns lead ethanoate paper black:

$$(CH_3 \cdot COO)_2Pb + H_2S \rightarrow PbS + 2CH_3 \cdot COOH$$
White Black

Lead ethanoate paper is made by dipping filter paper in lead ethanoate solution and then drying it.

Sulphur dioxide, SO_2

Sulphur dioxide may be prepared in the laboratory by heating sodium sulphite with dilute hydrochloric acid (Figure 12.26):

$$Na_2SO_3(s) + 2HCl(aq) \rightarrow 2NaCl(aq) + H_2O(l) + SO_2(g)$$

The gas is dried with concentrated sulphuric acid and collected by upward displacement of air; it is too soluble to collect over water.

An alternative method of preparing sulphur dioxide is to heat copper turnings with concentrated sulphuric acid:

$$Cu(s) + 2H_2SO_4(conc) \rightarrow CuSO_4(s) + 2H_2O(l) + SO_2(g)$$

The apparatus used is identical to that shown in Figure 12.26.

The manufacture of sulphur dioxide is described on page 210.

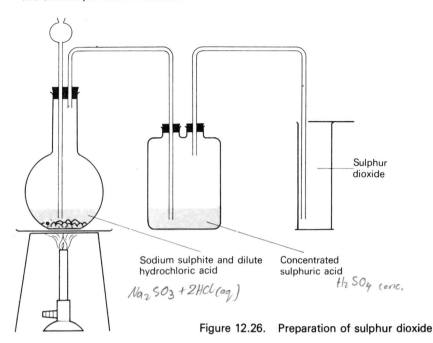

Figure 12.26. Preparation of sulphur dioxide

Physical properties

Sulphur dioxide is a colourless, poisonous gas with a very pungent, choking smell. It is well over twice as dense as air (the relative densities of sulphur dioxide and air being 32 and 14.4 respectively). The gas has a boiling point of −10 °C and so it is readily liquefied by pressure or by cooling with an ice/salt mixture. It is quite soluble in water; at room temperature, 1 volume of water dissolves about 70 volumes of the gas.

Reactions

1. Sulphur dioxide dissolves in water, giving an acidic solution known as sulphurous acid:

$$H_2O(l) + SO_2(g) \rightleftharpoons H_2SO_3(aq)$$

If sulphurous acid is being prepared, there is obviously no need to dry the sulphur dioxide and precautions have to be taken to prevent sucking back, i.e. an inverted funnel is fixed on to the end of the delivery tube (Figure 12.27).

Sulphurous acid is a fairly weak dibasic acid. The solution is unstable and, for example, is slowly oxidised by air to sulphuric acid:

$$2H_2SO_3(aq) + O_2(g) \rightarrow 2H_2SO_4(aq)$$

2. If excess sulphur dioxide is passed into sodium hydroxide solution, the acid salt, sodium hydrogensulphite, is obtained:

$$NaOH(aq) + SO_2(g) \rightarrow NaHSO_3(aq)$$

The normal salt, i.e. sodium sulphite, is made by treating the acid salt with an equivalent amount of sodium hydroxide solution:

$$NaHSO_3(aq) + NaOH(aq) \rightarrow Na_2SO_3(aq) + H_2O(l)$$

The other alkalis behave similarly.

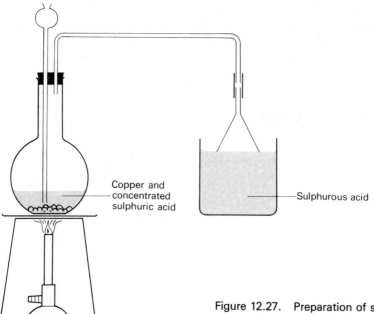

Figure 12.27. Preparation of sulphurous acid

3. Sulphur dioxide in aqueous soultion, i.e. sulphurous acid, is a powerful reducing agent:

$$2H_2O + SO_2 \rightarrow 4H^+ + SO_4^{2-} + 2e^-$$

(a) It reduces chlorine to hydrochloric acid:

$$2H_2O(l) + SO_2(g) + Cl_2(g) \rightarrow H_2SO_4(aq) + 2HCl(aq)$$

(b) It reduces iron(III) salts to iron(II) salts, e.g:

$$2FeCl_3(aq) + 2H_2O(l) + SO_2(g) \rightarrow 2FeCl_2(aq) + H_2SO_4(aq) + 2HCl(aq)$$
Yellow   Green

(c) It reduces acidified potassium manganate(VII) solution, the colour change being from purple to colourless. Note that it differs from hydrogen sulphide in that no sulphur is deposited; the sulphur dioxide is oxidised to sulphuric acid.

(d) It reduces acidified potassium dichromate(VI), the colour change being from orange to green.

4. Sulphur dioxide behaves as an oxidising agent in its reaction with hydrogen sulphide, a more powerful reducing agent. A trace of water must be present.

$$2H_2S(g) + SO_2(g) \rightarrow 3S(s) + 2H_2O(l)$$

Test for sulphur dioxide

The gas is detected by its pungent, choking smell and its property of turning filter paper, which has been dipped in acidified potassium dichromate(VI), from orange to green.

Uses of sulphur dioxide

Large quantities of sulphur dioxide are used in the manufacture of sulphuric acid. It is also important for bleaching wood pulp, wool, silk, and straw; it is much milder than chlorine. Its use as a food preservative depends on the fact that very low concentrations of it prevent fermentation. A further use is in the manufacture of calcium hydrogensulphite which is an important chemical in the paper making industry.

Sulphur(VI) oxide, SO_3

Preparation

Sulphur(VI) oxide is prepared both on a laboratory and industrial scale by the 'Contact' process. The apparatus used in the laboratory preparation is illustrated by Figure 12.28. Dry sulphur dioxide and oxygen are passed over a heated catalyst and the resultant sulphur(VI) oxide is condensed as a white crystalline solid by cooling it with ice.

$$2SO_2(g) + O_2(g) \rightarrow 2SO_3(s)$$

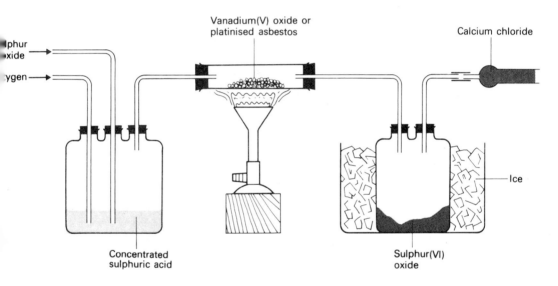

Figure 12.28. Preparation of sulphur(VI) oxide

The catalyst may be either platinised asbestos or vanadium(V) oxide whilst the operating temperature is 450 °C. The reasons behind the choice of these conditions have already been discussed (page 131).

Properties

Sulphur(VI) oxide is a very volatile white solid which is usually stored in sealed tubes. It reacts vigorously with water giving sulphuric acid; it is therefore the anhydride of sulphuric acid. *Anhydrides* are substances which react with water to give acids.

$$H_2O(l) + SO_3(s) \rightarrow H_2SO_4(aq)$$

Sulphur(VI) oxide dissolves in concentrated sulphuric acid giving *oleum* or fuming sulphuric acid:

$$H_2SO_4(conc) + SO_3(s) \rightarrow H_2S_2O_7(l)$$

Sulphuric acid, H_2SO_4

Manufacture (See also page 131)

Sulphuric acid is manufactured almost exclusively by the 'Contact' process. Sulphur dioxide is prepared by burning sulphur in air or by heating iron(II) disulphide (iron pyrites), lead(II) sulphide (galena), or zinc sulphide (zinc blende) in air:

$$S(s) + O_2(g) \rightarrow SO_2(g)$$
$$4FeS_2(s) + 11O_2(g) \rightarrow 2Fe_2O_3(s) + 8SO_2(g)$$
$$2PbS(s) + 3O_2(g) \rightarrow 2PbO(s) + 2SO_2(g)$$
$$2ZnS(s) + 3O_2(g) \rightarrow 2ZnO(s) + 2SO_2(g)$$

The sulphur dioxide is mixed with air, passed through an electrostatic precipitator to remove dust, washed with water, and then dried by bubbling it through concentrated sulphuric acid. This thorough purification is necessary because the catalyst, in the next stage, is poisoned by dust, traces of arsenic(III) oxide (As_2O_3), and moisture. The purified sulphur dioxide/air mixture is passed over a catalyst at about 450 °C. The catalyst may be either vanadium(V) oxide or platinum but the former is generally preferred because it is much cheaper than platinum.

About 99% of the sulphur dioxide is converted to sulphur(VI) oxide and this is dissolved in concentrated sulphuric acid to yield oleum. The oleum is treated with water to give concentrated, i.e. 98%, sulphuric acid.

$$2SO_2(g) + O_2(g) \rightarrow 2SO_3(g)$$
$$H_2SO_4(conc) + SO_3(g) \rightarrow H_2S_2O_7(l)$$
$$H_2S_2O_7(l) + H_2O(l) \rightarrow 2H_2SO_4(conc)$$

It is impracticable to dissolve the sulphur(VI) oxide directly in water because the heat evolved causes the solution to boil and a fine mist of sulphuric acid to be formed.

Since sulphur dioxide is very poisonous, the 1% of it which is unchanged in the process is absorbed in calcium hydroxide solution. The nitrogen and excess oxygen are returned to the atmosphere.

Properties

Sulphuric acid is a colourless liquid which boils at 330 °C. The pure compound is covalent but it fully dissociates into ions in aqueous solutions (see below). Concentrated sulphuric acid can behave as a strong acid, a dehydrating agent, and as an oxidising agent.

1. Sulphuric acid is fully dissociated in solution and so it is a strong dibasic acid:

$$H_2SO_4(l) + H_2O(l) \rightarrow H_3O^+(aq) + HSO_4^-(aq)$$
$$HSO_4^-(aq) + H_2O(l) \rightarrow H_3O^+(aq) + SO_4^{2-}(aq)$$

The dilute solution reacts with many metals giving a sulphate and hydrogen, e.g:

$$Zn(s) + H_2SO_4(aq) \rightarrow ZnSO_4(aq) + H_2(g)$$

It forms salts with bases and liberates carbon dioxide from carbonates, e.g:

$$CuO(s) + H_2SO_4(aq) \rightarrow CuSO_4(aq) + H_2O(l)$$
$$Na_2CO_3(s) + H_2SO_4(aq) \rightarrow Na_2SO_4(aq) + H_2O(l) + CO_2(g)$$

Concentrated sulphuric acid displaces weaker and more volatile acids from their salts. For example, nitric acid and hydrogen chloride are prepared in the laboratory by heating sodium nitrate and sodium chloride respectively with concentrated sulphuric acid:

i.e.
$$NaNO_3(s) + H_2SO_4(conc) \rightarrow NaHSO_4(s) + HNO_3(l)$$
$$NO_3^- + H_2SO_4(conc) \rightarrow HSO_4^-(s) + HNO_3(l)$$
$$NaCl(s) + H_2SO_4(conc) \rightarrow NaHSO_4(s) + HCl(g)$$
i.e.
$$Cl^-(s) + H_2SO_4(conc) \rightarrow HSO_4^-(s) + HCl(g)$$

2. Sulphuric acid has a strong affinity for water. The reaction is very exothermic and water added to the concentrated acid is likely to turn to steam and cause spattering of the acid. Concentrated sulphuric acid should therefore always be diluted by careful addition of the acid to water with continuous stirring.

Concentrated sulphuric acid is used to dry many gases and to remove water or the elements of water (i.e. one oxygen and two hydrogen atoms) from a number of compounds.

   (a) Copper(II) sulphate pentahydrate crystals change from blue to white, when added to concentrated sulphuric acid, due to the formation of the anhydrous salt:

$$CuSO_4 \cdot 5H_2O \xrightleftharpoons{H_2SO_4} CuSO_4 + 5H_2O$$

   (b) Carbon monoxide may be prepared by dehydration of methanoic (formic) acid with concentrated sulphuric acid (see page 181):

$$HCOOH \xrightarrow{H_2SO_4} H_2O + CO$$

   (c) Sugar is dehydrated slowly by cold, and rapidly by hot, concentrated sulphuric acid, giving a black mass of carbon:

$$C_{12}H_{22}O_{11} \xrightarrow{H_2SO_4} 12C + 11H_2O$$

   A rapid reaction occurs if the concentrated acid is added to sugar moistened with water: the heat of reaction between the acid and water accelerates the dehydration.

   (d) The dehydration of ethanol to ethene by the hot concentrated acid is discussed on page 230.

3. Hot concentrated sulphuric acid behaves as an oxidising agent and, in these reactions, it is itself generally reduced to sulphur dioxide:

$$H_2SO_4 + 2H^+ + 2e^- \rightarrow 2H_2O + SO_2$$

   (a) It reacts with most metals, the reaction with copper being used for the laboratory preparation of sulphur dioxide (page 206):

$$Cu(s) + 2H_2SO_4(conc) \rightarrow CuSO_4(s) + 2H_2O(l) + SO_2(g)$$

   Note that hydrogen is never a product using these conditions.

   (b) It oxidises some non-metals to their oxides, e.g:

$$C(s) + 2H_2SO_4(conc) \rightarrow 2H_2O(l) + CO_2(g) + 2SO_2(g)$$

(c) It oxidises hydrogen bromide to bromine:

$$H_2SO_4(conc) + 2HBr(g) \rightarrow Br_2(l) + 2H_2O(l) + SO_2(g)$$

Test for sulphuric acid and sulphates in solution
Sulphuric acid gives the normal reactions associated with strong acids, e.g. it turns blue litmus red, liberates carbon dioxide from carbonates, etc. The sulphate ion is detected by the white precipitate it gives on addition of barium chloride solution in the presence of dilute hydrochloric acid, e.g:

$$BaCl_2(aq) + H_2SO_4(aq) \rightarrow BaSO_4(s) + 2HCl(aq)$$
i.e. $$Ba^{2+}(aq) + SO_4^{2-}(aq) \rightarrow BaSO_4(s)$$

The hydrochloric acid decomposes carbonates and sulphites and so prevents their precipitation.

Uses of sulphuric acid
Roughly one-third of all sulphuric acid produced is used to make fertilisers. The acid also finds application in a wide range of other processes such as the manufacture of paints, detergents, dyestuffs, plastics, oil, and petrol.

CHLORINE, BROMINE, AND IODINE – GROUP VII ELEMENTS

The Group VII elements are often known as the *halogens*. They all have seven electrons in their outer shell, exhibit a valency of one, and exist as diatomic molecules. The first four members are fluorine, chlorine, bromine, and iodine and mention has already been made of their relative reactivities (page 41). Fluorine is beyond the scope of this text and, in the case of bromine and iodine, discussion will be limited to their hydrides.

CHLORINE

Preparation

Chlorine is prepared in the laboratory by oxidation of concentrated hydrochloric acid. When potassium manganate(VII) is used as the oxidising agent, reaction occurs at room temperature:

$$2KMnO_4(s) + 16HCl(conc) \rightarrow 2KCl(aq) + 2MnCl_2(aq) + 8H_2O(l) + 5Cl_2(g)$$

but heat is required with manganese(IV) oxide or lead(IV) oxide:

$$MnO_2(s) + 4HCl(conc) \rightarrow MnCl_2(aq) + 2H_2O(l) + Cl_2(g)$$
$$PbO_2(s) + 4HCl(conc) \rightarrow PbCl_2(aq) + 2H_2O(l) + Cl_2(g)$$

The apparatus is illustrated in Figure 12.29. The gas is washed with water to remove hydrogen chloride, dried with concentrated sulphuric acid, and collected by upward displacement of air.

Chlorine is obtained industrially by the electrolysis of various chlorides, e.g. molten sodium chloride or brine (see pages 88–89).

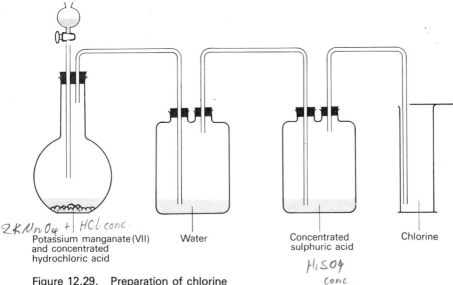

2KMnO₄ + HCl conc.
Potassium manganate(VII) and concentrated hydrochloric acid   Water   Concentrated sulphuric acid   Chlorine

H₂SO₄ conc.

Figure 12.29. Preparation of chlorine

Physical properties

Chlorine is a poisonous, yellow-green gas with a pungent, choking smell. It is slightly soluble in water: at room temperature, one volume of water dissolves a little over two volumes of chlorine. The gas is readily liquefied by pressure at room temperature and is transported in the liquid state in steel cylinders.

Reactions of chlorine

1. Chlorine reacts directly with many metals on heating, e.g:

$$Mg(s) + Cl_2(g) \rightarrow MgCl_2(s)$$

The reactions with aluminium and iron are discussed on pages 153 and 166 respectively.

Note that, since chlorine is an oxidising agent, metals with more than one valency react with chlorine to give the chloride with the metal in its highest valency state, e.g. iron reacts to give iron(III) chloride.

2. The reaction between chlorine and hydrogen has already been discussed (page 41). Chlorine also reacts with a number of compounds containing hydrogen; a few examples are given.

   (a) Chlorine oxidises hydrogen sulphide to sulphur:

   $$H_2S(g) + Cl_2(g) \rightarrow S(s) + 2HCl(g)$$

   (b) Filter paper dipped in warm turpentine ($C_{10}H_{16}$) inflames, when put in a gas jar of chlorine, and clouds of soot are produced:

   $$C_{10}H_{16}(l) + 8Cl_2(g) \rightarrow 10C(s) + 16HCl(g)$$

   (c) Chlorine reacts explosively with ethyne (see page 233).

3. Chlorine reacts with many non-metals. The reaction with phosphorus is discussed on page 200.

4. When chlorine is passed into water, some simply dissolves but a little reacts giving hydrochloric acid and chloric(I) acid (hypochlorous acid):

$$H_2O(l) + Cl_2(g) \rightarrow HCl(aq) + HOCl(aq)$$

The solution has bleaching properties. Chlorine bleaches moist litmus paper and this can be used as a test for the gas.

5. Chlorine reacts with alkali, the products depending on the conditions. With cold dilute sodium or potassium hydroxide solution, a chloride and chlorate(I) are formed, e.g:

$$2NaOH(aq) + Cl_2(g) \rightarrow NaCl(aq) + NaOCl(aq) + H_2O(l)$$

The mixture is used extensively as a bleaching agent and as a disinfectant.

Passage of chlorine into hot concentrated sodium or potassium hydroxide solution gives a mixture of chloride and chlorate(V), e.g:

$$6NaOH(aq) + 3Cl_2(g) \rightarrow 5NaCl(aq) + NaClO_3(aq) + 3H_2O(l)$$

Sodium and potassium chlorate(V) are powerful oxidising agents. The potassium salt is used in explosives, matches, and fireworks whilst the deliquescent sodium salt is used as a weed killer.

The formation of bleaching powder from chlorine and calcium hydroxide is discussed on page 149.

6. Chlorine oxidises bromides to bromine and iodides to iodine (see page 41).

Uses of chlorine

Chlorine is used to make hydrochloric acid and various chlorides. Very small concentrations of chlorine kill the bacteria in water and so it is used in water treatment. Large quantities are used to make organic chloro compounds for use as solvents, insecticides, and plastics. It is also used to bleach cotton, linen, and wood-pulp and for making other bleaches.

Hydrogen chloride, HCl

Preparation

Hydrogen chloride may be prepared in the laboratory by gently warming sodium chloride with concentrated sulphuric acid (Figure 12.30):

$$NaCl(s) + H_2SO_4(conc) \rightarrow NaHSO_4(s) + HCl(g)$$

Hydrogen chloride is made industrially by burning hydrogen in an atmosphere of chlorine. Both of these gases are obtained as by-products from the manufacture of sodium hydroxide by the electrolysis of brine (page 89).

Physical properties

Hydrogen chloride is a colourless gas which fumes in moist air and has a choking smell. It is very soluble in water, one volume of water at room temperature dissolving about 400 volumes of the gas. The high solubility enables the gas to be used in the fountain experiment (page 192).

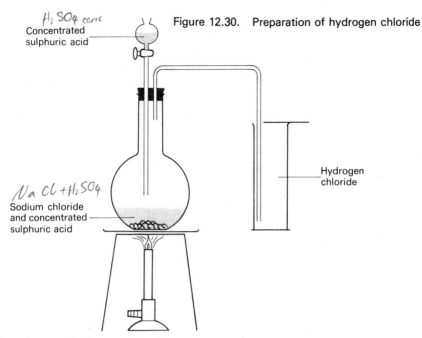

Figure 12.30. Preparation of hydrogen chloride

Reactions of hydrogen chloride
1. The gas does not burn or support combustion.

2. Some metals will form their chloride when heated in a stream of hydrogen chloride, e.g:

$$Mg(s) + 2HCl(g) \rightarrow MgCl_2(s) + H_2(g)$$

Hydrogen chloride, in contrast to chlorine, yields the lower chloride when heated with a metal which can have more than one valency state, e.g:

$$Fe(s) + 2HCl(g) \rightarrow FeCl_2(s) + H_2(g)$$

3. Hydrogen chloride reacts with ammonia to give white fumes of ammonium chloride: this reaction is used as a test for the gas.

$$HCl(g) + NH_3(g) \rightarrow NH_4Cl(s)$$

Hydrochloric acid

Preparation

Hydrochloric acid is prepared by passing hydrogen chloride into water. It is necessary to fit an inverted funnel to the end of the delivery tube (see page 194) to prevent sucking back.

It should be noted that the hydrogen and chlorine atoms in hydrogen chloride are joined by a strong covalent bond. The gas dissolves in methylbenzene (toluene) and tetrachloromethane, etc. giving solutions which are not acidic, i.e. they do not affect dry litmus paper, or liberate carbon dioxide from carbonates, or pass an electric current. Dissociation only occurs in aqueous solutions because the resultant ions are stabilised by their hydration:

$$H_2\ddot{O}(l) + HCl(g) \rightarrow H_3\dot{O}^+(aq) + Cl^-(aq)$$

Reactions of hydrochloric acid

Hydrochloric acid is a strong monobasic acid. It reacts with the metals above hydrogen in the electrochemical and activity series giving the metal chloride and hydrogen, e.g:

$$Fe(s) + 2HCl(aq) \rightarrow FeCl_2(aq) + H_2(g)$$

It reacts with bases giving salts, e.g:

$$CuO(s) + 2HCl(aq) \rightarrow CuCl_2(aq) + H_2O(l)$$
i.e. $$O^{2-}(s) + 2H^+(aq) \rightarrow H_2O(l)$$

and liberates carbon dioxide from carbonates, e.g:

$$Na_2CO_3(s) + 2HCl(aq) \rightarrow 2NaCl(aq) + H_2O(l) + CO_2(g)$$
i.e. $$CO_3^{2-}(s) + 2H^+(aq) \rightarrow H_2O(l) + CO_2(g)$$

Concentrated hydrochloric acid is oxidised by strong oxidising agents and these reactions are utilised in the laboratory preparation of chlorine (page 212).

Test for hydrochloric acid and chloride ions in solution
Hydrochloric acid gives the usual reactions associated with strong acids, e.g. it turns blue litmus red and liberates carbon dioxide from carbonates. The chloride ion in solution is detected by the white precipitate it gives on addition of silver nitrate solution, i.e:

$$Ag^+(aq) + Cl^-(aq) \rightarrow AgCl(s)$$

The precipitate is insoluble in dilute nitric acid but soluble in ammonia solution due to the formation of diamminesilver chloride:

$$AgCl(s) + 2NH_3(aq) \rightarrow [Ag(NH_3)_2]^+Cl^-(aq)$$

Hydrogen bromide, HBr

Preparation
Hydrogen bromide is prepared in the laboratory by adding bromine to moist red phosphorus (Figure 12.31). The reaction takes place in two stages:

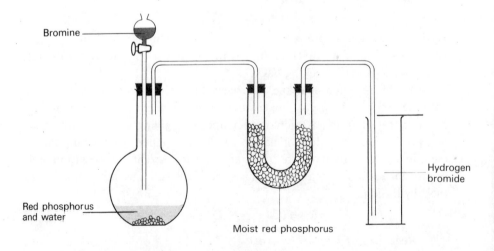

Figure 12.31. Preparation of hydrogen bromide

$$2P(s) + 3Br_2(l) \rightarrow 2PBr_3(l)$$
$$PBr_3(l) + 3H_2O(l) \rightarrow H_3PO_3(aq) + 3HBr(g)$$
Phosphonic acid

The gas is passed through moist red phosphorus, to remove bromine vapour, and then collected by upward displacement of air.

Physical properties

Hydrogen bromide is a colourless gas with a pungent smell. It is very soluble in water and it fumes in moist air.

Reactions of hydrogen bromide

1. Hydrogen bromide is readily soluble in water giving hydrobromic acid; an inverted funnel device (page 194) is needed on the end of the delivery tube to prevent sucking back.

$$H_2\ddot{O}(l) + HBr(g) \rightarrow H_3O^+(aq) + Br^-(aq)$$

   Hydrobromic acid is a strong acid.

2. The gas is similar in many ways to hydrogen chloride and it gives many analogous reactions, e.g. it reacts with bases and with some hot metals to give bromides and it forms white fumes of ammonium bromide when brought into contact with ammonia.

3. Hydrogen bromide is a more powerful reducing agent than hydrogen chloride. Unlike hydrogen chloride, it cannot be prepared by heating one of its salts with concentrated sulphuric acid because it reduces the acid:

$$H_2SO_4(conc) + 2HBr(g) \rightarrow 2H_2O(l) + SO_2(g) + Br_2(g)$$

Detection of the bromide ion

Bromides in solution give a pale yellow precipitate of silver bromide on addition of silver nitrate solution, e.g:

$$AgNO_3(aq) + HBr(aq) \rightarrow AgBr(s) + HNO_3(aq)$$
i.e. $$Ag^+(aq) + Br^-(aq) \rightarrow AgBr(s)$$

The precipitate is insoluble in dilute nitric acid but sparingly soluble in ammonia solution:

$$AgBr(s) + 2NH_3(aq) \rightarrow [Ag(NH_3)_2]^+(aq) + Br^-(aq)$$

Chlorine displaces bromine from solutions of bromides:

$$2Br^-(aq) + Cl_2(g) \rightarrow Br_2(aq) + 2Cl^-(aq)$$

If the mixture is shaken with tetrachloromethane, the latter forms a lower yellow-brown layer as the bromine dissolves in it.

Solid bromides are oxidised by a mixture of hot concentrated sulphuric acid and manganese(IV) oxide; brown fumes of bromine are evolved.

Hydrogen iodide

Preparation

Hydrogen iodide is prepared by adding water to a mixture of iodine and red phosphorus (Figure 12.32):

$$2P(s) + 3I_2(s) \rightarrow 2PI_3(s)$$
$$PI_3(s) + 3H_2O(l) \rightarrow H_3PO_3(aq) + 3HI(g)$$
Phosphoric acid

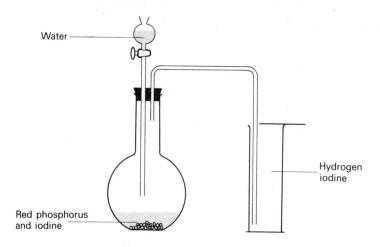

Figure 12.32. Preparation of hydrogen iodide

Physical properties

Hydrogen iodide is a colourless gas with a pungent smell. It fumes in moist air and is very soluble in water.

Reactions of hydrogen iodide

The gas is very soluble in water, giving a strongly acidic solution of hydriodic acid:

$$H_2\ddot{\underset{..}{O}}(l) + HI(g) \rightarrow H_3\underset{..}{O}^+(aq) + I^-(aq)$$

An inverted funnel is needed on the end of the delivery tube to prevent sucking back.

The reactions of hydrogen iodide are very similar to those of hydrogen chloride and hydrogen bromide, but it is a more powerful reducing agent. It is oxidised by most oxidising agents (see page 102).

Detection of iodide ions

Solutions containing iodide ions give a pale yellow precipitate of silver iodide when treated with silver nitrate solution, e.g:

$$AgNO_3(aq) + HI(aq) \rightarrow AgI(s) + HNO_3(aq)$$
i.e.
$$Ag^+(aq) + I^-(aq) \rightarrow AgI(s)$$

The precipitate is insoluble in dilute nitric acid and in ammonia solution.

Iodine is formed when chlorine is passed into a solution of an iodide:

$$2I^-(aq) + Cl_2(g) \rightarrow I_2(aq) \text{ or } (s) + 2Cl^-(aq)$$

The iodine dissolves in tetrachloromethane giving a violet lower layer.

Solid iodides give violet fumes of iodine when warmed with concentrated sulphuric acid and manganese(IV) oxide.

Questions

1 Carbon has atomic number 6 and relative atomic mass 12.
  (a) Draw a labelled diagram of a carbon atom.
  (b) Explain how carbon attains the noble gas electronic structure in the formation of tetrafluoromethane, CF_4 and state how this type of bonding differs from that in sodium fluoride. (Atomic numbers: $Na = 11, F = 9$.)
  (c) Describe the arrangement of the carbon atoms in the diamond and graphite lattices and hence explain the physical properties of these allotropes.
  (d) Briefly explain, without experimental details, how it could be shown that graphite and the black solid obtained from dehydration of sugar are the same chemical substance.

2 Carbon dioxide may be prepared by the action of hydrochloric acid on marble (calcium carbonate).
  (a) Give the equation for the reaction.
  (b) Why is hydrochloric acid used rather than sulphuric acid?
  (c) (i) Apart from water vapour, what other impurity is present in the gas and how can it be removed?
  (ii) How can the gas be dried?
  (iii) Explain how the gas may be collected.
  When carbon dioxide is passed over heated zinc some of the gas reacts as shown by the equation:

$$Zn + CO_2 \rightarrow ZnO + CO$$

  (d) Sketch an apparatus by which, using this reaction, you could obtain a sample of carbon monoxide free from carbon dioxide.
  (e) State and explain a chemical test for carbon monoxide.
  (f) Name the products that would be formed if magnesium was used instead of zinc in the above reaction and suggest a reason for any difference in the behaviour noted.

[A.E.B. June 1976]

3 (a) Give the equation for the decomposition of ammonia into nitrogen and hydrogen.
  Calculate the change in volume which occurs when 100 cm³ of ammonia is completely decomposed. (Assume all the measurements are made at the same temperature and pressure.)
  (b) A pupil attempted to decompose ammonia by passing the gas over heated iron wool. Although much of the gas decomposed, a little remained unchanged.
  (i) What was the purpose of the iron wool?
  (ii) By what simple chemical test could unchanged ammonia be detected?
  (iii) Suggest why the decomposition was incomplete.
  (iv) State a chemical reaction that may be used to remove hydrogen from a mixture of nitrogen and hydrogen.
  (c) One stage in the manufacture of nitric acid involves the oxidation of ammonia by the reaction:

$$4NH_3 + 5O_2 \rightarrow 4NO + 6H_2O$$

  State the conditions under which this reaction is carried out and outline the further stages in the production of nitric acid. (No diagram is required.)
  (d) 0.1 mol of nitric acid is neutralised by ammonia. Calculate the mass in grams of the product contained in the resulting solution.

[A.E.B. June 1976]

4 Ammonia is manufactured by passing a mixture of nitrogen and hydrogen, at a pressure of about 250 atmospheres, over a catalyst at 500 °C. Under these conditions about 15% of the gases are converted to ammonia.

(a) Write an equation for the reaction and state whether it is endothermic or exothermic.
(b) Why is a catalyst used?
(c) What happens to the gases which are not converted to ammonia?
(d) The yield of ammonia is increased if the process is carried out at:
   (i) even higher pressures,
   (ii) lower temperatures.
   Explain why these conditions are not used in industry.
(e) Explain why industrial plants for manufacturing ammonia used to be situated near coalfields but the plants built more recently are situated near oil refineries.
(f) If ammonia is passed over heated copper(II) oxide, the oxide is reduced to copper and the other products of the reaction are nitrogen and water.
Starting with dry ammonia coming through a delivery tube, draw a diagram of apparatus which could be used to carry out the above experiment and collect the three products.

[J.M.B.]

5 (a) Draw a fully labelled diagram and give an equation to show how you would prepare a dry sample of ammonia in the laboratory starting from a named ammonium salt and a named alkali.
Giving equations and reaction conditions, outline how nitric acid is manufactured from ammonia.
Nitrogen is necessary for all plant growth.
(b) Name a plant which can assimilate nitrogen from the atmosphere.
(c) Give the chemical name of a nitrogen containing compound which is used as a fertiliser.

[J.M.B.]

6 (a) In the manufacture of nitric acid, ammonia is first oxidised to nitrogen oxide according to the equation:

$$4NH_3 + 5O_2 \rightarrow 4NO + 6H_2O \quad \Delta H = -1088 \text{ kJ}$$

(i) Calculate the energy change involved in the conversion of one kilogram of ammonia by this reaction.
(ii) State the conditions under which the oxidation is carried out.
(iii) How is the nitrogen oxide further oxidised and then converted to nitric acid?

(b) Give two large scale uses for nitric acid.
(c) Describe two reactions in which nitric acid acts as an oxidising agent and state why you regard them as oxidations.

[A.E.B. Nov 1975]

7 A concentrated acid, A, is added to copper turnings and water in a flask and a brown gas, B, is seen to fill the flask.
In a second experiment the air is removed by filling the flask with nitrogen before the acid is added. It is found that a reaction occurs between the copper and the acid A, and a colourless gas, C, is formed.
A fresh portion of the acid, A, is diluted, neutralised with potassium hydroxide solution and the solution carefully evaporated to dryness to give a white solid, D. On heating D a new white solid, E, and a gas, F, are formed. F is found to ignite a glowing splint. When a mixture of ammonium chloride and E is heated, two volatile products are formed; one of these is easily condensed to a colourless liquid, G, while the other colourless product, H, remains gaseous. H does not support combustion.
Explain these results, identify A, B, C, D, E, F, G, H, and give equations for the reactions.

[J.M.B.]

8 (a) Draw a labelled diagram of the apparatus used in the laboratory to convert sulphur dioxide to sulphur(VI) oxide, SO_3, and to collect the product. Write an equation for the reaction.

(b) Describe what you would observe if sulphur(VI) oxide were cautiously treated with water and thus explain why this method of formation of sulphuric acid from sulphur(VI) oxide is unsuitable for a large scale process. What is the industrial alternative?

(c) Pure sulphuric acid is a liquid which boils at 330 °C at standard atmospheric pressure. What type of bonding is present in the pure liquid? How do you therefore justify the word 'acid' in its name?

(d) Name the gas given off when concentrated sulphuric acid is added to sodium chloride at room temperature and name the product formed when this gas is passed into water. Write an equation for one of these reactions.
[J.M.B.]

9 Identify the three gases X, Y and Z and the other substances indicated by letters. Write an equation for each reaction.

(a) When gas X was burned in an excess of air, two gases were produced. One of these condensed to a colourless liquid A which turned blue cobalt chloride paper pink. The other reacted with water to form an acidic solution.
When X was passed into iron(III) chloride solution, the yellow liquid turned pale green and a pale yellow precipitate B was formed.

(b) When Y was passed over heated copper(II) oxide, the black solid turned brown. Y burned on igniting in air, forming a single product C which gave a white precipitate with limewater.

(c) The gas Z combined with oxygen in the presence of a hot catalyst forming D which condensed to a white solid on cooling. D reacted violently with water, forming a liquid which reacted with zinc oxide forming a colourless solution.
[J.M.B.]

10 (a) Name and describe the appearance of one crystalline allotrope of sulphur. Outline how you would obtain a sample of the allotrope starting from powdered roll sulphur.

(b) Sulphur crystals melt easily; graphite crystals do not. Briefly relate this dissimilar behaviour to the differences between the crystal structures of these substances.

(c) The gas carbonyl sulphide (formula COS), burns in air, forming carbon dioxide and sulphur dioxide only.
(i) Calculate the mass of one litre of carbonyl sulphide at room temperature and pressure. (One mole of gas occupies 24 litres at room temperature and pressure.)
(ii) Write the equation for the combustion of the gas in oxygen.
(iii) Deduce the volume of oxygen required to burn 100 cm³ of carbonyl sulphide. (Assume all measurements are made at the same temperature and pressure.)

(d) Briefly state and explain one chemical test that would enable you to identify sulphur dioxide in a mixture of sulphur dioxide and carbon dioxide.
[A.E.B. Nov 1975]

11 Sulphur dioxide can be prepared by the reaction between hydrochloric acid and sodium sulphite.

(a) State the conditions necessary to give a steady reaction.
(b) Explain how you would collect the gas. (No diagram is required.)
(c) Give the equation for the reaction.
(d) State one distinctive test for the gas.
(e) When sulphur dioxide is passed into chlorine water the following reaction takes place:
$$2H_2O + SO_2 + Cl_2 \rightarrow 4H^+ + SO_4^{2-} + 2Cl^-$$
(i) Interpret this equation in terms of oxidation and reduction.
(ii) Describe and explain chemical tests by which you would confirm that the products of the reaction are those indicated in the equation. [A.E.B. Nov 1974]

12 In the 'Contact' process, pure sulphur dioxide is converted to gaseous sulphur(VI) oxide which, on further treatment, gives concentrated sulphuric acid.
- (a) State the conditions used in the conversion to sulphur(VI) oxide, and the equation for the reaction that takes place.
- (b) Why must the sulphur dioxide be purified before use?
- (c) The conversion is exothermic. What would be the effect of increasing the temperature at which the reaction is carried out?
- (d) How is concentrated sulphuric acid obtained from sulphur(VI) oxide? Explain why gaseous sulphur(VI) oxide is not dissolved in water directly.
  Give reactions (one in each case) in which sulphuric acid reacts as:
  (i) an acid,
  (ii) an oxidising agent,
  (iii) a dehydrating agent. [A.E.B. June 1974]

13 This question concerns the preparation in the laboratory of hydrogen sulphide from iron(II) sulphide and hydrochloric acid. No diagrams are required in your answers.
- (a) How can the iron(II) sulphide be made in the laboratory?
- (b) Give the equation for the reaction by which hydrogen sulphide is prepared.
- (c) Hydrogen sulphide made by this reaction usually contains hydrogen as an impurity. Suggest the reason for this.
- (d) Why is it desirable to pass the hydrogen sulphide through water before collecting it?
- (e) Explain how you would collect the gas.
- (f) When hydrogen sulphide is passed through a solution containing copper(II) ions the following reaction takes place:

$$Cu^{2+} + H_2S \rightarrow CuS + 2H^+$$

  (i) State what you would expect to see.
  (ii) Calculate the volume of hydrogen sulphide (measured at room temperature and pressure) required to react with a solution containing 0.1 mol of copper(II) ions. Also calculate the number of grams of copper(II) sulphide that would be formed by this reaction.
- (g) Describe and explain one experiment in which hydrogen sulphide is a reagent in an oxidation-reduction reaction.
  (1 mol of gas occupies 24.0 litres at room temperature and pressure.)
  [A.E.B. Nov 1976]

14 A grey-black compound, A, reacted with dilute hydrochloric acid to liberate a gas, B, which had an unpleasant smell and which blackened lead ethanoate paper. A white crystalline compound, C, reacted with warm dilute hydrochloric acid to liberate a gas, D, which had a sharp pungent smell and which turned an orange potassium dichromate(VI) paper green.
(i) What do you think the compounds, A, B, C and D might be? Explain the reactions and observations.
(ii) Describe the effect of each of the two gases B and D on iron(III) chloride solution.
(iii) What would be the effect of mixing the two gases B and D together?
[J.M.B.]

15 (a) When dry chlorine is passed through hot iron wool in a combustion tube, the iron glows strongly and black crystals are deposited in the cooler parts of the tube.
(i) Name the product of, and write the equation for, this reaction.
(ii) Name three members of the halogen family other than chlorine.
(iii) State whether each of the halogens named in (ii) would react more or less vigorously than would chlorine.
(b) (i) Describe the reaction of chlorine with hydrogen under any one stated set of conditions.
(ii) Compare the vigour of the reaction of hydrogen with chlorine with that of the reactions of hydrogen with the other halogens you have named.

(c) Name two halogens which react with potassium bromide solution. Write an equation for one of the reactions which occur.

[J.M.B.]

16 (a) Chlorine can be prepared by the reaction between manganese(IV) oxide and hydrochloric acid.
(i) State the conditions necessary for obtaining a reasonable quantity of chlorine by this method.
(ii) Give the equation for the reaction.
(iii) Name the principal impurity in the gas (apart from water vapour) and state how you would remove it.
(iv) Sketch and label the apparatus you would use to prepare and collect a reasonably pure sample of chlorine.
(b) Describe and explain the reactions that take place when chlorine is passed into aqueous solutions of:
(i) hydrogen sulphide,
(ii) iron(II) chloride,
(iii) potassium iodide.
(c) Explain the nature of the chemical bond between the two atoms in a molecule of chlorine. (The electronic configuration of a chlorine atom is 2.8.7.)

[A.E.B. Nov 1975]

17 (a) Chlorine and bromine are members of the same chemical family of elements. The atomic number of chlorine is 17.
(i) What name is given to this family of elements? Name one other member and state its physical appearance at room temperature.
(ii) Give electronic diagrams of a chlorine atom and of a hydrogen chloride molecule.
(iii) Why do the elements in this chemical family show similarities in their chemical reactions?
(iv) State the conditions under which each of the elements chlorine and bromine reacts with hydrogen. Indicate how these are related to the relative reactivities of the two elements.
(b) Hydrogen chloride dissolves readily in water. The change is exothermic and the solution produced is strongly acidic.
(i) What precautions would you take when dissolving hydrogen chloride in water?
(ii) What is meant by the term exothermic?
(iii) Explain why the solution is strongly acidic.

[A.E.B. June 1976]

18 (a) Chlorine may be prepared by oxidising hydrochloric acid.
(i) Name one suitable oxidising agent.
(ii) Write the equation for the reaction and say why you regard this to be an oxidation reaction.
(iii) State how you would react the oxidising agent with hydrochloric acid so as to produce a reasonable quantity of chlorine for use in the laboratory (a diagram is not required).
(iv) Why is it desirable to wash the chlorine with a little water before collecting it? Why would it be incorrect to use sodium hydroxide solution to wash the gas?
(v) State how you would collect the chlorine. Why is the method of collection suitable?
(vi) Give one chemical test for chlorine.
(b) Describe one experiment you have seen in which chlorine reacts with a named metallic element.
(c) Warming a gaseous oxide of chlorine, Cl_xO_y causes it to decompose:

$$2Cl_xO_y \rightarrow xCl_2 + yO_2$$

In an experiment 20 cm³ of the oxide gave 20 cm³ of chlorine and 10 cm³ of oxygen (all gas volumes being measured at the same temperature and pressure). Calculate the formula of this oxide showing your reasoning clearly.

[A.E.B. June 1977]

19 (a) Draw a fully labelled diagram for the laboratory preparation and collection of a sample of hydrogen chloride. Give an equation for the reaction.
How would you adapt your apparatus to make a solution of hydrogen chloride in water?

(b) Describe what you would see, and name what is formed, when dilute hydrochloric acid reacts with:
 (i) silver nitrate solution,
 (ii) sodium carbonate.
Write an equation for each of the reactions.

(c) When concentrated hydrochloric acid is heated with manganese(IV) oxide (MnO_2), chlorine is formed. Write the equation for the reaction and state the function of the manganese(IV) oxide.

[J.M.B.]

13 Organic Chemistry

INTRODUCTION

Originally, organic chemistry referred to the chemistry of carbon compounds present in living things, both animal and vegetable. However, the definition has been extended and the compounds of carbon, other than carbon monoxide, carbon dioxide, the disulphide, and the metal carbonates and carbides, are now all classed as being organic.

There are far more compounds of carbon than of all the other elements put together. The main reasons for this are the valency of four and the ability of carbon atoms to join up with one another to form stable chains. The carbon–carbon bonds are stable because carbon atoms are small and so the electrons being shared in the bonds are close to the nuclei and tightly held.

Even though there are so many organic compounds, relatively few other elements are involved. Most organic compounds contain carbon and hydrogen and there are, in fact, thousands of compounds which consist solely of these elements. The only other elements which are fairly consistently encountered are oxygen, nitrogen, and the halogens.

The majority of organic compounds are covalent and so they have low melting and boiling points. They tend to have limited solubility in water but are readily soluble in solvents such as benzene and ethoxyethane (ether). Most organic compounds are flammable and those with a high proportion of carbon to hydrogen burn with smoky, luminous flames.

Fortunately, the vast number of organic compounds are not completely unrelated; there are a number of homologous series. A *homologous series* is one in which each member differs from adjacent members by a CH_2 unit. For example, the first four members of the alkane series (page 227) have the formulae CH_4, C_2H_6, C_3H_8, and C_4H_{10} respectively. The members of a homologous series are prepared in a similar manner and have similar chemical properties. The physical properties change steadily down the series, e.g. the melting and boiling points increase gradually as the relative molecular mass increases.

EMPIRICAL, MOLECULAR, AND STRUCTURAL FORMULAE

The *structural formula* of a compound shows how the atoms are joined together in the molecule. Before this can be deduced, it is necessary to determine, in turn, the empirical and molecular formulae. The *empirical formula* shows the simplest whole number ratio of the atoms of each element in the molecule, whilst the *molecular formula* shows the actual number of atoms of each element present. The meaning of the various formulae is illustrated by the following example. The compound with the structural formula

$$\begin{array}{c} \text{H} \quad \text{H} \\ | \quad\quad | \\ \text{H}-\text{C}-\text{C}-\text{H} \\ | \quad\quad | \\ \text{H} \quad \text{H} \end{array}$$

has a molecular formula of C_2H_6 and an empirical formula of CH_3.

The empirical formula of a compound may be calculated once the percentage composition is known.

Hydrocarbons, i.e. compounds consisting of carbon and hydrogen only, can be analysed by burning a known mass of them in a stream of dry oxygen. The hydrogen present is converted into water and this can be absorbed in weighed tubes of calcium chloride. The carbon is converted into carbon dioxide and this can be absorbed in weighed bulbs containing potassium hydroxide solution. Once the mass of water and carbon dioxide formed from a given mass of the hydrocarbon is known, the percentage composition of the compound can be calculated. The calculation is illustrated by the following example.

Example If 1.16 g of a hydrocarbon gave, on ignition in oxygen, 3.65 g of carbon dioxide and 1.49 g of water, calculate its percentage composition.

$$C + O_2 \rightarrow CO_2$$
$$12 \quad\quad\quad 44$$

44 g of carbon dioxide contain 12 g of carbon

$\therefore$ 3.65 g of carbon dioxide contain $\frac{12}{44} \times 3.65 = 0.995$ g of carbon

$$\% \text{ carbon in the compound} = \frac{0.995}{1.16} \times 100 = 85.7\%$$

$$2H_2 + O_2 \rightarrow 2H_2O$$
$$4 \quad\quad\quad 36$$

36 g of water contain 4 g of hydrogen

$\therefore$ 1.49 g of water contain $\frac{4}{36} \times 1.49 = 0.166$ g of hydrogen

$$\% \text{ hydrogen in the compound} = \frac{0.166}{1.16} \times 100 = 14.3\%$$

The empirical formula of the compound can now be calculated because, dividing the percentage of each element by its relative atomic mass will give the fractional ratio of atoms present. Dividing each ratio by the smallest will give the simple ratio.

Element	C	H
Percentage	85.7	14.3
Percentage/relative atomic mass	$\frac{85.7}{12}$	$\frac{14.3}{1}$
Fractional ratio of atoms	7.14 :	14.3
Simple ratio	$\frac{7.14}{7.14}$:	$\frac{14.3}{7.14}$
i.e.	1 :	2

The empirical formula is therefore CH_2, i.e. for every carbon atom in the molecule there are two hydrogen atoms.

The molecular formula of the compound will be either the same as the empirical formula or a simple multiple of it. The molecular formula is found by multiplying the empirical formula by the number of times its relative molecular mass goes into the actual relative molecular mass of the compound. The relative molecular mass of a gas or vapour can be determined from its relative density using the relationship:

relative molecular mass = 2 × relative density (see page 63)

These calculations may be illustrated by assuming that, in the above example, the relative density of the hydrocarbon was found to be 21.

Relative molecular mass of the hydrocarbon = 2 × relative density
= 2 × 21
= 42

Relative molecular mass of the empirical formula, i.e. CH_2 = 12 + (2 × 1)
= 14

Molecular formula
$= (CH_2) \times \frac{\text{relative molecular mass of the hydrocarbon}}{\text{relative molecular mass of the empirical formula}}$
$= (CH_2) \times \frac{42}{14}$
$= (CH_2) \times 3$
$= C_3H_6$

The elucidation of structural formulae often used to be long and arduous. However, there are now several important instrumental methods which make the task easier but the details are beyond the scope of this text.

The simple chemistry of a few homologous series will now be considered.

ALKANES

The alkanes are classed as *saturated hydrocarbons*; this means that they contain only carbon and hydrogen atoms and they have no double or triple bonds. They are represented by a general formula C_nH_{2n+2} where n is any number from one upwards. The first four members of the series are given over.

Compound	Molecular formula	Structural formula	Boiling point °C
Methane	CH_4	H—C(H)(H)—H	−161
Ethane	C_2H_6	H—C(H)(H)—C(H)(H)—H	−88
Propane	C_3H_8	H—C(H)(H)—C(H)(H)—C(H)(H)—H	−42
Butane	C_4H_{10}	H—C(H)(H)—C(H)(H)—C(H)(H)—C(H)(H)—H	0

The physical properties alter gradually as the series is ascended as illustrated by the boiling points above. However, the members of the series undergo similar reactions and can be prepared by similar methods.

Preparation

Alkanes may be prepared by heating the sodium salts of carboxylic acids with soda-lime. Methane is prepared from sodium ethanoate using the apparatus shown in Figure 13.1.

$$CH_3 \cdot COONa(s) + NaOH(s) \rightarrow Na_2CO_3(s) + CH_4(g)$$

Soda-lime is a mixture of sodium hydroxide and calcium oxide and it is used in preference to sodium hydroxide alone because it is not deliquescent.

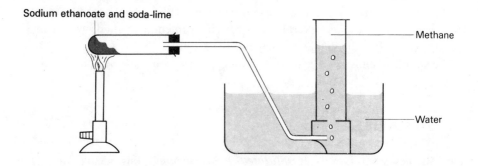

Figure 13.1. Preparation of methane

Properties

The first four members of the alkane series are colourless gases, very sparingly soluble in water but soluble in organic solvents. They are comparatively unreactive because they do not contain any double or triple bonds and carbon–carbon and carbon–hydrogen bonds are normally quite strong. Under normal conditions, alkanes do not react with concentrated acids, alkalis, or oxidising agents.

Methane reacts with chlorine under the influence of ultraviolet light to give a mixture of chlorinated products.

$$CH_4(g) + Cl_2(g) \rightarrow CH_3Cl(g) + HCl(g) \quad \text{chloromethane}$$
$$CH_3Cl(g) + Cl_2(g) \rightarrow CH_2Cl_2(l) + HCl(g) \quad \text{dichloromethane}$$
$$CH_2Cl_2(l) + Cl_2(g) \rightarrow CHCl_3(l) + HCl(g) \quad \text{trichloromethane}$$
$$CHCl_3(l) + Cl_2(g) \rightarrow CCl_4(l) + HCl(g) \quad \text{tetrachloromethane}$$

The other alkanes are not usually chlorinated because an even more complex mixture would be formed.

All alkanes burn in air, giving carbon dioxide and water, e.g:

$$CH_4(g) + 2O_2(g) \rightarrow CO_2(g) + 2H_2O(g)$$

Uses of alkanes

The alkanes are important fuels. Methane is the major component of natural gas. Propane and butane are sold in cylinders or cannisters for domestic use, butane being known as calor gas. Petrol, diesel oil, lubricating oil, and paraffin wax are all mixtures of alkanes.

Alkyl groups

The removal of a hydrogen atom from an alkane gives an alkyl group or radical. These groups have a valency of one and are named by removing the ending '-ane' from the alkane name and adding '-yl', e.g:

methane, $CH_4 \rightarrow$ methyl radical, CH_3-
ethane, $C_2H_6 \rightarrow$ ethyl radical, C_2H_5-

A knowledge of alkyl groups is necessary for naming branched chain alkanes. For example, the name 2-methylpropane for the compound

$$\overset{1}{C}H_3 - \overset{2}{C}H - \overset{3}{C}H_3$$
$$|$$
$$CH_3$$

indicates that a hydrogen atom on the second carbon of a propane molecule has been replaced by a methyl group

Structural isomerism

The fourth member of the alkane series has a molecular formula C_4H_{10}. Now these atoms can be arranged in two different ways; see over:

CH$_3$—CH$_2$—CH$_2$—CH$_3$ and   CH$_3$—CH—CH$_3$
                                        |
                                       CH$_3$

Butane   2-Methylpropane

This phenomenon of compounds having the same molecular formula but different structural formulae is known as *structural isomerism* and it is very common in organic chemistry. Structural isomers have different physical and chemical properties.

ALKENES

The alkenes are a homologous series of hydrocarbons with the general formula C_nH_{2n} where n is any number from two upwards. The first and the most important member of the series is ethene, C_2H_4 or $CH_2=CH_2$, whilst the second member is propene, C_3H_6, i.e. $CH_3 \cdot CH=CH_2$.

Alkenes are named after the alkane with the same number of carbon atoms, the alkane ending '-ane' being replaced by '-ene'. The '-ene' indicates the presence of a double bond between two carbon atoms.

Alkenes are reactive and, as most of their reactions involve addition across the double bond, they are classed as *unsaturated hydrocarbons*.

Preparation of ethene

Ethene may be prepared by dehydration of ethanol:

$$CH_3CH_2OH \rightarrow CH_2=CH_2 + H_2O$$

The dehydration can be brought about by heating the ethanol with excess concentrated sulphuric acid at 180 °C or by passing the alcohol vapour over aluminium oxide at about 350 °C (Figure 13.2).

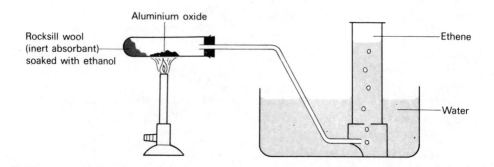

Figure 13.2. Preparation of ethene

In the latter method, the alcohol is absorbed in rocksill wool (an inert material) and it vaporises steadily as the aluminium oxide is heated. The ethene is collected over water whilst any unchanged ethanol dissolves in the water.

Ethene is manufactured in huge quantities by the 'cracking' of petroleum fractions (page 238).

Properties of ethene

Ethene is a colourless gas which is practically insoluble in water but soluble in organic solvents such as benzene, etc. It burns in air, giving carbon dioxide and water, but alkenes are not used as fuels – they are reactive and are used to make other products. Ethene, like the other alkenes, readily adds on a variety of reagents to give saturated products.

1. Hydrogen adds on to ethene, in the presence of a catalyst, to give ethane:

$$CH_2{=}CH_2(g) + H_2(g) \rightarrow CH_3 \cdot CH_3(g) \quad \text{(i.e.} \quad \overset{\overset{H}{|}}{CH_2}{-}\overset{\overset{H}{|}}{CH_2})$$

The reaction is known as catalytic hydrogenation. It occurs at room temperature when platinum or palladium is the catalyst, but temperatures of 200–300 °C are required with nickel as catalyst.

2. Addition of hydrogen chloride to ethene gives chloroethane:

$$CH_2{=}CH_2(g) + HCl(g) \rightarrow CH_3 \cdot CH_2Cl(g) \quad \left[\text{i.e.} \quad H{-}\overset{\overset{H}{|}}{\underset{\underset{H}{|}}{C}}{-}\overset{\overset{Cl}{|}}{\underset{\underset{H}{|}}{C}}{-}H \right]$$

3. Chlorine adds on to ethene to give 1,2-dichloroethane:

$$CH_2{=}CH_2(g) + Cl_2(g) \rightarrow CH_2Cl \cdot CH_2Cl(l)$$

The numbers in the name of the product indicate that there is a chlorine atom joined to each of the carbons. This distinguishes the compound from its structural isomer, $CH_3 \cdot CHCl_2$, i.e. 1,1-dichloroethane.

4. Bromine adds on to ethene to give 1,2-dibromoethane:

$$CH_2{=}CH_2(g) + Br_2(l) \rightarrow CH_2Br \cdot CH_2Br(l)$$

The brown colour of bromine disappears and a colourless, oily liquid is formed. This reaction is used as a test for unsaturation, i.e. for double and triple bonds. The bromine is used as a dilute solution in tetrachloromethane since, if excess bromine is added, no colour change is observed.

Uses of ethene

Ethene is used to manufacture ethanol, poly(ethene) ('polythene'), and a variety of other compounds.

ALKYNES

The alkynes are a homologous series with the general formula C_nH_{2n-2} where n may be any number from two upwards. Alkynes contain a carbon–carbon triple bond and so they are unsaturated hydrocarbons. The first, and most important, member of the series is ethyne (acetylene), $HC{\equiv}CH$.

Preparation of ethyne

Ethyne is made on a laboratory and on an industrial scale by the action of water on calcium dicarbide (Figure 13.3).

$$CaC_2(s) + 2H_2O(l) \rightarrow Ca(OH)_2(aq) + C_2H_2(g)$$

The calcium dicarbide is prepared by heating calcium oxide and coke at 2500 °C:

$$CaO(s) + 3C(s) \rightarrow CaC_2(s) + CO(g)$$

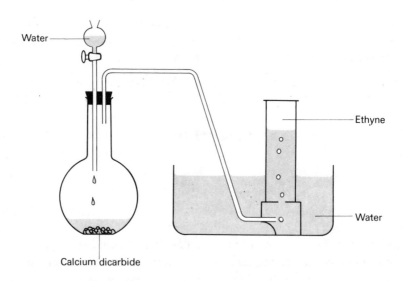

Figure 13.3. Preparation of ethyne

Properties of ethyne

Ethyne is a colourless gas, slightly soluble in water. It can form explosive mixtures with air. Ethyne burns in air with a sooty flame owing to the high proportion of carbon to hydrogen in the molecule. It is a reactive compound and readily undergoes addition reactions.
1. Catalytic hydrogenation occurs in two stages, giving ethene and then ethane:

$$CH{\equiv}CH(g) + H_2(g) \xrightarrow{\text{catalyst}} CH_2{=}CH_2(g) \xrightarrow{H_2,\ \text{catalyst}} CH_3{\cdot}CH_3(g)$$

i.e.

$$H-C\equiv C-H(g) + H_2(g) \xrightarrow{catalyst} H-\underset{\underset{H}{|}}{\overset{\overset{H}{|}}{C}}=\underset{\underset{H}{|}}{\overset{\overset{H}{|}}{C}}-H(g) \xrightarrow{H_2, catalyst} H-\underset{\underset{H}{|}}{\overset{\overset{H}{|}}{C}}-\underset{\underset{H}{|}}{\overset{\overset{H}{|}}{C}}-H(g)$$

The same catalysts and conditions are used as for the alkenes.

2. Hydrogen chloride adds on in two stages giving chloroethene and then 1,1-dichloroethane:

$$CH\equiv CH(g) + HCl(g) \rightarrow CH_2=CHCl(g) \xrightarrow{HCl} CH_3\cdot CHCl_2(l)$$

3. Chlorine and ethyne react explosively in the gas phase giving carbon and hydrogen chloride:

$$CH\equiv CH(g) + Cl_2(g) \rightarrow 2C(s) + 2HCl(g)$$

However, the chlorine adds on in two stages if it is diluted with nitrogen and the reaction is done in the presence of iron:

$$CH\equiv CH(g) + Cl_2(g) \xrightarrow{N_2/Fe} \underset{\text{1,2-dichloroethene}}{CHCl=CHCl} \xrightarrow{Cl_2/N_2/Fe} \underset{\text{1,1,2,2-tetrachloroethane}}{CHCl_2\cdot CHCl_2}$$

4. Bromine adds on to ethyne in two stages, the bromine colour disappearing as the colourless addition products are formed.

$$CH\equiv CH(g) + Br_2(l) \rightarrow \underset{\text{1,2-dibromoethene}}{CHBr=CHBr(l)} \xrightarrow{Br_2} \underset{\text{1,1,2,2-tetrabromoethane}}{CHBr_2\cdot CHBr_2(s)}$$

Uses of ethyne

Ethyne is used to make a number of polymers and synthetic fibres such as poly(chloroethene) (better known as polyvinyl chloride or P.V.C.), 'Orlon', 'Courtelle', etc. Ethyne burns in oxygen giving a flame with a temperature in excess of 2000 °C and so it is used in welding.

ALCOHOLS

Alcohols are formed when a hydroxyl group combines with an alkyl group. The general formula of the alcohols is $C_nH_{2n+1}OH$ where n is any whole number. The first two members are methanol, CH_3OH and ethanol, C_2H_5OH (alternatively written as $CH_3\cdot CH_2OH$). The alcohols are named after the alkane with the same number of carbon atoms, the end '-e' of the alkane name being replaced by '-ol'.

Ethanol is considerably more important than methanol.

Preparation of ethanol

Alcohols may be prepared in the laboratory by refluxing alkyl halides with a solution of an alkali. For example, ethanol is formed when bromoethane is refluxed with sodium hydroxide solution (Figure 13.4). When the reaction is complete, the mixture is distilled to give an aqueous solution of ethanol.

$$CH_3 \cdot CH_2Br(l) + NaOH(aq) \rightarrow CH_3 \cdot CH_2OH(aq) + NaBr(aq)$$

Ethanol may be manufactured from ethene or starch (or glucose) as described on page 238.

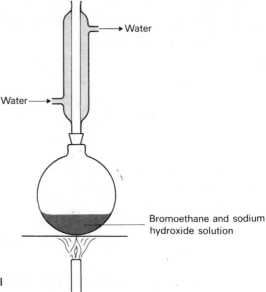

Figure 13.4. Preparation of ethanol

Properties of ethanol

Ethanol is a colourless liquid with a boiling point of 78 °C. It is soluble in water in all proportions. It undergoes a number of reactions, a few of which are given below.

1. Sodium reacts briskly with ethanol, giving sodium ethoxide and hydrogen. The reaction is analogous to that between sodium and water.

$$2Na + 2C_2H_5OH \rightarrow 2C_2H_5ONa + H_2$$
$$2Na + 2HOH \rightarrow 2NaOH + H_2$$

The sodium ethoxide remains as a white solid when the excess ethanol is evaporated.

2. Alcohols react with organic acids, giving esters and water, the process being known as *esterification*. These reactions are reversible and are catalysed by a few drops of concentrated sulphuric acid. Ethanol reacts with ethanoic acid to give ethyl ethanoate:

$$CH_3 \cdot COOH(l) + C_2H_5OH(l) \rightleftharpoons CH_3 \cdot COOC_2H_5(l) + H_2O(l)$$

i.e. $CH_3 \cdot C\begin{smallmatrix}\diagup O \\ \diagdown O-H\end{smallmatrix} + CH_3 \cdot CH_2OH \rightleftharpoons CH_3 \cdot C\begin{smallmatrix}\diagup O \\ \diagdown O \cdot CH_2 \cdot CH_3\end{smallmatrix} + H_2O$

Esters have fruity smells.

3. Phosphorus pentachloride reacts vigorously with ethanol, the products being chloroethane, phosphorus trichloride oxide, and hydrogen chloride:

$$PCl_5(s) + C_2H_5OH(l) \rightarrow POCl_3(l) + C_2H_5Cl(g) + HCl(g)$$

Phosphorus pentachloride reacts with all compounds containing hydroxyl groups – the hydroxyl group is replaced by a chlorine atom and white fumes of hydrogen chloride are evolved. This is used as a test for hydroxyl groups.

4. Ethanol is dehydrated to ethene by heating it with concentrated sulphuric acid at 180 °C or by passing its vapour over hot aluminium oxide (see page 230).

$$C_2H_5OH(l) \xrightarrow{-H_2O} C_2H_4(g)$$

5. Ethanol is oxidised to ethanoic acid by refluxing it with acidified potassium manganate(VII) or potassium dichromate(VI) – see the preparation of ethanoic acid below.

6. Ethanol burns with a blue flame:

$$C_2H_5OH(l) + 3O_2(g) \rightarrow 2CO_2(g) + 3H_2O(g)$$

Uses of ethanol

Ethanol is used in alcoholic drinks, in some petrols, as a solvent, and in the manufacture of a wide variety of other products.

CARBOXYLIC ACIDS

The carboxylic acids have the general formula $C_nH_{2n+1}COOH$ where n is any whole number but the first member simply has the formula HCOOH. The structure of the carboxyl group is $-C{\overset{=O}{\underset{O-H}{}}}$. The carboxylic acids are named after the alkane with the same number of carbon atoms, the end '-e' of the alkane name being replaced by '-oic acid'. The first two members are: methanoic acid (formic acid) HCOOH and ethanoic acid (acetic acid) $CH_3 \cdot COOH$.

Ethanoic acid is the most important member of the series.

Preparation of ethanoic acid

Ethanoic acid is prepared by oxidation of ethanol. On a laboratory scale, the ethanol is refluxed with acidified potassium manganate(VII) or potassium dichromate(VI). After the oxidation is complete, the mixture is distilled and an aqueous solution of the acid is obtained. The equation for the reaction is complicated and so a simplified version is given, [O] representing oxygen from the oxidising agent.

$$CH_3 \cdot CH_2OH(l) + 2[O] \rightarrow CH_3 \cdot COOH(aq) + H_2O(l)$$

Properties of ethanoic acid

Ethanoic acid is a colourless liquid with a pungent smell. It boils at 118 °C and freezes at 17 °C. On cold days, it may exist as colourless crystals and for this reason the pure acid is often known as glacial ethanoic acid. It is miscible with water in all proportions. It is a weak acid, i.e. it is only partially dissociated into ions in its aqueous solutions:

$$CH_3 \cdot COOH(l) + H_2O(l) \rightleftharpoons CH_3 \cdot COO^-(aq) + H_3O^+(aq)$$

Ethanoic acid neutralises sodium hydroxide solution giving sodium ethanoate and water:

$$CH_3 \cdot COOH(aq) + NaOH(aq) \rightarrow CH_3 \cdot COONa(aq) + H_2O(l)$$

It liberates carbon dioxide from carbonates, e.g:

$$Na_2CO_3(s) + 2CH_3 \cdot COOH(aq) \rightarrow 2CH_3 \cdot COONa(aq) + H_2O(l) + CO_2(g)$$

The reaction with alcohols to give esters is described on page 234.

Uses of ethanoic acid

The acid is widely used for making esters. It is also used in photographic processing and in food preservation – vinegar is a dilute solution of ethanoic acid.

SOURCES OF ORGANIC COMPOUNDS

Coal

Coal used to be a major source of organic compounds but natural gas and crude petroleum are now much more important in this respect.

Coal is the product of decayed vegetation from a few hundred million years ago. It may be decomposed by heating it between about 600 °C and 1500 °C, in the absence of air, giving solid, liquid, and gaseous products.

1. *Solid products*

    The low temperature decomposition is used to make smokeless fuel whilst the higher temperatures are used to make coke. Smokeless fuel is used mainly for domestic use but coke is utilised industrially in the manufacture of producer gas (page 180) and water gas (page 180), and in the extraction of various metals, e.g. iron and zinc.

2. *Liquid products*

    These may be separated into ammoniacal liquor and coal tar. The ammoniacal liquor consists of ammonia and benzene, C_6H_6, the former substance being used to make fertilisers, whilst the latter is used to make dyes, detergents, etc. The coal tar contains a large number of organic compounds and it is fractionally distilled to give such products as methylbenzene (toluene), phenol, naphthalene, and anthracene. A residue of pitch remains and this is used for road surfacing.

3. *Gaseous products*

The main gaseous products are hydrogen, carbon monoxide, methane, and ethane. The mixture is passed with a little air over hydrated iron(III) oxide to remove hydrogen sulphide by the reaction:

$$2H_2S(g) + O_2(g) \rightarrow 2S(s) + 2H_2O(l)$$

The resultant coal gas is used as a fuel.

Natural gas and crude petroleum

Natural gas and crude petroleum are probably the remains of marine plant and animal matter from many millions of years ago. They generally occur together but some wells produce gas only.

Natural gas consists mainly of methane, with progressively smaller quantities of ethane, propane, and butane. It is being used increasingly throughout the world as a replacement for coal gas. Its controlled oxidation is important industrially, e.g. it may be oxidised to methanol for use as a solvent, to methanal, HCHO, for producing plastics, and for making carbon black for use as a filler in car tyres. The carbon black is made by burning methane in a limited supply of air:

$$CH_4(g) + O_2(g) \rightarrow C(s) + 2H_2O(g)$$

Some hydrogen is made industrially by heating natural gas at 1200 °C in the absence of air:

$$CH_4(g) \rightarrow C(s) + 2H_2(g)$$

The composition of crude petroleum varies according to the part of the world from which the sample is obtained. It is a viscous, dark coloured liquid which consists mainly of alkanes with up to about forty carbon atoms, but some longer chain alkanes, sand, water, and sulphur compounds are present. The water and sand are separated and then the crude petroleum is subjected to fractional distillation. Various fractions are collected and it should be noted that the boiling point range and the composition of the fractions vary from refinery to refinery. An example of the type of result obtained is given below. There is some overlap of the chain length in the various fractions due to the different boiling points of structural isomers.

Name of fraction	Carbon atoms per molecule	Boiling range	Uses
Petroleum ethers, ligroin	C_5-C_7	Up to 60 °C	Solvents
Petrol or gasoline	C_6-C_{10}	50–200 °C	Motor fuel
Kerosene or paraffin oil	$C_{11}-C_{16}$	200–280 °C	Jet fuel and cracking into petrol, etc.
Gas oil	$C_{13}-C_{20}$	280–350 °C	Diesel fuel, domestic fuel and cracking
Lubricating oil	$C_{20}-C_{30}$	350–400 °C	Lubricants, paraffin wax, vaseline, etc.
Residue, i.e. bitumen	Over C_{30}	Above 400 °C	Asphalt for roads

The rough fractions outlined above may be redistilled, depending upon the purpose for which they are going to be used, to obtain fractions with narrower boiling ranges. The petrol or gasoline fraction is refined to remove sulphur compounds, since combustion of these would give sulphur dioxide which is corrosive and has a pungent smell. Further treatment causes rearrangement to give branching of the carbon chains and a better performance petrol.

The demand for kerosene and gas oil has greatly increased in recent years as the use of jet and diesel engines has multiplied. Nevertheless, petrol is still the major requirement. Now the petrol or gasoline fraction only accounts for about 20% of crude petroleum. It has therefore been necessary to develop a process for converting the less useful long chain hydrocarbons into shorter chain ones. The process is known as *cracking* and it involves passing the vapour of the high boiling fractions over a catalyst such as aluminium oxide at a temperature of about 500 °C. The long chain alkanes break down to give a mixture of hydrogen and short chain alkanes and alkenes. The alkenes are very important starting materials for preparing polymers, alcohols, etc.

Since all our crude petroleum is imported or obtained from off-shore wells, the refineries are situated on the coast or rivers in order to reduce transport costs.

SOME INDUSTRIAL PROCESSES

1. Manufacture of starch and glucose

Starch, a constituent of all green plants, has a formula of $(C_6H_{10}O_5)_n$ where n is a large number. It is obtained industrially from wheat, potatoes, rice, etc., which are ground up, mashed with water, and then sieved. The starch passes through the sieve as a very fine suspension whilst the fibrous material is retained. The starch granules, which slowly settle out from the filtrate, are collected and dried with hot air.

Starch is hydrolysed to glucose by heating it with dilute hydrochloric acid under pressure:

$$(C_6H_{10}O_5)_n + nH_2O \rightarrow nC_6H_{12}O_6$$

Starch may be detected by the dark blue colouration it gives with iodine solution.

2. Manufacture of ethanol

(a) Ethanol is used in large quantities in alcoholic drinks and, in this connection, it is always made by fermentation processes. Wheat, barley, potatoes, etc., are mashed with hot water and then heated with malt, i.e. germinated barley, at 50 °C for one hour. Malt contains an enzyme, diastase, (*an enzyme is an organic catalyst formed by living cells*) which hydrolyses the starch to maltose:

$$(C_6H_{10}O_5)_n(s) + \frac{n}{2}H_2O(l) \xrightarrow{\text{diastase}} \frac{n}{2}C_{12}H_{22}O_{11}(aq)$$

The mixture is then cooled to 30 °C and fermented with yeast for a few days. Yeast contains several enzymes; one, known as maltase, converts the maltose into glucose whilst another, known as zymase, converts the glucose into ethanol.

$$C_{12}H_{22}O_{11}(aq) + H_2O(l) \xrightarrow{\text{maltase}} 2C_6H_{12}O_6(aq)$$

$$C_6H_{12}O_6(aq) \xrightarrow{\text{zymase}} 2C_2H_5OH(aq) + 2CO_2(g)$$

The resultant solution, which contains 6—10% of ethanol, is concentrated by distillation. The carbon dioxide, produced during the fermentation, is a useful by-product.

(b) Ethanol for industrial use is made by treating ethene with fairly concentrated sulphuric acid under slight pressure. The resultant ethyl hydrogensulphate is hydrolysed by boiling it with water.

$$H_2SO_4(aq) + CH_2{=}CH_2(g) \rightarrow CH_3 \cdot CH_2 \cdot OSO_3H(aq)$$
$$CH_3 \cdot CH_2 \cdot OSO_3H(aq) + H_2O(l) \rightarrow CH_3 \cdot CH_2OH(aq) + H_2SO_4(aq)$$

3. Saponification of fats

Fats are solid esters derived from long chain carboxylic acids and glycerol, $CH_2OH \cdot CHOH \cdot CH_2OH$, a trihydroxy alcohol. They are hydrolysed by refluxing them with sodium or potassium hydroxide solution, the products being glycerol and a salt of the long chain acid, e.g:

$$\begin{array}{l} CH_2 \cdot OOC \cdot C_{17}H_{35} \\ | \\ CH \cdot OOC \cdot C_{17}H_{35}(s) + 3NaOH(aq) \rightarrow \\ | \\ CH_2 \cdot OOC \cdot C_{17}H_{35} \\ \text{Glyceryl stearate} \end{array} \begin{array}{l} CH_2OH \\ | \\ CHOH(aq) + 3C_{17}H_{35} \cdot COONa(aq) \\ | \\ CH_2OH \\ \text{Glycerol} \qquad \text{Sodium stearate} \end{array}$$

The sodium and potassium salts of the long chain acids are used as soaps and this accounts for the hydrolysis process being known as saponification. Addition of concentrated sodium chloride solution to the reaction mixture causes the soap to collect as a layer on the surface. The lower aqueous layer is run off to recover the sodium chloride and separate the glycerol.

Note that, for the sake of simplicity, the common names rather than the systematic names have been used for the fat and its hydrolysis products.

4. Polymerisation

Polymers are very large molecules which are formed by many small molecules combining together. The small molecule used to make a polymer is generally known as the monomer.

(a) Many ethene molecules can be made to join up to give poly(ethene) ('polythene') by, for example, heating with a little oxygen under pressure.

$$n(CH_2{=}CH_2) \rightarrow {+\!\!}CH_2{-}CH_2{+\!\!}_n \quad \text{where} \quad n \text{ is a large number.}$$

Poly(ethene) is used for making containers, piping, and plastic bags, etc. Its average relative molecular mass is about 16 000.

(b) Poly(chloroethene) ('polyvinyl chloride' or 'P.V.C.') results from the polymerisation of chloroethene:

$$n\mathrm{CH_2{=}CH} \rightarrow {+}\mathrm{CH_2{-}CH}{+}_n$$
$$\qquad\quad |\qquad\qquad\quad |$$
$$\qquad\quad \mathrm{Cl}\qquad\qquad\ \mathrm{Cl}$$

Poly(chloroethene) is used for making insulators, piping, and thin sheets for various purposes.

(c) Terylene is an example of a polyester. It can be made by heating ethane-1,2-diol, $CH_2OH \cdot CH_2OH$, with a dicarboxylic acid which for simplicity will be represented as HOOC—☐—COOH:

HOOC—☐—CO{OH ⋯ H}OCH$_2 \cdot$CH$_2$O{H ⋯ HO}OC—☐—COOH
$$\downarrow$$
HOOC—☐—COOCH$_2 \cdot$CH$_2$OOC—☐—COOH + 2H$_2$O

Each end of the chain will react alternately with more alcohol and acid molecules to give a polymer with a relative molecular mass in the region of 15 000.

Terylene is used for making a wide variety of clothing, etc.

Questions

1. 0.282 g of a liquid A was found to contain 0.036 g of carbon, 0.006 g of hydrogen and the rest bromine.
    0.470 g of A occupied 112 cm³ when vaporised at 273 °C and 1 atmosphere pressure.
    (Relative atomic masses: H = 1, C = 12, Br = 80. Molar volume of a gas at s.t.p. is 22.4 litres.)

    (a) Calculate the moles of carbon, hydrogen and bromine present in 0.282 g of A.
    (b) Using the results from (a), state the empirical formula of A.
    (c) Calculate the volume that 0.470 g of the vapour would have occupied if measured at s.t.p.
    (d) Calculate the mass of A which would occupy 22.4 litres at s.t.p. and hence state the relative molecular mass of A.
    (e) What is the molecular formula of A?
    (f) Liquid A is one of two isomeric compounds. Name the reagents from which one of these compounds can be made in the laboratory, and write the structural formula for this compound and the equation for the reaction.
    [J.M.B.]

2. (a) Draw diagrams to show the structural formulae of:
    (i) ethane,
    (ii) ethene,
    (iii) ethyne.

    (b) Under what conditions does ethene react with:
    (i) hydrogen,
    (ii) hydrogen chloride?
    In each case name the product and write an equation.

Organic Chemistry 241

- (c) Under what conditions does ethyne react with an excess of:
  - (i) hydrogen,
  - (ii) bromine?

  In each case name the product and write an equation.
- (d) When ethyne reacts with hydrogen chloride, a compound is formed which will undergo polymerisation. Draw a diagram to illustrate the structure of a section of the polymer which is formed.

[J.M.B.]

3 (a) Briefly explain in terms of the electronic theory the bonding between the atoms in a molecule of methane, CH_4.
   (b) Methane is said to be a saturated hydrocarbon; name one unsaturated hydrocarbon. How does the structure of an unsaturated hydrocarbon differ from that of a saturated hydrocarbon?
   Compare the reactivities of methane and the unsaturated hydrocarbon with bromine, and indicate the structures of the products formed.
   (c) State (without describing the apparatus used):
   (i) how an unsaturated hydrocarbon can be obtained from ethanol in the laboratory,
   (ii) how an unsaturated hydrocarbon can be converted to a polymer.
   (d) Why is it dangerous to use an appliance burning methane (natural gas) in a badly ventilated room?

[A.E.B. June 1977]

4 (a) By drawing their structural formulae show the differences in structure between ethane and ethene. Explain how the bond between a carbon atom and a hydrogen atom in these compounds is formed.
   (b) By naming the reagents, stating the conditions, and writing an equation for each reaction, describe how ethene could be converted into:
   (i) ethane,
   (ii) chloroethane,
   (iii) 1,2-dibromoethane,
   (iv) ethanol.

[J.M.B.]

5 (a) Describe how you would obtain an aqueous solution of ethanol from a sugar, for example glucose, $C_6H_{12}O_6$. State without experimental details, the method you would use to obtain a small sample of reasonably pure ethanol from the aqueous solution and indicate the physical bases of this preparation.
   (b) Dehydration of an alcohol, E (not ethanol), results in the formation of a gaseous hydrocarbon, F, having the empirical formula CH_2. F rapidly decolourises bromine water.
   Under suitable conditions, 100 cm³ of F reacts with 100 cm³ of hydrogen, giving 100 cm³ of G and no other product. The relative molecular mass of G is 44.
   (i) Suggest the identities and molecular formulae of E, F, and G.
   (ii) What reaction takes place when F decolourises bromine water?
   (iii) Briefly state how you would carry out in the laboratory the dehydration of E to F.

[A.E.B. Nov 1976]

6 Outline an example of each of the following processes either on a laboratory or industrial scale. Give equations along with essential conditions and the names of the products formed.
   (i) Chlorination.
   (ii) Saponification.
   (iii) Esterification.
   (iv) Dehydration.
   (v) Polymerisation.

7 (a) Name four fractions obtained by the fractional distillation of petroleum and give one major use for each fraction.

(b) At present, most of our ethanol is made from ethene. How is this done on a large scale?

(c) In future, it may be necessary to make ethene from ethanol. How could this be done industrially?

(d) Ethene and bromine were allowed to react together at a temperature high enough to ensure that all the substances involved in the reaction were gaseous.
   (i) Write the equation for the reaction.
   (ii) Give the structural formula of the product.
   (iii) State what would be seen during the course of the reaction.
   (iv) If 50 cm³ of ethene and 50 cm³ of bromine vapour were used, what volume of gaseous product would be formed?

[J.M.B.]

8 Give the equations and conditions showing how the following substances may be obtained:
   (i) an acid from an alcohol,
   (ii) an ester from an acid,
   (iii) soap from a fat,
   (iv) a polymer from an alkene,
   (v) an alcohol from an alkene.

14 Practical Exercises

EXPERIMENTS ON CHAPTER 1

EXPERIMENT 1

The Effect of Heating Substances in Air

Place copper(II) carbonate in a clean dry test-tube to a depth of about 1 cm and then weigh the tube and its contents. Heat the tube over a small bunsen flame until no further change is apparent, then allow the tube to cool and reweigh it. Record your observations and classify the process as a physical or chemical change.

Perform similar experiments using, in turn, each of the following substances: sodium hydrogencarbonate, zinc oxide, copper(II) sulphate crystals, and sodium chloride. Finally, use magnesium powder but this time do the experiment in a silica crucible, on a pipe clay triangle, and leave the lid partly off while the heating is in progress. A larger flame will be necessary in this last case.

EXPERIMENT 2

Separation of a Mixture of Potassium Dichromate(VI) and Sand

Weigh out 5 g of the mixture and put it into a 250 cm³ beaker. Add 20 cm³ of distilled water and stir the mixture until all the orange particles of potassium dichromate(VI) have dissolved. Pour the mixture, with the aid of a glass rod (Figure 14.1), into a *weighed* filter paper in a funnel.

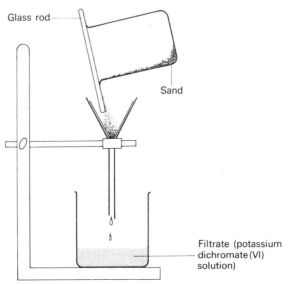

Figure 14.1. Separation of sand from potassium dichromate(VI) solution

Wash all the particles of sand out of the beaker into the filter paper with the minimum quantity of water. Now, using very small portions of water, wash the filter paper until all traces of orange have been removed. Let the filter paper drain then place it on a watch glass and put it in an oven to dry. When the paper and contents are dry, weigh them and hence find the mass of sand.

Gently boil the filtrate (potassium dichromate(VI) solution) until its volume is reduced by about two-thirds and then transfer it to a *weighed* evaporating basin with the aid of a little water. Place the evaporating basin on a boiling water-bath and evaporate the solution to dryness. Weigh the evaporating basin containing the dry residue and hence find the mass of potassium dichromate(VI).

Record your results as follows and then calculate the percentage of sand in the mixture.

Mass of filter paper + sand	=	g
Mass of filter paper	=	g
Mass of sand	=	g
Mass of basin + potassium dichromate(VI)	=	g
Mass of basin	=	g
Mass of potassium dichromate(VI)	=	g

Explain why, if you repeated the experiment, you would not necessarily obtain the same answer even if both experiments were carried out very carefully.

EXPERIMENT 3

Separation of the Pigments in Inks

Place 25 cm^3 of the solvent mixture (made from 3 volumes of butan-1-ol to 1 volume of ethanol, and 1 volume of 2 M ammonia solution) in a litre beaker and then cover the beaker with a watch glass.

Handling it only by the edges, take an oblong piece of filter paper (about 25 × 12 cm) and place it on a sheet of clean paper. Now, with the aid of a melting point tube drawn to a fine jet, make a small dot of ink about 1 cm up from the edge of one of the long sides of the paper. Repeat the process with different coloured inks along the same edge; the dots should be about 3 cm apart and no more than about 3 mm in diameter. The colour of the samples can be written in pencil along the top edge. Allow the inks to dry and then secure the filter paper in the form of a cylinder using paper clips or, preferably, plastic clips. Place the paper in the beaker containing the solvent (Figure 1.4) and then replace the watch glass. Note that the ink samples must be above the solvent level.

Leave the beaker and contents *perfectly still* until the solvent has risen about two-thirds of the way up the paper. Remove the paper from the beaker and allow the solvent to evaporate. Examine the dry filter paper and note the colour of the pigments making up each ink.

EXPERIMENT 4

Examination of the Colouring Matter in Grass

Place 25 cm³ of methylbenzene (toluene) in a clean litre beaker and cover the latter with a watch glass.

Cut up a dozen or so lengths of grass into small pieces and then grind them up with 5 cm³ of propanone using a mortar and pestle. Decant the resultant deep green solution into a test tube. Place a small spot of the solution on a piece of filter paper as in the previous experiment. Make the filter paper into a cylinder with the aid of clips and then put it in the beaker containing the methylbenzene. Allow the chromatogram to develop as before.

Remove the filter paper from the beaker, allow the solvent to evaporate, and then note the colour and order of the spots. Three spots should be apparent; in ascending order, these are due to carotene, chlorophyll, and xanthophyll respectively.

EXPERIMENT 5

Separation of Iodine from Salt by Sublimation

Weigh out 2 g of a mixture, containing about 90% salt and 10% iodine, into a 100 cm³ beaker. Place the beaker on a tripod and gauze and then clamp a round bottomed flask, containing cold water, on top of the beaker (Figure 14.2). Gently heat the beaker with a *small flame* until all the iodine has vaporised and condensed on the cold flask. Allow the apparatus to cool, weigh the residual salt, and hence calculate the percentage of salt in the mixture.

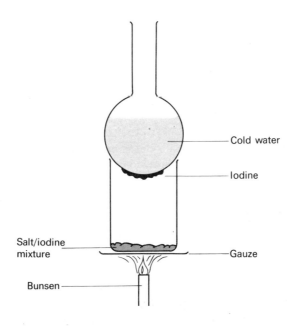

Figure 14.2. Separation of iodine from salt by sublimation

Since iodine vapour is poisonous, the experiment should be done in a fume cupboard or in a well ventilated room.

EXPERIMENT 6

Separation of Ammonium Chloride from Salt by Sublimation

The method is identical to that used in Experiment 5 but stronger heating is required to sublime the ammonium chloride. The sublimation is complete when no more white fumes are seen coming from the solid in the bottom of the beaker.

EXPERIMENT 7

Separation of a Solvent from a Non-Volatile Solute by Distillation

Dissolve about 10 g of potassium dichromate(VI) in 150 cm^3 of water and pour the resultant solution into a 250 cm^3 round bottomed flask. Add two or three small pieces of unglazed porcelain and then set up the apparatus for distillation as shown in Figure 1.2. Turn on the water so that a steady stream is passing through the condenser and then light the bunsen. Continue the heating until about three-quarters of the water has distilled over and note the temperature of the water vapour leaving the flask. Pour the remaining solution into a small beaker and observe the appearance of crystals as the concentrated solution cools.

EXPERIMENT 8

Purification of a Solid by Recrystallisation

This experiment may be done using a mixture consisting of 95% benzenecarboxylic acid (benzoic acid) and 5% phenylethanamide (acetanilide).

Determine the melting point of the contaminated benzenecarboxylic acid as follows. Grind up a small amount of the mixture and introduce a little into a melting point tube to a depth of about 3 mm. Place the tube in a melting point apparatus, raise the temperature slowly, and note the range of temperature between which the first trace of liquid and a clear solution is obtained.

Now weigh out 5 g of the mixture into a 250 cm^3 conical flask, add about 40 cm^3 of water, and then heat the mixture. Add water, a little at a time, to the boiling mixture until a clear solution is obtained, i.e. until no crystals or trace of oil is left. Allow the solution to cool to room temperature and then filter off the crystals using a Buchner funnel and flask (Figure 14.3). When no more liquid can be sucked off the crystals, spread them out on a piece of filter paper and leave them to dry in the air. Finally, weigh the dry crystals and determine their melting point.

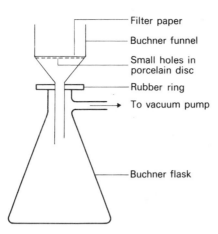

Figure 14.3. Filtration using a Buchner funnel and flask

EXPERIMENT 9

The Effect of Heat on a Mixture of Iron and Sulphur

Thoroughly mix 21 g of iron filings (or preferably iron powder) with 12 g of powdered sulphur. Divide the mixture into two portions, the one portion being about twice as large as the other. Place the larger portion in a test-tube and heat the bottom of the tube in a small flame. When the mixture starts to glow, remove the flame. The glow will spread throughout the mixture, without further heating, as the sulphur and iron combine to give iron(II) sulphide.

$$\text{Iron + sulphur} \rightarrow \text{iron(II) sulphide}$$

Allow the tube and contents to cool, put on a pair of safety spectacles, and then carefully break the tube using a mortar and pestle. Separate the glass from the grey-black iron(II) sulphide.

Perform the following tests on the mixture and on the iron(II) sulphide.

(a) Place one end of a magnet in the sample, raise the magnet and tap it gently. Does separation occur?

(b) Put approximately 1 cm³ of the sample in a test tube and add a few cm³ of dilute hydrochloric acid. Effervescence occurs in each case. Cautiously smell each gas and then place a finger over the end of the tube for a few seconds before holding the mouth of the tube to a flame. Record your observations in each case.

EXPERIMENTS ON CHAPTER 2

EXPERIMENT 10

Verification of Charles' Law

Draw about 50 cm³ of air into a graduated 100 cm³ glass syringe. Seal the end of the syringe with a vaccine cap or a piece of rubber tube and glass rod. Now clamp the syringe in a litre beaker containing cold water and a thermometer (Figure 14.4). Leave the apparatus to stand for about five minutes to equilibrate and then note the volume of the air and the temperature of the water. Next, raise the temperature of the water by about 10 °C using a very small bunsen flame. Stirring the water continuously with the thermometer, maintain the temperature as steady as possible and, when equilibrium has been reached, record the new temperature and volume. Repeat the procedure at 10 °C intervals up to the boiling point of the water.

Plot graphs of (a) volume against temperature in degrees Celsius, and (b) volume against temperature in degrees Kelvin.

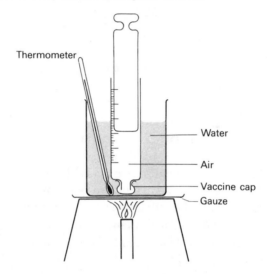

Figure 14.4. Volume–temperature relationships for gases

EXPERIMENT 11

The Action of Heat on Ammonium Chloride

Heat a little ammonium chloride in a test-tube and note the appearance of white solid in the cooler parts of the tube. The ammonium chloride sublimes by a process known as thermal dissociation, i.e. it decomposes on heating but recombines on cooling.

$$\text{Ammonium chloride} \underset{\text{cool}}{\overset{\text{heat}}{\rightleftharpoons}} \text{ammonia} + \text{hydrogen chloride}$$

Now, in a second experiment, put a little ammonium chloride in the end of a test-tube and then clamp the tube horizontally. Place a piece of moist, neutral litmus paper, then a loose plug of Rocksill wool (or glass wool), and then another piece of moist, neutral litmus paper in the tube as illustrated in Figure 14.5. Heat the ammonium chloride and note the colour changes of the litmus papers. Explain your observations in terms of the fact that the Rocksill wool acts as a diffusion plug. Ammonia forms an alkaline solution which turns litmus blue, whilst hydrogen chloride gives an acidic solution which turns litmus red.

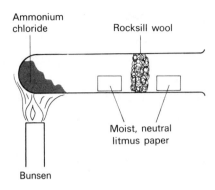

Figure 14.5. Action of heat on ammonium chloride

EXPERIMENTS ON CHAPTER 4

EXPERIMENT 12

To Construct a Solubility Curve for Potassium Chlorate

This experiment is best done by about three students each determining the solubility at a particular temperature and then averaging their results. Different temperatures are alloted to different groups and then the results are pooled in order to construct the solubility curve.

Place 35 cm^3 of distilled water in a boiling tube and add the quantity of powdered potassium chlorate indicated in the table below.

Temperature	Mass of KClO$_3$	Temperature	Mass of KClO$_3$
20 °C	3 g	55 °C	9 g
30 °C	5 g	60 °C	10 g
35 °C	5.5 g	65 °C	11 g
40 °C	6 g	70 °C	12.5 g
45 °C	7 g	75 °C	13.5 g
50 °C	8 g	80 °C	15 g

Heat up the tube and contents in a beaker of water about 10 °C above the temperature at which the solubility is being determined and stir until all the crystals have dissolved. Now let the tube and the bath cool until the desired temperature is reached. Quickly decant off as much as possible of the resultant solution into a weighed evaporating basin without letting any crystals escape from the tube. Weigh the basin containing the solution and then evaporate the solution to dryness on a boiling water-bath. Weigh the basin and dry potassium chlorate and so find the mass of potassium chlorate in the known mass of solution at the given temperature. Calculate the solubility of the potassium chlorate at that temperature and, when all the results are complete, plot a graph of solubility against temperature.

The results are recorded and calculated as in the example below.

Mass of basin	= 68.32 g
Mass of basin + $KClO_3$ solution	= 91.48 g
Mass of $KClO_3$ solution	= 23.16 g
Mass of basin + dry $KClO_3$	= 70.09 g
Mass of $KClO_3$	= 1.77 g

Thus, 23.16 g of the saturated solution at 25 °C contain 1.77 g of $KClO_3$ i.e. 21.39 g of water (i.e. 23.16–1.77) at 25 °C dissolve 1.77 g of $KClO_3$

$\therefore$ 1000 g of water at 25 °C dissolve $\frac{1.77}{21.39} \times 1000$ = 82.7 g of $KClO_3$

Hence the solubility of potassium chlorate at 25 °C is 82.7 g per kg of water.

EXPERIMENT 13

Preparation of a Super-Saturated Solution

Half fill a boiling-tube with sodium thiosulphate crystals, $Na_2S_2O_3 \cdot 5H_2O$, and then add about 2 cm³ of water. Warm the tube gently until all the crystals have dissolved. When a colourless solution has been obtained, cool the tube and contents under a cold tap, being careful to keep the tube perfectly still. No crystals form even when the solution is cold; a super-saturated solution has been formed. Now add a very small crystal of sodium thiosulphate to the solution and note that crystallisation immediately commences. The reason for the phrase 'in the presence of excess solute' in the definition of a saturated solution (page 53) should now be clearly apparent.

Note that crystallisation of the super-saturated solution can also be induced by shaking it or by scratching the tube under the solution with a glass rod—this provides nuclei for the crystals to form.

EXPERIMENT 14

The Hardness of Water

(a) Pass carbon dioxide through a few cm³ of limewater in a test-tube. The limewater turns milky as a fine suspension of calcium carbonate is formed. Continue passing the carbon dioxide until the precipitate redissolves. The resultant solution contains calcium hydrogencarbonate and this is one of the causes of temporary hardness. Carefully boil the solution for a few minutes, shaking the tube all the time, and note that the white solid reappears as the temporary hardness (calcium hydrogencarbonate) is destroyed.

(b) (i) Using a measuring cylinder, place 100 cm³ of temporary hard water in a stoppered 250 cm³ bottle. Add 0.5 cm³ of soap solution from a burette, replace the stopper, and shake the mixture vigorously. Repeat the additions and shakings until a lather is formed which lasts for two minutes. Record your observations and the volume of soap solution used.

Practical Exercises

(ii) Place 100 cm³ of temporary hard water in a 250 cm³ conical flask, boil the water for 5 minutes, cool it, and then filter it into a clean 250 cm³ bottle. Titrate the water with the soap solution as above. Record the result, explain any difference with (b)(i), and give the equation involved in the change.

(c) (i) Titrate 100 cm³ of permanently hard water with soap solution and record your observations and the volume of soap solution used.

(ii) Place 100 cm³ of permanently hard water in a conical flask and shake it with about 1 g of sodium carbonate. Filter the mixture into a clean bottle and titrate the solution with soap. Record the volume of soap solution used, explain any difference with (c)(i), and give the relevant equations.

(d) (i) Titrate 100 cm³ of distilled water with soap solution and record the volume of soap used to produce a lather.

(ii) Place a fresh 100 cm³ sample of distilled water in the bottle and add about 1 g of sodium chloride. Shake the bottle until the salt has dissolved and then titrate the solution with soap. Record the volume of soap solution used.

(iii) Take a further 100 cm³ sample of distilled water, dissolve about 1 g of sodium sulphate in it, and then titrate the solution with soap. Record your result and state what conclusions may be drawn from experiments (d)(i), (ii), and (iii).

EXPERIMENTS ON CHAPTER 5

EXPERIMENT 15

To Demonstrate the Law of Constant Composition

Prepare samples of copper(II) oxide by the three methods below.

(a) Dissolve about 4 g of copper(II) sulphate crystals in 100 cm³ of water in a 250 cm³ beaker and then add sodium hydroxide solution, with stirring, until no more precipitate is formed.

$$CuSO_4(aq) + 2NaOH(aq) \rightarrow Cu(OH)_2(s) + Na_2SO_4(aq)$$

Heat the beaker on a tripod and gauze until all the blue copper(II) hydroxide has decomposed into black copper(II) oxide:

$$Cu(OH)_2(s) \rightarrow CuO(s) + H_2O(l)$$

Filter off the copper(II) oxide and wash it several times on the paper with distilled water until the washings no longer turn litmus paper blue. Scrape off the precipitate into a crucible, warm gently at first and then more strongly until all the water appears to have been removed. Let the crucible cool and then weigh it. Reheat the crucible for a few minutes, allow it to cool, and then reweigh it. Continue this procedure until two consecutive weighings are the same and then store the crucible and contents in a desiccator to keep it dry.

(b) Heat about half a crucible of copper(II) carbonate to constant mass as above and then store it in a desiccator.

$$CuCO_3(s) \rightarrow CuO(s) + CO_2(g)$$

(c) Place about 2 g of copper turnings in a crucible and add concentrated nitric acid to it dropwise, in a fume cupboard, until all the copper has dissolved.

$$Cu(s) + 4HNO_3(conc) \rightarrow Cu(NO_3)_2(aq) + 2H_2O(l) + 2NO_2(g)$$

Gently heat the crucible containing the copper(II) nitrate solution, on a tripod and gauze, until all traces of liquid have disappeared and then heat the residue strongly until no more brown fumes are evolved.

$$2Cu(NO_3)_2(s) \rightarrow 2CuO(s) + 4NO_2(g) + O_2(g)$$

Store the copper(II) oxide in a desiccator.

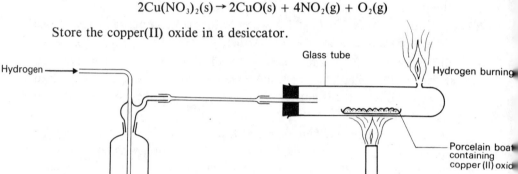

Figure 14.6. Analysis of copper(II) oxide

Analysis of the Copper(II) Oxide Samples

Set up the apparatus illustrated in Figure 14.6. Make sure that the tube is tilted slightly towards the jet to prevent any water formed in the reaction from running back and cracking the glass. Weigh a procelain boat empty and then again with 1 to 1.5 g of copper(II) oxide in it. Put the boat in the glass tube and then gently turn on the hydrogen. Allow the hydrogen to pass through the apparatus for one minute before lighting it at the jet. (**This time must be allowed for the expulsion of air from the apparatus because oxygen-hydrogen mixtures are explosive.**) Heat the tube under the boat until all the solid in the boat has turned copper colour.

$$CuO(s) + H_2(g) \rightarrow Cu(s) + H_2O(g)$$

Now turn off the bunsen but leave the current of hydrogen on until the tube is cool so that the hot copper does not recombine with oxygen. Weigh the boat and copper. Repeat the heating etc. as above until constant mass is attained.

Repeat the above process with each sample of copper(II) oxide and in each case calculate the percentage of copper present. The calculation is illustrated on page 60.

EXPERIMENT 16

To Demonstrate the Law of Multiple Proportions

This law can be illustrated by analysing either copper(I) and copper(II) oxides or lead(II) and lead(IV) oxides by heating them in a current of hydrogen as in the previous experiment. In each case the results may be used to calculate the mass of oxygen which combines with 1 g of the metal. The masses of oxygen combined with 1 g of the metal in its two oxides are found, within experimental error, to be in a simple ratio.

EXPERIMENT 17

To Find the Molar Ratio of Lead(II) and Iodide Ions which React Together

The reaction between lead(II) nitrate and potassium iodide may be represented as:

lead(II) nitrate(aq) + potassium iodide(aq) → lead(II) iodide(s) + potassium nitrate(aq)

Confirm this reaction by mixing a little of the two solutions.

Now obtain 6 identical test-tubes, place them in a rack and run in to each 5.0 cm³ of molar potassium iodide solution from a burette. From a different burette add 1.0, 1.5, 2.0, 2.5, 3.0, and 3.5 cm³ respectively of molar lead(II) nitrate solution to the 6 tubes. Thoroughly mix the contents of each tube and then allow them to settle for 10 minutes before measuring the depth of each precipitate. Plot a graph of depth of precipitate against volume of lead(II) nitrate solution added.

After a certain volume of lead(II) nitrate solution has been added, the depth of precipitate remains constant. The first volume which gives this depth of precipitate is the volume of molar lead(II) nitrate solution which is required to react with 5.0 cm³ of molar potassium iodide solution.

Since 1000 cm³ of M KI solution contains 1 mole of I^- ions

5 cm³ of M KI solution contains $\frac{1}{1000} \times 5 = 0.005$ mole of I^- ions.

From the volume of lead(II) nitrate solution used, work out the moles of Pb^{2+} ions which took part in the reaction and hence write the ionic equation for reaction between Pb^{2+} and I^- ions.

EXPERIMENT 18

Determination of the number of Molecules of Water of Crystallisation in $MgSO_4 \cdot xH_2O$

Weigh a clean crucible and lid then half fill the crucible with magnesium sulphate crystals and reweigh. Place the crucible on a pipe-clay triangle and move the lid slightly to one side. Heat the crucible gently with a small bunsen flame and, when all the water appears to have been driven off, heat strongly for 5 minutes. Allow the crucible to cool in a desiccator and finally reweigh it along with the lid and anhydrous magnesium sulphate. Record your results as follows.

	Mass of crucible + lid	=	g
	Mass of crucible + lid + magnesium sulphate crystals	=	g
∴	Mass of magnesium sulphate crystals	=	g
	Mass of crucible + lid + anhydrous magnesium sulphate	=	g
∴	Mass of anhydrous magnesium sulphate	=	g
∴	Mass of water of crystallisation	=	g

Divide the mass of *anhydrous* magnesium sulphate by its relative molecular mass to find the moles of magnesium sulphate present. Find the number of moles of water similarly. Divide the moles of water by the moles of magnesium sulphate to give the simple whole number ratio of moles of water : moles of magnesium sulphate, and hence the formula of the crystals.

A specimen calculation using barium chloride crystals is given on page 58.

EXPERIMENTS ON CHAPTER 6

EXPERIMENT 19

Preparation of Zinc Sulphate Crystals from Zinc

In this experiment a salt is made from reaction between an acid and a metal.

Place about 50 cm^3 of dilute (approximately 2 M) sulphuric acid in a 250 cm^3 beaker and add a few pieces of granulated zinc.

$$Zn(s) + H_2SO_4(aq) \rightarrow ZnSO_4(aq) + H_2(g)$$

If the reaction is slow, add a few drops of copper(II) sulphate solution and *gently* warm the mixture. If all the zinc dissolves, add more. When the evolution of hydrogen ceases, filter the solution to remove excess zinc and traces of impurity such as carbon. Place the filtrate in an evaporating basin and evaporate the solution, on a water-bath, to about half of the original volume. Cool a little of the solution in a test tube under a tap, shaking well. If crystals form, allow the remaining solution in the basin to cool slowly. However, if no crystals are produced, further evaporation is required. Filter off the crystals and wash them with a *small* quantity of cold distilled water to remove traces of acid. Finally, dry the crystals with filter paper; they are zinc sulphate heptahydrate, $ZnSO_4 \cdot 7H_2O$.

Magnesium sulphate and iron(II) sulphate crystals may be prepared in a similar manner.

EXPERIMENT 20

Preparation of Copper(II) Sulphate Crystals from Copper(II) Oxide

This experiment involves the preparation of a salt from an acid and an insoluble oxide.

Place about 50 cm^3 of approximately 2 M sulphuric acid in a 250 cm^3 beaker, add a little copper(II) oxide and simmer the mixture gently, with stirring. The black solid dissolves to give a blue solution.

$$CuO(s) + H_2SO_4(aq) \rightarrow CuSO_4(aq) + H_2O(l)$$

Add further small samples of copper(II) oxide to the boiling solution until a permanent residue of the oxide is obtained, i.e. until all the acid is used up. Remove the excess copper(II) oxide by filtration of the hot solution and then obtain the copper(II) sulphate crystals in a similar manner to that described for zinc sulphate in the previous experiment. The blue crystals are copper(II) sulphate pentahydrate, $CuSO_4 \cdot 5H_2O$.

EXPERIMENT 21

Preparation of Sodium Sulphate

In this experiment, a salt is made from an acid and a soluble base. Before performing the experiment, read pages 76 to 79 on volumetric analysis.

Pipette 25 cm^3 of approximately 2 M sodium hydroxide solution into a 250 cm^3 conical flask. Add two or three drops of screened methyl orange and then titrate with approximately 1 M sulphuric acid from a burette until the indicator changes colour from green to blue-grey. If too much acid is added the solution will turn red. Wash out the conical flask and do a further titration to check your result.

$$H_2SO_4(aq) + 2NaOH(aq) \rightarrow Na_2SO_4(aq) + 2H_2O(l)$$

Now mix the equivalent quantities of acid and alkali in the absence of the indicator. Concentrate the resultant solution and obtain the crystals in a similar manner to that described for zinc sulphate in Experiment 19. The crystals obtained are sodium sulphate decahydrate, $Na_2SO_4 \cdot 10H_2O$.

EXPERIMENT 22

Preparation of Sodium Sulphate

This experiment entails the preparation of a salt from an acid and a soluble carbonate. The method is similar to that of the previous experiment.

Titrate 25 cm^3 portions of an approximately 1 M solution of sodium carbonate with molar sulphuric acid solution, using screened methyl orange as indicator, until concordant results are obtained.

$$H_2SO_4(aq) + Na_2CO_3(aq) \rightarrow Na_2SO_4(aq) + H_2O(l) + CO_2(g)$$

Then mix the reacting volumes together in the absence of the indicator and isolate the crystals as in the previous experiments.

EXPERIMENT 23

Preparation of Lead(II) Nitrate

In this experiment a salt is made from an acid and an insoluble carbonate.

Put about 50 cm³ of approximately 2 M nitric acid in a 250 cm³ beaker and add lead(II) carbonate a little at a time. Carbon dioxide is evolved and lead(II) nitrate is produced:

$$PbCO_3(s) + 2HNO_3(aq) \rightarrow Pb(NO_3)_2(aq) + H_2O(l) + CO_2(g)$$

Continue addition of the carbonate until effervescence ceases, i.e. until all the acid has been used up. Filter off the excess carbonate, concentrate the filtrate, and then obtain the crystalline product as in the previous experiments. Lead(II) nitrate does not have any water of crystallisation.

EXPERIMENT 24

Preparation of Insoluble Salts

Insoluble salts are prepared by mixing two solutions chosen so that one product is soluble whilst the other is insoluble.

For each pair of compounds below, dissolve a few crystals in distilled water, mix the two solutions, and note the result. If required, the precipitates may be filtered off, washed with water to remove traces of the soluble product, and then dried.
(a) Barium chloride and potassium sulphate.
(b) Lead(II) nitrate and potassium iodide.
(c) Copper(II) sulphate and sodium carbonate.
(d) Iron(III) chloride and sodium hydroxide.

Write the relevant equation for each reaction.

EXPERIMENTS ON CHAPTER 7

Experiment 25

To Find what Type of Substance Conducts Electricity and the Conditions Required

The apparatus is arranged as in Figure 14.7, the purpose of the bulb being to indicate when the circuit is complete.

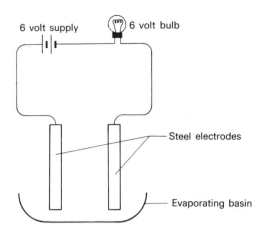

Figure 14.7. Apparatus for testing electrical conductivity

(a) Insert the electrodes, in turn, in each of the following and note whether the bulb lights up: sulphur, zinc, charcoal, and iron.

(b) Insert the electrodes, in turn, in solid samples of sugar, sodium chloride, naphthalene ($C_{10}H_8$), copper(II) sulphate, and lead(II) bromide. Repeat the tests with aqueous solutions of sugar, sodium chloride, and copper(II) sulphate.

(c) Investigate the effect of melting the naphthalene and the lead(II) bromide. In each case use a small flame. Note that naphthalene melts at 80 °C and is flammable. The experiment with lead(II) bromide must be done in a fume cupboard (lead(II) bromide vapour is poisonous).

Answer the following questions from your results.
(i) What type of elements conduct electricity?
(ii) Do compounds in the solid state conduct?
(iii) Does dissolving a compound in water have any effect on its ability to conduct an electric current? If so, what type of compound is affected?
(iv) Does melting the compound have any effect? If so, what type of compound is affected?

EXPERIMENT 26

To Find What Happens When Solutions Conduct Electricity

Ionic substances conduct electricity when they are molten or are dissolved in water. The question arises, 'does water play any other part besides allowing the ions to move?' This may be answered by examining the products produced at the cathode (negative electrode) and anode (positive electrode) on electrolysis of approximately 0.5 M solutions of:

(a) sulphuric acid,   (b) hydrochloric acid,
(c) potassium iodide,   (d) copper(II) chloride.

The apparatus required for the electrolyses is illustrated in Figure 14.8.

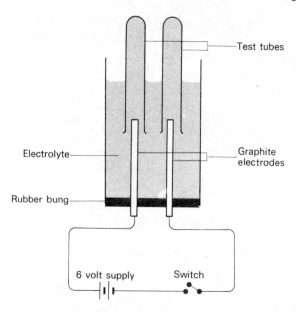

Figure 14.8. Collection of gases evolved during electrolysis

If a gas is evolved, examine its properties—note its colour, if any, test whether it 'pops' when the tube is held to a flame (i.e. hydrogen), whether it relights a glowing splint (i.e. oxygen), or whether it bleaches moist litmus (i.e. chlorine).

Record your results as follows:

Electrolyte	Observation at cathode	Observation at anode	Suggested products

How do the types of product at the cathode and anode differ?

EXPERIMENT 27

Investigation of the Displacement Reaction between Zinc and Copper(II) Sulphate

Warm about 6 g of copper(II) sulphate crystals with 50 cm³ of distilled water in a 250 cm³ beaker until all the crystals have dissolved. Now accurately weigh out about 1 g of zinc foil and add it to the warm copper(II) sulphate solution. Copper is deposited on the foil as zinc goes into solution as zinc sulphate. Stir the mixture to dislodge the copper deposited on the zinc. Repeat the stirring frequently during the next ten minutes by which time reaction should be complete, i.e. all the zinc should have dissolved. The solution should still be blue because excess copper(II) sulphate was used.

Practical Exercises

Filter off the copper using a pre-weighed filter paper. Wash the copper and filter paper with distilled water until all traces of blue have disappeared and then dry the paper and its contents in an oven. Weigh the dry filter paper and contents and hence find the mass of copper displaced by the known mass of zinc. Divide the masses of zinc used and copper deposited by their respective relative atomic masses to give the molar ratio of zinc atoms and copper ions taking part in the reaction. The results should confirm the equation:

$$Zn(s) + Cu^{2+}(aq) \rightarrow Cu(s) + Zn^{2+}(aq)$$

EXPERIMENTS ON CHAPTER 8

EXPERIMENT 28

Oxidation – Reduction Reactions between Metals and Metallic Ions

It was seen in the previous experiment that zinc reduced Cu^{2+} ions to copper and was itself oxidised to Zn^{2+} ions. Several metals and metallic ions will now be studied in order to obtain an activity series. The tests require approximately molar solutions of magnesium, copper(II), and zinc sulphates and lead(II) nitrate and thin strips of magnesium, zinc, copper, and lead.

Take three test-tubes and put about 3 cm^3 of zinc sulphate solution in one, copper(II) sulphate in another, and lead(II) nitrate solution in the final one. Add about 2.5 cm of clean magnesium ribbon to each and note whether there is any sign of displacement occurring. Record your results. Repeat the procedure using each metal in turn with solutions of the salts of the other three metals.

The metal which is best at displacing the ions of the other metals, is the strongest reducing agent, i.e. the most reactive. On the basis of your results, arrange the metals in order of decreasing reducing power.

EXPERIMENT 29

Detection of Oxidising and Reducing Agents

(a) (i) Make about 5 cm^3 of a dilute solution of iron(II) sulphate in a test-tube. Add dilute (about 2 M) sodium hydroxide solution, and note the colour of the precipitated iron(II) hydroxide.

(ii) Add dilute sodium hydroxide solution to a dilute solution of iron(III) chloride and note the colour of the precipitated iron(III) hydroxide.

(iii) Mix equal volumes of dilute solutions of iron(II) sulphate and sulphuric acid and then add a few drops of concentrated nitric acid (an oxidising agent). Boil the mixture gently for about two minutes and then make it alkaline with sodium hydroxide solution. The result is the same as that obtained in test (ii) and so it is apparent that oxidising agents convert iron(II) compounds into iron(III) compounds. The reaction may be used to detect oxidising agents.

(b) (i) Mix equal volumes of dilute sulphuric acid and dilute (approximately 1%) potassium manganate(VII), add a little sodium nitrite solution (a reducing agent), and note the colour of the resultant solution.

(ii) Mix equal volumes of dilute sulphuric acid and dilute potassium dichromate(VI). Add a little sodium nitrite solution and note what happens.

Distinct changes occur in the above two reactions and so these tests may be used to detect reducing agents.

(c) Using the tests with acidified solutions of iron(II) sulphate, potassium manganate(VII), and potassium dichromate(VI) outlined above, classify the following as oxidising or reducing agents: sodium chlorate(I) (sodium hypochlorite) solution, sodium sulphite, hydrogen peroxide, and potassium iodide. Record all your observations. All three tests must be done with each substance.

EXPERIMENTS ON CHAPTER 9

EXPERIMENT 30

The Reaction of Various Elements with Oxygen

Five gas jars of oxygen are required in this experiment; they may be obtained either from a cylinder of oxygen or by the catalytic decomposition of hydrogen peroxide (page 113). The oxygen is collected over water and a little of the latter (a depth of about 1 cm) should be left in each gas jar.

(a) Ignite about 4 cm of magnesium ribbon in a deflagrating spoon and then plunge it in a gas jar of oxygen (Figure 9.10). When reaction has ceased, push the spoon, containing the magnesium oxide, down into the water, swirl the mixture round, and then test with litmus paper.

(b) Wrap some iron wire round a deflagrating spoon, heat the wire in a flame until it glows, and then plunge it into a jar of oxygen. After reaction has ceased, proceed as in (a).

(c) Do three separate experiments using wood charcoal (carbon), yellow phosphorus, and sulphur respectively. In each case use an amount of the element equivalent to about half of the size of a small pea. Ignite the element and allow it to burn in the oxygen. After reaction has ceased, withdraw the deflagrating spoon, fit the gas jar with a lid and then shake the jar thoroughly to dissolve the gaseous oxide. Test the resulting solution with litmus paper.

Record your observations for each experiment and state what conclusions may be drawn.

EXPERIMENT 31

Preparation and Reactions of Hydrogen Peroxide

Cool about 25 cm³ of dilute sulphuric acid in a small conical flask by immersion in an ice and salt mixture. Slowly add, whilst stirring, moist barium peroxide to the acid until the mixture is only weakly acidic.

$$BaO_2(s) + H_2SO_4(aq) \rightarrow BaSO_4(s) + H_2O_2(aq)$$

Filter off the barium sulphate to obtain a dilute solution of hydrogen peroxide.

Perform the following tests with the hydrogen peroxide and explain the results.

(a) Add a little of the hydrogen peroxide to a solution of potassium iodide acidified with dilute sulphuric acid.

(b) Add hydrogen peroxide to each of the following mixed with an equal volume of dilute sulphuric acid: (i) potassium manganate(VII) solution; (ii) potassium dichromate(VI) solution.

(c) Add hydrogen peroxide to a little lead(II) sulphide prepared by passing hydrogen sulphide through lead(II) nitrate solution.

EXPERIMENTS ON CHAPTER 10

EXPERIMENT 32

Determination of the Heat of Combustion of an Alcohol

Perform the experiment described on page 122. Any of the common alcohols may be used, e.g. methanol, ethanol, or propanol.

EXPERIMENT 33

Determination of the Heat of Neutralisation of Hydrochloric Acid with Sodium Hydroxide Solution

Accurately measure out 50 cm³ of 2 M hydrochloric acid into a polystyrene beaker and then take its temperature. Now measure the temperature of a 2 M sodium hydroxide solution. Rapidly add 50 cm³ of the sodium hydroxide solution to the acid, stir the mixture with the thermometer, and note the maximum temperature attained.

Record your results and calculate the heat of neutralisation as illustrated in the example below.

Temperature of hydrochloric acid = 20.5 °C

Temperature of sodium hydroxide solution = 19.7 °C

Maximum temperature attained
on mixing 50 cm³ of acid and alkali = 33.2 °C

The average initial temperature of the reactants = $\frac{20.5 + 19.7}{2}$ = 20.1 °C

Temperature rise = 33.2−20.1 = 13.1 °C

Within the limits of the accuracy of the experiment with the equipment used, it may be assumed that the dilute solutions of the acid and alkali have the same density and heat capacity as water.

Now, heat evolved = mass × heat capacity × temperature rise
= 100 × 4.2 × 13.1 J
= 5502 J

The equation for the reaction is:

$$HCl(aq) + NaOH(aq) \rightarrow NaCl(aq) + H_2O(l)$$

and so 1 mole of acid reacts with 1 mole of alkali to give 1 mole of water. But, in the experiment 50 cm³ each of 2 M acid and alkali were used and these contain $\frac{50}{1000} \times 2$ = 0.1 mole of acid and alkali respectively.

The heat evolved in the formation of one mole of water will therefore be 5502 × 10 = 55 020 J or 55 kJ.

Hence, the heat of neutralisation of hydrochloric acid with sodium hydroxide is −55 kJ mole⁻¹ of water produced. The answer is a little below the true value due to small heat losses to the atmosphere and the approximations made in the calculation.

EXPERIMENT 34

To Show the Effect of Concentration on the Rate of a Reaction

A suitable reaction for study is that between sodium thiosulphate solution and nitric acid:

$$Na_2S_2O_3(aq) + 2HNO_3(aq) \rightarrow 2NaNO_3(aq) + S(s) + H_2SO_3(aq)$$

Two series of experiments are performed, the concentration of nitric acid being constant in each series whilst the concentration of sodium thiosulphate is varied.

Prepare five 100 cm³ conical flasks containing 50 cm³ samples of sodium thiosulphate solution with the following concentrations: 1 M, 0.5 M, 0.25 M,

0.125 M, and 0.0625 M respectively. To the first flask add 5 cm^3 of 2 M nitric acid from a small measuring cylinder, start a stop-clock, and swirl the flask to mix the solutions. Place the flask on a piece of paper with a cross marked on it and note the time taken for the precipitate of sulphur to appear and obscure the cross, when viewed from above. Repeat the process with the other flasks.

Prepare a second series of flasks again containing 50 cm^3 samples of sodium thiosulphate with the respective concentrations as above. Repeat the previous procedure but this time use 0.2 M nitric acid. Tabulate your results as below.

Concentration of thiosulphate	Time in seconds	Rate, i.e. $\frac{1}{time}$

For both experiments, plot a graph of $\frac{1}{time}$ against concentration of sodium thiosulphate. Note that the graphs should pass through the origin since, when the concentration is zero, the rate must be zero.

EXPERIMENT 35

Examination of the Effect of Temperature on Rate of Reaction

Perform the first part of the previous experiment with both reagents at room temperature and then again with them about 20 °C higher. Plot rate against concentration at both temperatures on the same graph.

EXPERIMENT 36

Effect of Concentration on the Rate of Decomposition of Hydrogen Peroxide

Add 10 cm^3 of 100 volume hydrogen peroxide to 90 cm^3 of water to give a nominal 10 volume solution. Place this solution in one burette and water in another. In five boiling tubes, add the volumes of hydrogen peroxide and water stated below.

cm^3 of 10 volume peroxide	cm^3 of water	Volume concentration of resultant solution
12.5	12.5	5
10.0	15.0	4
7.5	17.5	3
5.0	20.0	2
2.5	22.5	1

Set up the apparatus shown in Figure 14.9. Clamp one of the boiling tubes in the beaker of cold water (this is necessary because the reaction is exothermic) and add exactly 1 g of manganese(IV) oxide powder. Immediately fit the bung and tube, start a stop-clock, and note the time taken for 20 cm³ of oxygen to be evolved.

$$2H_2O_2(aq) \xrightarrow{MnO_2 \text{ catalyst}} 2H_2O(l) + O_2(g)$$

Repeat the process with the other four solutions, each time using exactly 1 g of the catalyst.

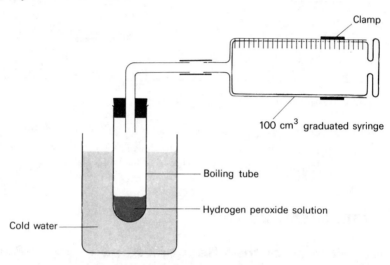

Figure 14.9. Measuring the rate of decomposition of hydrogen peroxide

Plot graphs of (a) concentration of hydrogen peroxide against time, and (b) concentration against $\frac{1}{time}$ i.e. the rate of reaction.

It should be noted that the range of concentrations which will give satisfactory results is rather limited. Thus, if the solution is too concentrated, the time is too short whereas, if the solution is too dilute, insufficient oxygen is evolved. Also the particle size of the catalyst affects the results markedly. It may be necessary therefore to adjust the conditions slightly.

EXPERIMENTS ON CHAPTER 11

EXPERIMENT 37

Reactions of Metal Ions in Solution

Dissolve a quantity of magnesium nitrate, equivalent to the size of a small pea, in about 25 cm³ of distilled water. Pour 3–4 cm³ of this solution into each of five test-tubes. To one of the samples, add dropwise, and with shaking, an approximately 2 M solution of sodium hydroxide, until no further change is apparent. Follow the same procedure with the other four samples using 2 M solutions of aqueous ammonia, sodium carbonate, hydrochloric acid, and sulphuric acid respectively. Repeat the above five tests in turn with solutions of calcium nitrate, zinc nitrate, copper(II) nitrate, lead(II) nitrate, iron(II) sulphate, and iron(III) chloride. Record your results in tabular fashion as illustrated below. Note that some of the precipitates may be soluble in excess sodium hydroxide or ammonia solution.

Metal ion	Reaction with				
	NaOH(aq)	NH_3(aq)	Na_2CO_3(aq)	HCl(aq)	H_2SO_4(aq)
Mg^{2+}	White precipitate insoluble in excess reagent	White precipitate insoluble in excess reagent	White precipitate	No apparent change	No apparent change

EXPERIMENT 38
Flame Tests

Dip a platinum or nichrome wire into concentrated hydrochloric acid and then hold it just above the blue part of a bunsen flame, the air-hole being open (Figure 14.10). Repeat this procedure until the wire is clean. Now dip the wire into the acid and then into a little solid sodium chloride on a watch glass. Note the colour that the sodium ions impart to the flame. Clean the wire thoroughly as before.

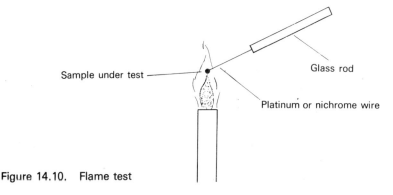

Figure 14.10. Flame test

Perform the test, in turn, with samples of potassium, calcium, copper(II), and lead(II) salts. The potassium flame is still visible when viewed through blue glass.

EXPERIMENTS ON CHAPTER 12

EXPERIMENT 39
Preparation of Ammonia and Examination of its Properties

(a) Heat, in dry test-tubes, ammonium chloride with (i) sodium hydroxide; (ii) calcium hydroxide. In each case test the gas evolved with moist red litmus paper.

Repeat the experiments using ammonium sulphate instead of ammonium chloride. Write an equation for each of the reactions.

(b) Heat a mixture of calcium hydroxide and ammonium chloride and collect three test-tubes of ammonia as shown in Figure 14.11. Cork each tube.

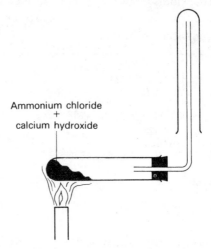

Figure 14.11. Preparation of ammonia

(i) Invert a test-tube of ammonia in water contained in an evaporating basin.

(ii) Invert a tube of the gas in a basin containing dilute sulphuric acid.

(iii) To a test-tube of the gas, add a little water, shake well, and then add a drop of litmus solution.

EXPERIMENT 40

A Study of the Behaviour of some Chlorides and Oxides in Water

(a) Put distilled water in a test-tube to a depth of about 3 cm and note its temperature. Add an amount of magnesium chloride equivalent to half the size of a small pea, shake the tube well, and note if there has been any significant change in the temperature. Determine the approximate pH of the solution using universal indicator paper and compare the result with that obtained with distilled water alone. Repeat the procedure using, in turn, aluminium and sodium chlorides, phosphorus pentachloride, and two drops of tetrachloromethane.

If the chloride is soluble, decide, on the basis of your results, whether it simply dissolves or if reaction occurs. Draw any relevant conclusions.

(b) Using a similar procedure to that above, investigate if there is any reaction between the following oxides and distilled water: calcium, copper(II), and magnesium oxides, and phosphorus(V) oxide respectively. If an oxide is insoluble in water, test its solubility in dilute sulphuric acid, warming the mixture if necessary.

As a result of your observations, classify the oxides as being basic or acidic.

EXPERIMENT 41

The Detection of some Acid Radicals

Perform the tests below, record your observations, and write an equation for each reaction.

(a) (i) Add a small quantity of a carbonate to a few cm³ of dilute hydrochloric acid in a test-tube. Pour the evolved gas into a test-tube containing a little limewater then close the tube and shake it. Repeat the experiment using sodium hydrogencarbonate instead of a carbonate. Does the test distinguish between the carbonate and hydrogencarbonate?

(ii) Make dilute solutions of sodium carbonate and sodium hydrogencarbonate and add to each a solution of a magnesium salt, e.g. the sulphate or chloride. An immediate change is apparent with the carbonate but the corresponding change with hydrogencarbonate does not occur until the mixture is boiled; confirm this.

(b) Make a dilute solution of a sulphate, e.g. sulphuric acid or sodium, potassium, magnesium, aluminium, zinc, or iron(II) sulphate. Acidify the solution with dilute hydrochloric acid and then add barium chloride solution.

(c) To about 2 cm³ of a nitrate solution add a similar volume of iron(II) sulphate solution and of dilute sulphuric acid. Shake the mixture and then, holding the tube at an angle, *carefully and slowly* pour about 1 cm³ of concentrated sulphuric acid down the side of the tube.

(d) (i) Make dilute solutions of potassium chloride, bromide, and iodide, acidify each with dilute nitric acid, and then add silver nitrate solution. A silver halide is precipitated in each case. Now, note the effect, if any, of making each solution just alkaline with ammonia solution.

(ii) Acidify dilute solutions of potassium bromide and iodide with dilute hydrochloric acid, add approximately 1 cm³ of tetrachloromethane to each and then a few drops of sodium chlorate(I) (sodium hypochlorite). Shake both tubes thoroughly and then observe the colour of the dense organic layer.

General Questions

1 (a) Explain the term diffusion, as applied to gases, and why this phenomenon occurs.
 (b) Describe an experiment in which a dense, coloured vapour diffuses.
 (c) Describe how you could show that hydrogen diffuses faster than air.
 (d) At room temperature and pressure, one litre of nitrogen has a mass of 1.17 g whilst one litre of gas X has a mass of 4.14 g.
  (i) Calculate the volume occupied by one mole of nitrogen under these conditions.
  (ii) Calculate the relative molecular mass of gas X.

2 (a) (i) State the law of constant composition.
  (ii) Give the equations for two reactions in which lead(II) oxide is formed by thermal decomposition.
  (iii) Draw a labelled diagram showing the apparatus which can be used to reduce a sample of lead(II) oxide to lead.
  (iv) State two precautions which have to be taken in the reduction procedure.
  (v) State the measurements you would make in the experiment and explain how they would be used to verify the law.
 (b) Two oxides of cobalt (Co, relative atomic mass 59) gave the following results on analysis:
  oxide A: 2 g of cobalt were combined with 0.543 g of oxygen;
  oxide B: 3 g of cobalt were combined with 1.220 g of oxygen.
  (i) Calculate how many grams of oxygen were combined with 1 mole of cobalt in each oxide.
  (ii) State the law of multiple proportions and show how the above results verify it.
  (iii) Deduce the formulae of the two oxides.

3 (a) State three differences between mixtures and compounds.
 (b) Describe experiments by which you could obtain:
  (i) nitrogen from a mixture of nitrogen, carbon dioxide, and oxygen;
  (ii) sodium chloride from a mixture of sodium chloride, soot (carbon), and ammonium chloride.

4 (a) A salt hydrate with a relative molecular mass of 286 was found to contain 62.94% water. Calculate the number of molecules of water of crystallisation per molecule of the salt.
 (b) Calculate the solubility of copper(II) sulphate crystals at 25 °C given that 0.63 g of them dissolve in 18 g of water at this temperature.
 (c) 25 cm^3 of 0.9 M sodium hydroxide solution were found to be neutralised by 21.2 cm^3 of sulphuric acid solution. Calculate the molarity of the acid solution.

5 Describe how you would determine:
 (a) the molarity of dilute sulphuric acid using sodium hydroxide solution of known molarity.
 (b) the percentage of copper in a sample of copper(II) oxide.
  Your answers should include:
  (i) a description of the approximate experimental procedure and the precautions you would take,
  (ii) statements of the measurements you would record, together with an

outline of the necessary calculations,
(iii) the equations for the reactions that take place.

[A.E.B. Nov 1975]

6 (a) Draw a diagram of an apparatus suitable for the electrolysis of copper(II) sulphate solution using platinum electrodes, and for the collection of the products. Give the names and polarities of the electrodes, the names of the products, and the equations for the electrode reactions.
After passing a current for, say, ten minutes, what would be the effect of reversing the current and passing it in the opposite direction for about 20 minutes?

(b) In the example given above, electricity is being used to bring about a chemical change. Describe a way in which a chemical change can be used to release electrical energy. Carefully indicate the reactions which occur and where they take place.

[J.M.B.]

7 (a) How does
(i) the motion of the molecules,
(ii) the spacing between the molecules,
differ in the three states of matter, viz. solid, liquid, and vapour?

(b) How and why does atmospheric pressure affect the boiling point of a liquid?

(c) (i) An ionic solid $(X^+Y^-)_n$ may have its crystal lattice broken down by two different physical processes. Give the names of these two processes and describe, for each of them, how the necessary energy to break down the crystal lattice is obtained.
(ii) For each of the processes named in (c) (i), state with a reason, whether you would expect that process always to be endothermic.

[J.M.B.]

8 5 cm³ of 1.0 M hydrochloric acid were measured into a conical flask. 1 g of powdered zinc (excess) was added to the acid and hydrogen was evolved. The equation for the reaction is

$$Zn + 2HCl \rightarrow ZnCl_2 + H_2$$

(a) Describe an experiment by which you could study the rate of the reaction.

(b) Sketch a graph indicating the changes in volume of hydrogen observed during the experiment, continuing the graph for some minutes after the reaction has finished. Mark this graph A. Label your axes clearly and explain the reasons for the shape of your graph.

(c) Using the same axes as for (b), sketch the graph you would expect to obtain if the experiment was repeated using 1 g of granulated zinc.
Mark this graph B. Explain the reasons why graphs A and B have different shapes.

(d) Calculate the volume of hydrogen gas which should theoretically be evolved when the reaction is completed at room temperature and pressure.
(1 mole of a gas at room temperature and pressure occupies 24 litres.)

(e) Copper acts as a catalyst for the reaction. If 0.2 g of copper powder were added to the initial reaction mixture, what effect would this have on the total volume of hydrogen evolved?

[J.M.B.]

9 (a) Write three different equations (one in each case) for reactions in which the rate is increased by:
(i) ultraviolet light,
(ii) raising the temperature,
(iii) raising the pressure.
Explain why the changes in rate occur.

(b) Explain how the products from the reaction between sulphuric acid and sodium hydroxide can be varied.

10 (a) Describe in detail how, starting from copper(II) oxide, you would make a pure sample of copper(II) sulphate crystals. Write an equation for the reaction.

(b) Calculate the maximum mass of copper(II) sulphate crystals, $CuSO_4 \cdot 5H_2O$, which could be obtained from 16.0 g of copper(II) oxide.
Give two reasons why the maximum mass is never obtained.
(Relative atomic masses: H = 1; O = 16; S = 32; Cu = 64.)

(c) When copper(II) sulphate crystals are heated gently, a white solid is formed. When the white solid is heated strongly for a long time, a black solid remains. Name the white and black solids and write an equation for each reaction.

[J.M.B.]

11 (a) Explain how and why chlorine forms different types of bonds with different elements.

(b) State and explain two characteristic properties of (i) covalent, (ii) ionic compounds.

(c) Explain why magnesium and calcium (atomic numbers 12 and 20 respectively) have similar properties. Give equations for two reactions which illustrate the similarity.

12 (a) Give an account of the experiments you would carry out to demonstrate the conditions under which iron rusts.

(b) Explain why zinc protects iron from corrosion and suggest why copper-plating would not have the same effect.

(c) Why does aluminium corrode much less rapidly than iron when exposed to the same atmospheric conditions?

(d) Give three important uses for aluminium, each illustrating a different physical property of the metal.

[A.E.B. Nov 1974]

13 (a) (i) Outline, with essential details, how you would prepare a pure dry sample of zinc sulphate heptahydrate, $ZnSO_4 \cdot 7H_2O$, starting from zinc carbonate and dilute sulphuric acid.
(ii) Calculate the maximum theoretical mass of the crystals which could be obtained from 40 cm³ of 2 M sulphuric acid.

(b) (i) Write equations showing what happens at the electrodes when an electric current is passed through dilute sulphuric acid using platinum electrodes.
(ii) Calculate the volume of the gas, measured at s.t.p., which would be liberated at the anode by the passage of 0.2 Faraday.
(Molar volume = 22.4 litres at s.t.p.)

(c) In a titration, 25.0 cm³ of a sodium hydroxide solution were found to be neutralised by 24.3 cm³ of 0.057 M sulphuric acid. Calculate the molarity of the sodium hydroxide and hence its concentration in g/$^{-1}$.

14 (a) Name the product at each electrode and write equations for the reactions which occur during the electrolysis of copper(II) sulphate solution using platinum electrodes.

(b) Calculate the mass of each product of electrolysis if the current were stopped after the passage of 0.01 Faraday. (Cu = 64; O = 16.)

(c) Describe what would be seen at each electrode if the direction of the current were then reversed and the current allowed to flow until a further 0.02 faraday had passed. Write an equation for any reaction which occurs in (c) but which does not occur in (a).

(d) Outline a method of obtaining copper from copper(II) sulphate in which electrolysis is not used.

[J.M.B.]

General Questions 271

15 Describe and explain, with equations where relevant, the observations which could be made if the following were heated:
(i) a mixture of iodine and common salt,
(ii) sodium hydrogencarbonate,
(iii) copper(II) nitrate crystals,
(iv) dinitrogen tetraoxide. [J.M.B.]

16 Explain the following, giving equations for the reactions involved.
(a) Addition of 25 cm^3 of 2 M sulphuric acid to 50 cm^3 of 1 M potassium hydroxide gives a solution from which a pure crystalline product can be isolated. However, if 25 cm^3 of 2 M sulphuric acid is added to 50 cm^3 of 0.5 M potassium hydroxide, a different product is obtained.
(b) A reaction occurs when copper is added to silver nitrate solution.
(c) A white precipitate is formed when aqueous ammonia is slowly added to zinc sulphate solution but this precipitate redissolves if excess of the reagent is added.
(d) Despite its high position in the electrochemical and activity series, aluminium needs to be strongly heated in steam before it reacts with it.

17 Consider the following changes:
(a) copper(II) oxide to copper.
(b) ammonium chloride to ammonia,
(c) iron(III) chloride to iron(II) chloride,
(d) sodium sulphite to sulphur dioxide,
(e) potassium iodide to iodine.

In each case:
(i) name a reagent which will bring about this change,
(ii) state whether the substance underlined is oxidised, reduced, or undergoes neither of these processes,
(iii) write an equation for the reaction. [J.M.B.]

18 (a) The sentences below describe how a salt can be prepared. Name the salt which would be prepared following these instructions and give a reason why each of the words or phrases printed in italics is essential for the preparation of a pure crystalline salt.
An excess of slightly impure zinc carbonate was added to *dilute* sulphuric acid. When the reaction was complete, the mixture was *filtered*. The filtrate was *partially evaporated*, allowed to cool and crystallise. *Before crystallisation was complete, the liquid was poured off* and the crystals were dried *on filter paper rather than by warming*.
(b) Describe what would be seen on adding solutions of:
(i) barium chloride,
(ii) sodium hydroxide,
to separate solutions of the salt named in (a) until no further reaction occurred.
(c) Give an equation for, or name the products of, each reaction. [J.M.B.]

19 Describe briefly how you would separate a pure sample of the first named substance from the impurity in each of the following mixtures.
(i) Magnesium sulphate from a mixture of powdered glass.
(ii) A solution of an orange dye which, together with a blue dye, forms black ink.
(iii) Copper filings from a mixture with magnesium filings.
(iv) Oxygen from a mixture with chlorine.
(v) Ammonium chloride from a mixture with sodium chloride. [J.M.B.]

20 (a) Describe how you would obtain samples of:
  (i) nitrogen from ammonia,
  (ii) copper from copper(II) sulphate crystals.
  Interpret the reactions involved in terms of oxidation-reduction.

(b) Explain the following.
  (i) An aqueous solution of ammonia is alkaline, but dry ammonia has no effect on dry red litmus paper.
  (ii) Addition of a little aqueous ammonia to a solution containing copper(II) ions causes the formation of a pale blue precipitate, but when excess ammonia is added, a different product is obtained.
  [A.E.B. June 1974]

21 Describe and explain experiments by which you could obtain:
  (i) hydrogen from hydrogen contaminated with hydrogen sulphide,
  (ii) copper from a mixture of copper powder and zinc oxide,
  (iii) pure water from copper(II) sulphate solution,
  (iv) sulphur from a mixture of sulphur and carbon.
  [A.E.B. Nov 1974]

22 Suppose you were given a solution suspected of containing potassium sulphate.
  (i) Describe how you would confirm the presence of potassium ions and sulphate ions in the solution.
  (ii) State how you would decide whether the solution was saturated at the prevailing temperature.
  (iii) Assuming that the solution was saturated, describe how you would use it to determine the solubility of potassium sulphate at that temperature. State how you would calculate the solubility from your results.

23 The atomic numbers of some elements are given below.

Element:	hydrogen	carbon	sodium	chlorine
Atomic number	1	6	11	17

(a) Give the electron structures of carbon, sodium, and chlorine atoms.
(b) State how sodium and chlorine form ions and why these ions are stable.
(c) Give a diagram showing how the atoms are bonded in one named covalent compound.
(d) Briefly explain the following.
  (i) Crystals of sodium chloride have a regular shape.
  (ii) Sodium chloride in the solid state does not conduct electricity, but it will conduct when molten and it is decomposed by the electric current.
  (iii) Molecules of hydrogen chloride dissolve in water by an exothermic process.
  (iv) The melting point of diamond is extremely high.
  [A.E.B. Nov 1976]

24 The halogens have the following boiling points: fluorine (−188 °C), chlorine (−34 °C), bromine (+59 °C), and iodine (+184 °C). The heats of dissociation of the molecules are:

$$F_2 \rightarrow 2F \quad \Delta H = +318 \text{ kJ mol}^{-1}$$
$$Cl_2 \rightarrow 2Cl \quad \Delta H = +234 \text{ kJ mol}^{-1}$$
$$Br_2 \rightarrow 2Br \quad \Delta H = +188 \text{ kJ mol}^{-1}$$
$$I_2 \rightarrow 2I \quad \Delta H = +146 \text{ kJ mol}^{-1}$$

(a) State how the stabilities of the halogen molecules vary.
(b) At 1700 °C one halogen only is completely dissociated into atoms. Name this element.
(c) Both the boiling points and the heats of dissociation show a progressive change through the family of halogens. Show that this pattern of change is also followed in their reactivity with:
  (i) hydrogen,

General Questions

(ii) metals,
(iii) in the displacement of halogens from halides.
(d) How does chlorine react with sodium hydroxide solution?

[J.M.B.]

25     $_{11}N$ $_{12}Mg$ $_{13}Al$ $_{14}Si$ $_{15}P$ $_{16}S$ $_{17}Cl$ $_{18}Ar$

The above is a list of elements and their atomic numbers. By reference to the structure of the atoms, explain why,
(i) magnesium and sulphur both have a valency of two,
(ii) argon is chemically unreactive,
(iii) atoms of chlorine can have different masses,
(iv) sodium forms compounds which are usually electrolytes whilst silicon forms compounds which are usually non-electrolytes,
(v) an aluminium ion, Al^{3+}, is smaller than an aluminium atom but a phosphide ion, P^{3-}, is bigger than a phosphorus atom.

[J.M.B.]

26   Explain the following observations, with the aid of equations where possible.
(a) When dilute nitric acid is added to metallic copper and the reagents are warmed, a colourless gas is evolved which turns brown near the mouth of the tube. A blue solution is formed.
(b) When carbon dioxide is bubbled through limewater, a white precipitate forms at first but, on passing excess carbon dioxide, the solution clears. On boiling this clear solution, the white precipitate forms once more.
(c) Iron filings and sulphur were mixed together and the resulting grey/yellow mixture was split into two portions A and B.
To a sample of A was added dilute hydrochloric acid and a gas was evolved which extinguished a lighted splint with a squeaky pop.
Portion B was heated strongly and allowed to cool. Upon addition of dilute hydrochloric acid to the solid remaining, a gas was given off which had an obnoxious odour and which burned with a blue flame when a lighted splint was applied.
(d) When copper(II) sulphate solution is electrolysed using platinum electrodes, oxygen gas is evolved at the anode and the solution gradually becomes less blue. If, however, copper electrodes are used in place of platinum, no oxygen is liberated and the blue colour remains constant in intensity.

[J.M.B.]

27 (a) Describe how you would prepare a pure, dry sample of ammonium chloride from dilute hydrochloric acid and any other necessary reagent. Indicate the precautions needed to obtain the best possible yield of the product. Give an equation for the reaction.
(b) Describe experiments, one in each case, to show that a solution of ammonium chloride contains:
(i) ammonium ions,
(ii) chloride ions.
(c) Calculate the theoretical maximum mass of ammonium chloride which could be obtained by starting with 50 cm³ of 2 M (2 N) hydrochloric acid.
(Relative atomic masses: H = 1; N = 14; Cl = 35.5.)

[J.M.B.]

28 (a) Describe the essential features of an industrial process by which chlorine and sodium hydroxide are manufactured electrolytically. Name the electrolyte, give the polarity of the electrodes, state the materials of which they are made, and write equations for the discharge of the ions at the electrodes. Your account should state clearly how and where each product is obtained and should indicate how these two products are kept separate.
(b) Name the products and write an equation for a reaction between chlorine and sodium hydroxide solution.
(c) State two large-scale uses of chlorine.

[J.M.B.]

29 The formation of sulphur(VI) oxide can be represented by the equation
$$2SO_2 + O_2 \rightleftharpoons 2SO_3; \quad \Delta H = -94.5 \text{ kJ mol}^{-1}$$
(a) What would be the effect on the rate of reaction of sulphur dioxide with oxygen of increasing:
  (i) the temperature,
  (ii) the pressure?
(b) Describe one way in each case to illustrate sulphuric acid reacting as:
  (i) an acid,
  (ii) a dehydrating agent,
  (iii) an oxidising agent.
State the conditions, describe what would be observed, and write an equation for each reaction.
[J.M.B.]

30 (a) Draw a fully labelled diagram for the laboratory preparation and collection of carbon dioxide. Give the equation.
(b) State, giving reagents and reaction conditions, how carbon dioxide can be reduced to carbon monoxide. Give an equation for the reaction.
(c) Describe carefully how, by a chemical test, you would distinguish between solid samples of hydrated sodium carbonate and sodium hydrogencarbonate. Give the result of the test on each substance.
(d) When carbon monoxide and chlorine are passed over a heated catalyst, gaseous carbonyl chloride is formed.
$$CO(g) + Cl_2(g) \rightarrow COCl_2(g)$$
(i) If 20 cm³ of carbon monoxide were reacted with 20 cm³ of chlorine, what would be the maximum volume of carbonyl chloride formed if all measurements were made at the same temperature and pressure?
(ii) Using lines to represent covalent bonds, give the probable structural formula of carbonyl chloride.
[J.M.B.]

31 (a) (i) Name two reagents which are mixed together in the laboratory preparation of hydrogen chloride.
  (ii) State the conditions necessary for the reaction.
  (iii) Write an equation for the reaction.
(b) Describe experiments, one in each case, to demonstrate that hydrogen chloride contains:
  (i) hydrogen,
  (ii) chlorine.
(c) At high temperatures hydrogen and chlorine are in equilibrium with hydrogen chloride as represented by the equation
$$H_2(g) + Cl_2(g) \rightleftharpoons 2HCl(g); \quad \Delta H = -185 \text{ kJ}.$$
State and explain the effect on the position of equilibrium of:
  (i) lowering the temperature at constant pressure,
  (ii) adding a catalyst at constant temperature and pressure,
  (iii) increasing the pressure at constant temperature.
[J.M.B.]

32 Suggest a name for each substance indicated by a letter and write an equation for each reaction described below.
(a) When a white crystalline powder, A, was warmed with sodium hydroxide solution, a gas B was evolved which turned damp red litmus paper blue. On addition of a barium chloride solution to a solution of A in dilute hydrochloric acid a white precipitate C was seen.
(b) When a dull grey metal, D, was added to cold water hydrogen was evolved and a white suspension, E, was formed. When the reaction mixture was filtered and carbon dioxide was passed through the filtrate a second white suspension, F, was seen.

General Questions 275

(c) When a red-orange powder, *G*, was warmed with an excess of dilute nitric acid, a brown precipitate, *H*, was formed. When *H* was warmed with concentrated hydrochloric acid, a gas, *I*, was evolved which bleached moist litmus paper.

[J.M.B.]

33   Describe what you would see and explain the reactions which take place during the following experiments.

(a) Dry chlorine is passed through a combustion tube containing heated iron wool.
(b) A piece of calcium is placed in cold water.
(c) Ammonium chloride is gently heated in a test tube.
(d) Sodium hydroxide solution is added to a solution of zinc sulphate until no further change occurs.
(e) Concentrated sulphuric acid is warmed with potassium nitrate in a test tube.

[A.E.B. Nov 1974]

34   Identify each substance indicated by a letter and write an equation for each reaction described below.

(a) An orange-red solid *A* decomposed on heating giving oxygen and a solid residue which was yellow when cold. When another sample of *A* was warmed with an excess of dilute nitric acid, a brown precipitate *B* and a clear colourless solution *C* were formed.
(b) When chlorine was passed into a solution of a gas *D*, a pale yellow precipitate *E* was formed. On igniting a dry sample of *E* in air, it burned forming a gas *F*.
(c) When an excess of zinc was added to a blue solution *G*, a red-brown precipitate *H* was formed and the solution became colourless. The addition of silver nitrate solution to this colourless solution resulted in the formation of a white precipitate *J*.

[J.M.B.]

35   (a) When 1.0 g of powdered manganese(IV) oxide was shaken with 25 cm^3 of hydrogen peroxide solution, a reaction took place which was completed in 1 minute. State three different ways in which the reaction could be slowed down.

(b) It is said that copper(II) oxide will catalyse the decomposition of hydrogen peroxide. Outline the experiments which you would make to test this statement. Include in your answer details of the measurements you would make and the results you would expect if the oxide were a catalyst.

(c) Explain why catalysts are important in manufacturing processes. Illustrate your answer with two examples from industrial chemistry, naming the catalyst in, and writing an equation for, each reaction.

[J.M.B.]

36   Give the names of substances indicated by letters and write an equation for each reaction which occurs.

(a) A mixture of manganese(IV) oxide (manganese dioxide) and copper(II) sulphate was stirred in water. The mixture was filtered and a residue, *A*, was left in the filter paper.
After washing, *A* was added to hydrogen peroxide solution and a gas, *B*, was evolved. When sodium hydroxide solution was added to the filtrate from the first stage, a blue solid, *C*, was precipitated.

(b) A mixture of powdered sulphur and powdered magnesium was treated with an excess of dilute hydrochloric acid. When the reaction was complete, the mixture was filtered.
The residue was washed and then heated in air when it burned forming a gas, *D*. The filtrate, *E*, on addition of lead(II) nitrate solution, gave a white precipitate, *F*.

(c) On heating a white solid, *G*, nitrogen dioxide and oxygen were given off and a yellow solid, *H*, which turned white on cooling, remained. On the addition of dilute sulphuric acid to *H*, a colourless solution, *J*, was formed.

[J.M.B.]

37 Explain the following statements.
(a) Aqueous hydrogen peroxide should be stored in a cool, dark place.
(b) It is dangerous to use an appliance burning natural gas (methane, CH_4) in a badly ventilated room.
(c) Although atmospheric oxygen is used by living organisms, the proportion of this gas in air is fairly constant.
(d) It would be unwise to use an iron vessel for the storage of copper(II) sulphate solution.
(e) Sodium chloride crystals have a regular shape.

[A.E.B. Nov 1975]

38 (a) Explain the reaction between iron(III) oxide and aluminium in terms of oxidation and reduction, giving the relevant half equations.
(b) A gas contains 12.1% carbon, 16.2% oxygen, and 71.7% chlorine by mass and its density at room temperature and pressure is 4.125 g per litre. Calculate:
(i) the empirical formula,
(ii) its relative molecular mass,
(iii) the molecular formula.
(1 mole of gas at room temperature and pressure occupies 24 litres.)
(c) Give one equation in each case for sulphuric acid acting as
(i) an acid,
(ii) a dehydrating agent,
(iii) an oxidising agent.

39 Explain the meaning of each of the following terms, illustrating your answer with suitable examples in each case.
(i) Thermal decomposition.
(ii) Chemical equilibrium.
(iii) Isotopes.
(iv) Structural isomers.
(v) Hydrolysis.

[J.M.B.]

40 (a) Explain the following, giving equations for the reactions involved.
(i) Hydrogen chloride reacts with water forming a solution that is strongly acidic but the solution obtained by dissolving ethanoic (acetic) acid in water has weak acidic properties.
(ii) Copper reacts with hot concentrated sulphuric acid but is unaffected by the dilute acid.
(iii) By evaporating the solution obtained on neutralising sodium hydroxide solution with dilute sulphuric acid, it is possible to isolate a crystalline solid. By changing the relative volumes of the two reagents a different solid product can be obtained.
(b) Scandium (symbol Sc) reacts with hydrochloric acid releasing hydrogen:

$$2Sc + 6HCl \rightarrow 2ScCl_3 + 3H_2$$

It was found that 3 grams of scandium reacted exactly with 20 cm³ of 10 M hydrochloric acid. Calculate the mass of scandium which reacts with 6 moles of hydrochloric acid. Then use the equation to deduce the relative atomic mass of scandium.

[A.E.B. June 1977]

41 Describe how the following conversions are carried out by naming the reagents if any, stating the conditions, and writing an equation for each reaction.
(i) Hydrogen sulphide to sulphur dioxide.
(ii) Sodium chloride to sodium.
(iii) Iron(III) oxide to iron.
(iv) Sugar to ethanol (ethyl alcohol).
(v) Nitrogen dioxide to nitric acid.

[J.M.B.]

General Questions

42 This question is concerned with sodium hydroxide and with its manufacture from salt (sodium chloride) by an electrolytic process.
- (a) (i) Sodium chloride is said to be an *electrolyte*. What do you understand by this term?
  - (ii) In what form is salt used in this process?
  - (iii) Name the materials used for the electrodes.
  - (iv) Give the changes which take place at the cathode and outline how the sodium hydroxide is obtained.
- (b) For what reason is sodium hydroxide used in the manufacture of soaps from oils and fats?
- (c) Briefly describe how you would use sodium hydroxide solution to distinguish:
  - (i) ammonium chloride from sodium chloride,
  - (ii) an aqueous solution containing magnesium ions, Mg^{2+}, from one containing zinc ions, Zn^{2+}.

[A.E.B. June 1977]

43 In the following industrial processes, explain briefly, giving one relevant equation in each section, the use of;
- (i) electricity in the manufacture of sodium from sodium chloride,
- (ii) pressure in the manufacture of ammonia from nitrogen,
- (iii) heat in the formation of sulphur(VI) oxide from sulphur dioxide (this is an exothermic process),
- (iv) a catalyst in the manufacture of nitric acid starting from ammonia,
- (v) ethene in the formation of ethanol.

[J.M.B.]

44 Give one chemical test in each case by which you could distinguish between the two substances in each of the following pairs. Give the results of each test on *both* members of the pair.
- (i) Copper(II) oxide and manganese(IV) oxide (manganese dioxide).
- (ii) Ethane and ethene.
- (iii) Ammonium chloride and aluminium chloride.
- (iv) Ethanol (ethyl alcohol) and tetrachloromethane (carbon tetrachloride).
- (v) Solutions of magnesium chloride and magnesium nitrate.

[J.M.B.]

45 For each of the following mixtures, describe and explain what happens when it is heated.
- (a) Copper and concentrated sulphuric acid.
- (b) Silicon(IV) oxide and sodium sulphate.
- (c) Ammonia and copper(II) oxide.
- (d) Ethanol and concentrated sulphuric acid.
- (e) Ammonium sulphate and sodium hydroxide.

46
- (a) Describe, by means of a labelled diagram and an equation, how you would prepare and collect a dry sample of ammonia starting with ammonium sulphate.
- (b) Explain the reaction which occurs when ammonia dissolves in water causing the resulting solution to have a pH value greater than 7. Give an equation or equations.
- (c) Describe and explain with equations, the observations you would make when ammonia reacts with:
  - (i) copper(II) sulphate, and
  - (ii) hydrogen chloride.

[J.M.B.]

47
- (a) Explain, with examples, the following terms:
  - (i) structural isomerism,
  - (ii) unsaturated hydrocarbon.

(b) A gaseous hydrocarbon with relative density 14 was found to contain 85.7% carbon and 14.3% hydrogen by mass. Calculate:
  (i) the empirical formula,
  (ii) the relative molecular mass,
  (iii) the molecular formula of the gas.
Give the equations and conditions for the reaction of the hydrocarbon with hydrogen and with hydrogen bromide.

(c) Give the equation for the combustion of methane in air. What volume of oxygen would be required for 60 cm³ of methane to undergo complete combustion, all volumes being measured at the same temperature and pressure. Explain your answer.

48 From the information given below, identify A, B, C, and D explaining the observations and giving equations wherever possible.

(a) Addition of a colourless liquid A to copper causes the liberation of a colourless gas that turns brown on exposure to the atmosphere. The brown gas dissolves in aqueous sodium hydroxide.

(b) A white solid B reacts with water, evolving much heat. If enough water is added the resultant solid dissolves. When carbon dioxide is passed into this solution, a white precipitate is obtained.

(c) A black solid C dissolves in hydrochloric acid, evolving a foul-smelling gas which forms a black precipitate when bubbled into lead(II) nitrate solution. An aqueous solution of the gas slowly turns cloudy on standing.

(d) A pure colourless liquid D, when mixed with water, gives a non-conducting solution. On warming a mixture of D with ethanoic (acetic) acid and a few drops of concentrated sulphuric acid, a vapour with a fruity smell is obtained.

[A.E.B. June 1974]

49 (a) What do you understand by the term *allotrope*?

(b) Show how a knowledge of the crystal structures of diamond and graphite can be used to explain the difference in their hardness.

(c) The empirical formula of a gaseous hydrocarbon, X, is CH_2 and its density is the same as that of nitrogen under the same conditions. When X was passed into bromine water, the latter lost its colour and a sweet-smelling product was obtained. Under suitable conditions X can be converted to a white solid which also has the empirical formula CH_2. Identify X and explain these observations. State briefly how X may be obtained from ethanol.

[A.E.B. June 1975]

50 (a) State one compound of carbon which is a gas and one which is a high melting solid. Explain the difference in melting points of your compounds.

(b) A gaseous hydrocarbon was found to contain 82.76% carbon and 17.24% hydrogen. The mass of 140 cm³ of the gas at s.t.p. was found to be 0.363 g. Calculate the empirical and molecular formulae of the gas.
(Molar volume is 22.4 litres at s.t.p.)

(c) Explain how a solution of ethanol can be obtained, by fermentation, from glucose ($C_6H_{12}O_6$). State what you would see during the reaction and give the relevant equation.

RELATIVE ATOMIC MASSES (ATOMIC WEIGHTS)

Element	Symbol	Atomic number	Relative atomic mass	Element	Symbol	Atomic number	Relative atomic mass
Actinium	Ac	89	227.0	Mercury	Hg	80	200.6
Aluminium	Al	13	26.9	Molybdenum	Mo	42	95.9
Americium	Am	95	243.0	Neodymium	Nd	60	144.2
Antimony	Sb	51	121.8	Neon	Ne	10	20.2
Argon	Ar	18	39.9	Neptunium	Np	93	237.0
Astatine	At	85	210.0	Nickel	Ni	28	58.7
Arsenic	As	33	74.9	Niobium	Nb	41	92.9
Barium	Ba	56	137.3	Nitrogen	N	7	14.0
Berkelium	Bk	97	249.0	Osmium	Os	76	190.2
Beryllium	Be	4	9.0	Oxygen	O	8	16.0
Bismuth	Bi	83	209.0	Palladium	Pd	46	106.4
Boron	B	5	10.8	Phosphorus	P	15	31.0
Bromine	Br	35	79.9	Platinum	Pt	78	195.1
Cadmium	Cd	48	112.4	Plutonium	Pu	94	242.0
Caesium	Cs	55	132.9	Polonium	Po	84	210.0
Calcium	Ca	20	40.1	Potassium	K	19	39.1
Californium	Cf	98	251.0	Praseodymium	Pr	59	140.9
Carbon	C	6	12.0	Promethium	Pm	61	145.0
Cerium	Ce	58	140.1	Protactinium	Pa	91	231.0
Chlorine	Cl	17	35.5	Radium	Ra	88	226.1
Chromium	Cr	24	52.0	Radon	Rn	86	222.0
Cobalt	Co	27	58.9	Rhenium	Re	75	186.2
Copper	Cu	29	63.5	Rhodium	Rh	45	102.9
Curium	Cm	96	247.0	Rubidium	Rb	37	85.5
Dysprosium	Dy	66	162.5	Ruthenium	Ru	44	101.1
Einsteinium	Es	99	254.0	Samarium	Sm	62	150.4
Erbium	Er	68	167.3	Scandium	Sc	21	45.0
Europium	Eu	63	152.0	Selenium	Se	34	79.0
Fermium	Fm	100	253.0	Silicon	Si	14	28.1
Fluorine	F	9	19.0	Silver	Ag	47	107.9
Francium	Fr	87	223.0	Sodium	Na	11	23.0
Gadolinium	Gd	64	157.3	Strontium	Sr	38	87.6
Gallium	Ga	31	69.7	Sulphur	S	16	32.1
Germanium	Ge	32	72.6	Tantalum	Ta	73	180.9
Gold	Au	79	197.0	Technetium	Tc	43	99.0
Hafnium	Hf	72	178.5	Tellurium	Te	52	127.6
Helium	He	2	4.0	Terbium	Tb	65	158.9
Holmium	Ho	67	164.9	Thallium	Tl	81	204.4
Hydrogen	H	1	1.0	Thorium	Th	90	232.0
Indium	In	49	114.8	Thulium	Tm	69	168.9
Iodine	I	53	126.9	Tin	Sn	50	118.7
Iridium	Ir	77	192.2	Titanium	Ti	22	47.9
Iron	Fe	26	55.8	Tungsten	W	74	183.9
Krypton	Kr	36	83.8	Uranium	U	92	238.0
Lanthanum	La	57	138.9	Vanadium	V	23	50.9
Lead	Pb	82	207.2	Xenon	Xe	54	131.3
Lithium	Li	3	6.9	Ytterbium	Yb	70	173.0
Lutetium	Lu	71	175.0	Yttrium	Y	39	88.9
Magnesium	Mg	12	24.3	Zinc	Zn	30	65.4
Manganese	Mn	25	54.9	Zirconium	Zr	40	91.2

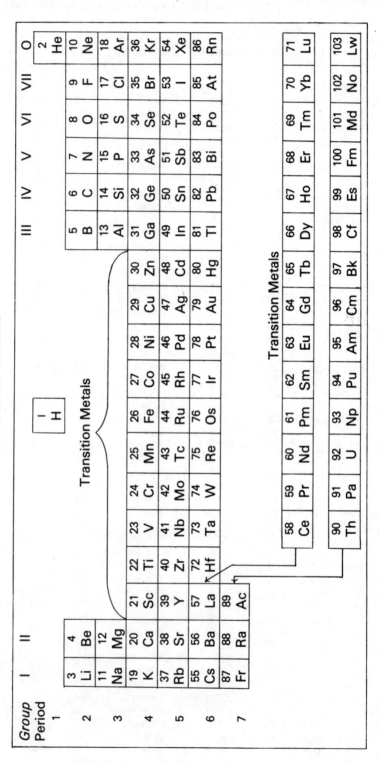

Answers to Numerical Questions

Chapter 2

4	557.3 cm^3
5	177.1 cm^3
6	1086 mm

Chapter 4

5 (b)	$x = 1$	

Chapter 5

2 (c) (v)	PbO$_2$	
(d) (ii)	PbO	
3	6.28 g	
4	15.53 g	
5	0.889 litre	
6	3.27 g	
7	2.419 litres	

Chapter 6

1 (a)	0.15 M
(b)	0.125 M; 7.875 gl^{-1}
2 (b)	0.0943 M
3 (a)	56; 39
4 (b)	0.0833 M
5 (c) (i)	0.2 M
(ii)	0.216 M
(iii)	100
(iv)	39

Chapter 7

1 (e)	60 cm^3
2 (d)	0.224 litre; 0.112 litre
(e)	16 minutes 5 seconds (965 s)
(f)	0.01 mole
3 (b)	4.5 g
4 (c)	0.05 mole

Chapter 9

4 (b)	1500 cm^3

Chapter 10

1 (b) (i)	−56.4 kJ
(ii)	−112.8 kJ

Chapter 11

4 (d) (i)	200 cm^3
(ii)	24.63 g
5 (c)	13.24 g Pb(NO$_3$)$_2$; 4.78 g PbO$_2$
8	699.2 kg
13	50.21 g
19	−4.34 kJ

Chapter 12

3 (a)	100 cm^3 increase
(d)	8.0 g
6 (a) (i)	−16000 kJ
10 (c) (i)	2.5 g
(iii)	150 cm^3
13 (f) (ii)	2.4 litre H$_2$S; 9.55 g CuS
18 (c)	Cl$_2$O

Chapter 13

1 (a)	0.003 mole; 0.006 mole; 0.003 mole
(b)	CH$_2$Br
(c)	56 cm^3
(d)	188
(e)	C$_2$H$_4$Br$_2$
7	50 cm^3

General Questions

1 (d) (i)	23.93 litres
(ii)	99
2 (b) (i)	Oxide A 16 g, Oxide B 24 g
(ii)	A = CoO, B = Co$_2$O$_3$
4 (a)	10
(b)	35.0 g kg^{-1} of water
(c)	0.53 M
8 (d)	0.06 litres
10 (b)	50 g
13 (a) (ii)	22.99 g
(b) (ii)	1.12 litres
(c)	0.1108 M, 4.432 gl^{-1}
14 (b)	0.32 g Cu
	0.08 g O$_2$

27 (c)   5.35 g
38 (b) (i) $COCl_2$
    (ii)   99
    (iii) $COCl_2$
40 (b)   90 g Sc
      45

47 (b) (i) CH_2
    (ii)   28
    (iii) C_2H_4
  (c)   120 cm³
50 (b) C_2H_5, C_4H_{10}

Logarithms

Logarithms

	0	1	2	3	4	5	6	7	8	9	1	2	3	4	5	6	7	8	9
10	0000	0043	0086	0128	0170						4	8	13	17	21	25	30	34	38
						0212	0253	0294	0334	0374	4	8	12	16	20	24	28	32	36
11	0414	0453	0492	0531	0569						4	8	12	15	19	23	27	31	35
						0607	0645	0682	0719	0755	4	7	11	15	18	22	26	30	33
12	0792	0828	0864	0899	0934						4	7	11	14	18	21	25	28	32
						0969	1004	1038	1072	1106	3	7	10	14	17	20	24	27	31
13	1139	1173	1206	1239	1271						3	7	10	13	16	20	23	26	30
						1303	1335	1367	1399	1430	3	6	9	13	16	19	22	25	28
14	1461	1492	1523	1553	1584						3	6	9	12	15	18	21	24	27
						1614	1644	1673	1703	1732	3	6	9	12	15	18	21	24	27
15	1761	1790	1818	1847	1875						3	6	9	11	14	17	20	23	26
						1903	1931	1959	1987	2014	3	6	8	11	14	17	19	22	25
16	2041	2068	2095	2122	2148						3	5	8	11	13	16	19	21	24
						2175	2201	2227	2253	2279	3	5	8	10	13	16	18	21	23
17	2304	2330	2355	2380	2405						3	5	8	10	13	15	18	20	23
						2430	2455	2480	2504	2529	2	5	7	10	12	15	17	20	22
18	2553	2577	2601	2625	2648						2	5	7	10	12	14	17	19	21
						2672	2695	2718	2742	2765	2	5	7	9	12	14	16	19	21
19	2788	2810	2833	2856	2878						2	5	7	9	11	14	16	18	20
						2900	2923	2945	2967	2989	2	4	7	9	11	13	15	18	20
20	3010	3032	3054	3075	3096	3118	3139	3160	3181	3201	2	4	6	8	11	13	15	17	19
21	3222	3243	3263	3284	3304	3324	3345	3365	3385	3404	2	4	6	8	10	12	14	16	18
22	3424	3444	3464	3483	3502	3522	3541	3560	3579	3598	2	4	6	8	10	12	14	15	17
23	3617	3636	3655	3674	3692	3711	3729	3747	3766	3784	2	4	6	7	9	11	13	15	17
24	3802	3820	3838	3856	3874	3892	3909	3927	3945	3962	2	4	5	7	9	11	12	14	16
25	3979	3997	4014	4031	4048	4065	4082	4099	4116	4133									
26	4150	4166	4183	4200	4216	4232	4249	4265	4281	4298	2	3	5	7	9	10	12	14	15
27	4314	4330	4346	4362	4378	4393	4409	4425	4440	4456	2	3	5	7	8	10	11	13	14
28	4472	4487	4502	4518	4533	4548	4564	4579	4594	4609	2	3	5	6	8	9	11	13	14
29	4624	4639	4654	4669	4683	4698	4713	4728	4742	4757	1	3	4	6	7	9	10	12	13
30	4771	4786	4800	4814	4829	4843	4857	4871	4886	4900	1	3	4	6	7	9	10	11	13
31	4914	4928	4942	4955	4969	4983	4997	5011	5024	5038	1	3	4	6	7	8	10	11	12
32	5051	5065	5079	5092	5105	5119	5132	5145	5159	5172	1	3	4	5	7	8	9	11	12
33	5185	5198	5211	5224	5237	5250	5263	5276	5289	5302	1	3	4	5	6	8	9	10	12
34	5315	5328	5340	5353	5366	5378	5391	5403	5416	5428	1	3	4	5	6	8	9	10	11
35	5441	5453	5465	5478	5490	5502	5514	5527	5539	5551	1	2	4	5	6	7	9	10	11
36	5563	5575	5587	5599	5611	5623	5635	5647	5658	5670	1	2	4	5	6	7	8	10	11
37	5682	5694	5705	5717	5729	5740	5752	5763	5775	5786	1	2	3	5	6	7	8	9	10
38	5798	5809	5821	5832	5843	5855	5866	5877	5888	5899	1	2	3	5	6	7	8	9	10
39	5911	5922	5933	5944	5955	5966	5977	5988	5999	6010	1	2	3	4	5	7	8	9	10
40	6021	6031	6042	6053	6064	6075	6085	6096	6107	6117	1	2	3	4	5	6	8	9	10
41	6128	6138	6149	6160	6170	6180	6191	6201	6212	6222	1	2	3	4	5	6	7	8	9
42	6232	6243	6253	6263	6274	6284	6294	6304	6314	6325	1	2	3	4	5	6	7	8	9
43	6335	6345	6355	6365	6375	6385	6395	6405	6415	6425	1	2	3	4	5	6	7	8	9
44	6435	6444	6454	6464	6474	6484	6493	6503	6513	6522	1	2	3	4	5	6	7	8	9
45	6532	6542	6551	6561	6571	6580	6590	6599	6609	6618	1	2	3	4	5	6	7	8	9
46	6628	6637	6646	6656	6665	6675	6684	6693	6702	6712	1	2	3	4	5	6	7	7	8
47	6721	6730	6739	6749	6758	6767	6776	6785	6794	6803	1	2	3	4	5	5	6	7	8
48	6812	6821	6830	6839	6848	6857	6866	6875	6884	6893	1	2	3	4	4	5	6	7	8
49	6902	6911	6920	6928	6937	6946	6955	6964	6972	6981	1	2	3	4	4	5	6	7	8

Logarithms

	0	1	2	3	4	5	6	7	8	9	1	2	3	4	5	6	7	8	9
50	6990	6998	7007	7016	7024	7033	7042	7050	7059	7067	1	2	3	3	4	5	6	7	8
51	7076	7084	7093	7101	7110	7118	7126	7135	7143	7152	1	2	3	3	4	5	6	7	8
52	7160	7168	7177	7185	7193	7202	7210	7218	7226	7235	1	2	2	3	4	5	6	7	7
53	7243	7251	7259	7267	7275	7284	7292	7300	7308	7316	1	2	2	3	4	5	6	6	7
54	7324	7332	7340	7348	7356	7364	7372	7380	7388	7396	1	2	2	3	4	5	6	6	7
55	7404	7412	7419	7427	7435	7443	7451	7459	7466	7474	1	2	2	3	4	5	5	6	7
56	7482	7490	7497	7505	7513	7520	7528	7536	7543	7551	1	2	2	3	4	5	5	6	7
57	7559	7566	7574	7582	7589	7597	7604	7612	7619	7627	1	2	2	3	4	5	5	6	7
58	7634	7642	7649	7657	7664	7672	7679	7686	7694	7701	1	1	2	3	4	4	5	6	7
59	7709	7716	7723	7731	7738	7745	7752	7760	7767	7774	1	1	2	3	4	4	5	6	7
60	7782	7789	7796	7803	7810	7818	7825	7832	7839	7846	1	1	2	3	4	4	5	6	6
61	7853	7860	7868	7875	7882	7889	7896	7903	7910	7917	1	1	2	3	4	4	5	6	6
62	7924	7931	7938	7945	7952	7959	7966	7973	7980	7987	1	1	2	3	3	4	5	6	6
63	7993	8000	8007	8014	8021	8028	8035	8041	8048	8055	1	1	2	3	3	4	5	5	6
64	8062	8069	8075	8082	8089	8096	8102	8109	8116	8122	1	1	2	3	3	4	5	5	6
65	8129	8136	8142	8149	8156	8162	8169	8176	8182	8189	1	1	2	3	3	4	5	5	6
66	8195	8202	8209	8215	8222	8228	8235	8241	8248	8254	1	1	2	3	3	4	5	5	6
67	8261	8267	8274	8280	8287	8293	8299	8306	8312	8319	1	1	2	3	3	4	5	5	6
68	8325	8331	8338	8344	8351	8357	8363	8370	8376	8382	1	1	2	3	3	4	4	5	6
69	8388	8395	8401	8407	8414	8420	8426	8432	8439	8445	1	1	2	2	3	4	4	5	6
70	8451	8457	8463	8470	8476	8482	8488	8494	8500	8506	1	1	2	2	3	4	4	5	6
71	8513	8519	8525	8531	8537	8543	8549	8555	8561	8567	1	1	2	2	3	4	4	5	5
72	8573	8579	8585	8591	8597	8603	8609	8615	8621	8627	1	1	2	2	3	4	4	5	5
73	8633	8639	8645	8651	8657	8663	8669	8675	8681	8686	1	1	2	2	3	4	4	5	5
74	8692	8698	8704	8710	8716	8722	8727	8733	8739	8745	1	1	2	2	3	3	4	5	5
75	8751	8756	8762	8768	8774	8779	8785	8791	8797	8802	1	1	2	2	3	3	4	5	5
76	8808	8814	8820	8825	8831	8837	8842	8848	8854	8859	1	1	2	2	3	3	4	5	5
77	8865	8871	8876	8882	8887	8893	8899	8904	8910	8915	1	1	2	2	3	3	4	4	5
78	8921	8927	8932	8938	8943	8949	8954	8960	8965	8971	1	1	2	2	3	3	4	4	5
79	8976	8982	8987	8993	8998	9004	9009	9015	9020	9025	1	1	2	2	3	3	4	4	5
80	9031	9036	9042	9047	9053	9058	9063	9069	9074	9079	1	1	2	2	3	3	4	4	5
81	9085	9090	9096	9101	9106	9112	9117	9122	9128	9133	1	1	2	2	3	3	4	4	5
82	9138	9143	9149	9154	9159	9165	9170	9175	9180	9186	1	1	2	2	3	3	4	4	5
83	9191	9196	9201	9206	9212	9217	9222	9227	9232	9238	1	1	2	2	3	3	4	4	5
84	9243	9248	9253	9258	9263	9269	9274	9279	9284	9289	1	1	2	2	3	3	4	4	5
85	9294	9299	9304	9309	9315	9320	9325	9330	9335	9340	1	1	2	2	3	3	4	4	5
86	9345	9350	9355	9360	9365	9370	9375	9380	9385	9390	1	1	2	2	3	3	4	4	5
87	9395	9400	9405	9410	9415	9420	9425	9430	9435	9440	0	1	1	2	2	3	3	4	4
88	9445	9450	9455	9460	9465	9469	9474	9479	9484	9489	0	1	1	2	2	3	3	4	4
89	9494	9499	9504	9509	9513	9518	9523	9528	9533	9538	0	1	1	2	2	3	3	4	4
90	9542	9547	9552	9557	9562	9566	9571	9576	9581	9586	0	1	1	2	2	3	3	4	4
91	9590	9595	9600	9605	9609	9614	9619	9624	9628	9633	0	1	1	2	2	3	3	4	4
92	9638	9643	9647	9652	9657	9661	9666	9671	9675	9680	0	1	1	2	2	3	3	4	4
93	9685	9689	9694	9699	9703	9708	9713	9717	9722	9727	0	1	1	2	2	3	3	4	4
94	9731	9736	9741	9745	9750	9754	9759	9763	9768	9773	0	1	1	2	2	3	3	4	4
95	9777	9782	9786	9791	9795	9800	9805	9809	9814	9818	0	1	1	2	2	3	3	4	4
96	9823	9827	9832	9836	9841	9845	9850	9854	9859	9863	0	1	1	2	2	3	3	4	4
97	9868	9872	9877	9881	9886	9890	9894	9899	9903	9908	0	1	1	2	2	3	3	4	4
98	9912	9917	9921	9926	9930	9934	9939	9943	9948	9952	0	1	1	2	2	3	3	4	4
99	9956	9961	9965	9969	9974	9978	9983	9987	9991	9996	0	1	1	2	2	3	3	3	4

Antilogarithms

	0	1	2	3	4	5	6	7	8	9	1	2	3	4	5	6	7	8	9
0.00	1000	1002	1005	1007	1009	1012	1014	1016	1019	1021	0	0	1	1	1	1	2	2	2
0.01	1023	1026	1028	1030	1033	1035	1038	1040	1042	1045	0	0	1	1	1	1	2	2	2
0.02	1047	1050	1052	1054	1057	1059	1062	1064	1067	1069	0	0	1	1	1	1	2	2	2
0.03	1072	1074	1076	1079	1081	1084	1086	1089	1091	1094	0	0	1	1	1	1	2	2	2
0.04	1096	1099	1102	1104	1107	1109	1112	1114	1117	1119	0	1	1	1	1	2	2	2	2
0.05	1122	1125	1127	1130	1132	1135	1138	1140	1143	1146	0	1	1	1	1	2	2	2	2
0.06	1148	1151	1153	1156	1159	1161	1164	1167	1169	1172	0	1	1	1	1	2	2	2	2
0.07	1175	1178	1180	1183	1186	1189	1191	1194	1197	1199	0	1	1	1	1	2	2	2	2
0.08	1202	1205	1208	1211	1213	1216	1219	1222	1225	1227	0	1	1	1	1	2	2	2	3
0.09	1230	1233	1236	1239	1242	1245	1247	1250	1253	1256	0	1	1	1	1	2	2	2	3
0.10	1259	1262	1265	1268	1271	1274	1276	1279	1282	1285	0	1	1	1	1	2	2	2	3
0.11	1288	1291	1294	1297	1300	1303	1306	1309	1312	1315	0	1	1	1	2	2	2	2	3
0.12	1318	1321	1324	1327	1330	1334	1337	1340	1343	1346	0	1	1	1	2	2	2	3	3
0.13	1349	1352	1355	1358	1361	1365	1368	1371	1374	1377	0	1	1	1	2	2	2	3	3
0.14	1380	1384	1387	1390	1393	1396	1400	1403	1406	1409	0	1	1	1	2	2	2	3	3
0.15	1413	1416	1419	1422	1426	1429	1432	1435	1439	1442	0	1	1	1	2	2	2	3	3
0.16	1445	1449	1452	1455	1459	1462	1466	1469	1472	1476	0	1	1	1	2	2	2	3	3
0.17	1479	1483	1486	1489	1493	1496	1500	1503	1507	1510	0	1	1	1	2	2	2	3	3
0.18	1514	1517	1521	1524	1528	1531	1535	1538	1542	1545	0	1	1	1	2	2	2	3	3
0.19	1549	1552	1556	1560	1563	1567	1570	1574	1578	1581	0	1	1	1	2	2	3	3	3
0.20	1585	1589	1592	1596	1600	1603	1607	1611	1614	1618	0	1	1	1	2	2	3	3	3
0.21	1622	1626	1629	1633	1637	1641	1644	1648	1652	1656	0	1	1	2	2	2	3	3	3
0.22	1660	1663	1667	1671	1675	1679	1683	1687	1690	1694	0	1	1	2	2	2	3	3	3
0.23	1698	1702	1706	1710	1714	1718	1722	1726	1730	1734	0	1	1	2	2	2	3	3	4
0.24	1738	1742	1746	1750	1754	1758	1762	1766	1770	1774	0	1	1	2	2	2	3	3	4
0.25	1778	1782	1786	1791	1795	1799	1803	1807	1811	1816	0	1	1	2	2	2	3	3	4
0.26	1820	1824	1828	1832	1837	1841	1845	1849	1854	1858	0	1	1	2	2	3	3	3	4
0.27	1862	1866	1871	1875	1879	1884	1888	1892	1897	1901	0	1	1	2	2	3	3	3	4
0.28	1905	1910	1914	1919	1923	1928	1932	1936	1941	1945	0	1	1	2	2	3	3	4	4
0.29	1950	1954	1959	1963	1968	1972	1977	1982	1986	1991	0	1	1	2	2	3	3	4	4
0.30	1995	2000	2004	2009	2014	2018	2023	2028	2032	2037	0	1	1	2	2	3	3	4	4
0.31	2042	2046	2051	2056	2061	2065	2070	2075	2080	2084	0	1	1	2	2	3	3	4	4
0.32	2089	2094	2099	2104	2109	2113	2118	2123	2128	2133	0	1	1	2	2	3	3	4	4
0.33	2138	2143	2148	2153	2158	2163	2168	2173	2178	2183	0	1	1	2	2	3	3	4	4
0.34	2188	2193	2198	2203	2208	2213	2218	2223	2228	2234	1	1	2	2	3	3	4	4	5
0.35	2239	2244	2249	2254	2259	2265	2270	2275	2280	2286	1	1	2	2	3	3	4	4	5
0.36	2291	2296	2301	2307	2312	2317	2323	2328	2333	2339	1	1	2	2	3	3	4	4	5
0.37	2344	2350	2355	2360	2366	2371	2377	2382	2388	2393	1	1	2	2	3	3	4	4	5
0.38	2399	2404	2410	2415	2421	2427	2432	2438	2443	2449	1	1	2	2	3	3	4	4	5
0.39	2455	2460	2466	2472	2477	2483	2489	2495	2500	2506	1	1	2	2	3	3	4	5	5
0.40	2512	2518	2523	2529	2535	2541	2547	2553	2559	2564	1	1	2	2	3	4	4	5	5
0.41	2570	2576	2582	2588	2594	2600	2606	2612	2618	2624	1	1	2	3	3	4	5	5	5
0.42	2630	2636	2642	2649	2655	2661	2667	2673	2679	2685	1	1	2	2	3	4	4	5	6
0.43	2692	2698	2704	2710	2716	2723	2729	2735	2742	2748	1	1	2	3	3	4	4	5	6
0.44	2754	2761	2767	2773	2780	2786	2793	2799	2805	2812	1	1	2	3	3	4	5	5	6
0.45	2818	2825	2831	2838	2844	2851	2858	2864	2871	2877	1	1	2	3	3	4	5	5	6
0.46	2884	2891	2897	2904	2911	2917	2924	2931	2938	2944	1	1	2	3	3	4	5	5	6
0.47	2951	2958	2965	2972	2979	2985	2992	2999	3006	3013	1	1	2	3	3	4	5	5	6
0.48	3020	3027	3034	3041	3048	3055	3062	3069	3076	3083	1	1	2	3	4	4	5	6	6
0.49	3090	3097	3105	3112	3119	3126	3133	3141	3148	3155	1	1	2	3	4	4	5	6	6

Antilogarithms

	0	1	2	3	4	5	6	7	8	9	1	2	3	4	5	6	7	8	9
0.50	3162	3170	3177	3184	3192	3199	3206	3214	3221	3228	1	1	2	3	4	4	5	6	7
0.51	3236	3243	3251	3258	3266	3273	3281	3289	3296	3304	1	2	2	3	4	5	5	6	7
0.52	3311	3319	3327	3334	3342	3350	3357	3365	3373	3381	1	2	2	3	4	5	5	6	7
0.53	3388	3396	3404	3412	3420	3428	3436	3443	3451	3459	1	2	2	3	4	5	6	6	7
0.54	3467	3475	3483	3491	3499	3508	3516	3524	3532	3540	1	2	2	3	4	5	6	6	7
0.55	3548	3556	3565	3573	3581	3589	3597	3606	3614	3622	1	2	2	3	4	5	6	7	7
0.56	3631	3639	3648	3656	3664	3673	3681	3690	3698	3707	1	2	3	3	4	5	6	7	8
0.57	3715	3724	3733	3741	3750	3758	3767	3776	3784	3793	1	2	3	3	4	5	6	7	8
0.58	3802	3811	3819	3828	3837	3846	3855	3864	3873	3882	1	2	3	4	4	5	6	7	8
0.59	3890	3899	3908	3917	3926	3936	3945	3954	3963	3972	1	2	3	4	5	5	6	7	8
0.60	3981	3990	3999	4009	4018	4027	4036	4046	4055	4064	1	2	3	4	5	6	6	7	8
0.61	4074	4083	4093	4102	4111	4121	4130	4140	4150	4159	1	2	3	4	5	6	7	8	9
0.62	4169	4178	4188	4198	4207	4217	4227	4236	4246	4256	1	2	3	4	5	6	7	8	9
0.63	4266	4276	4285	4295	4305	4315	4325	4335	4345	4355	1	2	3	4	5	6	7	8	9
0.64	4365	4375	4385	4395	4406	4416	4426	4436	4446	4457	1	2	3	4	5	6	7	8	9
0.65	4467	4477	4487	4498	4508	4519	4529	4539	4550	4560	1	2	3	4	5	6	7	8	9
0.66	4571	4581	4592	4603	4613	4624	4634	4645	4656	4667	1	2	3	4	5	6	7	9	10
0.67	4677	4688	4699	4710	4721	4732	4742	4753	4764	4775	1	2	3	4	5	7	8	9	10
0.68	4786	4797	4808	4819	4831	4842	4853	4864	4875	4887	1	2	3	4	6	7	8	9	10
0.69	4893	4909	4920	4932	4943	4955	4966	4977	4989	5000	1	2	3	5	6	7	8	9	10
0.70	5012	5023	5035	5047	5058	5070	5082	5093	5105	5117	1	2	4	5	6	7	8	9	11
0.71	5129	5140	5152	5164	5176	5188	5200	5212	5224	5236	1	2	4	5	6	7	8	10	11
0.72	5248	5260	5272	5284	5297	5309	5321	5333	5336	5358	1	2	4	5	6	7	9	10	11
0.73	5370	5383	5395	5408	5420	5433	5445	5458	5470	5483	1	3	4	5	6	8	9	10	11
0.74	5495	5508	5521	5534	5546	5559	5572	5585	5598	5610	1	3	4	5	6	8	9	10	12
0.75	5623	5636	5649	5662	5675	5689	5702	5715	5728	5741	1	3	4	5	7	8	9	10	12
0.76	5754	5768	5781	5794	5808	5821	5834	5848	5861	5875	1	3	4	5	7	8	9	11	12
0.77	5888	5902	5916	5929	5943	5957	5970	5984	5998	6012	1	3	4	5	7	8	10	11	12
0.78	6026	6039	6053	6067	6081	6095	6109	6124	6138	6152	1	3	4	6	7	8	10	11	13
0.79	6166	6180	6194	6209	6223	6237	6252	6266	6281	6295	1	3	4	6	7	9	10	11	13
0.80	6310	6324	6339	6353	6368	6383	6397	6412	6427	6442	1	3	4	6	7	9	10	12	13
0.81	6457	6471	6486	6501	6516	6531	6546	6561	6577	6592	2	3	5	6	8	9	11	12	14
0.82	6607	6622	6637	6653	6668	6683	6699	6714	6730	6745	2	3	5	6	8	9	11	12	14
0.83	6761	6776	6792	6808	6823	6839	6855	6871	6887	6902	2	3	5	6	8	9	11	13	14
0.84	6918	6934	6950	6966	6982	6998	7015	7031	7047	7063	2	3	5	6	8	10	11	13	15
0.85	7079	7096	7112	7129	7145	7161	7178	7194	7211	7228	2	3	5	7	8	10	12	13	15
0.86	7244	7261	7278	7295	7311	7328	7345	7362	7379	7396	2	3	5	7	8	10	12	13	15
0.87	7413	7430	7447	7464	7482	7499	7516	7534	7551	7568	2	3	5	7	9	10	12	14	16
0.88	7586	7603	7621	7638	7656	7674	7691	7709	7727	7745	2	4	5	7	9	11	12	14	16
0.89	7762	7780	7798	7816	7834	7852	7870	7889	7907	7925	2	4	5	7	9	11	13	14	16
0.90	7943	7962	7980	7998	8017	8035	8054	8072	8091	8110	2	4	6	7	9	11	13	15	17
0.91	8128	8147	8166	8185	8204	8222	8241	8260	8279	8299	2	4	6	8	9	11	13	15	17
0.92	8318	8337	8356	8375	8395	8414	8433	8453	8472	8492	2	4	6	8	10	12	14	15	17
0.93	8511	8531	8551	8570	8590	8610	8630	8650	8670	8690	2	4	6	8	10	12	14	16	18
0.94	8710	8730	8750	8770	8790	8810	8831	8851	8872	8892	2	4	6	8	10	12	14	16	18
0.95	8913	8933	8954	8974	8995	9016	9036	9057	9078	9099	2	4	6	8	10	12	15	17	19
0.96	9120	9141	9162	9183	9204	9226	9247	9268	9290	9311	2	4	6	8	11	13	15	17	19
0.97	9333	9354	9376	9397	9419	9441	9462	9484	9506	9528	2	4	7	9	11	13	15	17	20
0.98	9550	9572	9594	9616	9638	9661	9683	9705	9727	9750	2	4	7	9	11	13	16	18	20
0.99	9772	9795	9817	9840	9863	9886	9908	9931	9954	9977	2	5	7	9	11	14	16	18	20

Index

Absolute temperature, 21
Acid salts, 73
Acids, 69—70
    strong, 69
    weak, 69
Activity series, 135
Adsorption chromatography, 6
Agate, 187
Air, 44—47
    liquefaction and distillation, 112—113
    relative density, 182
    solubility in water, 47
Alcohols, 233—235
Alkali metals, 135
Alkaline earths, 143
Alkalis, 71, 72
Alkanes, 227—229
Alkenes, 230—231
Alkyl groups, 229
Alkynes, 232—233
Allotropy, 177
Alumina, 6
Aluminium, 3, 150—151
    chloride, 39, 152, 153—154
    compounds, 152—154
        test for, 154
    hydroxide, 152—153
    oxide, 6, 38, 150, 152, 230
    sulphate, 154
Amethyst, 187
Ammonia
    catalytic oxidation, 196
    manufacture, 130—131
    molecule, shape of, 36
    preparation, 191
    properties and reactions, 71, 139, 192—194
    solubility, 55
    structure, 32
Ammonia solution, 160, 169, 192, 194—195
Ammonium
    chloride, 10, 132, 188, 191, 195
    hydrogencarbonate, 139
    ion, 32
    nitrate, 198
    nitrite, 188
    phosphate, 200
    salts, 195—196
    thiocyanate, 167
Amphoteric behaviour, 38
Analysis, volumetric, 76—80
Anhydrides, 70, 209
Anode, 87
Anodising, 92
Association, 52
Atomic
    number, 27
    structure, 27
    theory, 15
Atoms, 15, 27
    electronic structure of, 28

Avogadro
    constant, 62, 63
    law, 61

Barium
    chloride, 212
    peroxide, 118
    sulphate, 212
Bases, 71—72
Basic salts, 74
Basicity, 69—70
Benene-1,2,3-triol, 45
Birkeland — Eyde process, 131—132
Bleaching powder, 149
Boiling
    points, 12, 24
    under reduced pressure, 24
Bosch process, 180
Boyle's law, 20
Brass, 168
Brine, 89, 139
Bromides, test for, 217
Bromine, 19, 40—42, 216, 231, 233
Bronze, 168
Brown ring test, 166
Brownian movement, 24
Burettes, 77
Butane, 228

Calcium, 51, 107, 111, 115, 143—144, 147
    carbonate, 129, 132, 149
    chlorate(I), 149
    chloride, 11, 45, 55, 147, 149, 191
    compounds, 147—150
        test for, 150
    dicarbide, 150, 232
    hydride, 111, 147
    hydrogencarbonate, 56, 57, 148
    hydrogensulphite, 209
    hydroxide, 57, 148—149, 191
    oxide, 11, 139, 148, 150, 163
    phosphate, 150, 199
    silicate, 163
    sulphate, 11, 56, 149
Calor gas, 229
Carbon, 101, 116, 177—180
    black, 179
    cycle, 180—181
    dioxide, 10, 45, 47, 55, 139, 148, 183—185
        cycle, 180—181
        molecule, shape of, 36
    disulphide, 180, 203
    monoxide, 101, 162, 180, 181—183
Carbonates, 74, 185—186
Carbonic acid, 184—185
Carboxylic acids, 235—236
Cast iron, 163
Catalysts, 128

Cathode, 87
Cells, 103—104
   Daniell, 103
Chalk, 149
Changes
   chemical, 1—3
   physical, 1—2
Charcoal
   animal, 179
   wood, 179
Charles' law, 20
Chemical
   changes, 1—3
   symbols, 4
Chilean saltpetre, 141
Chloric(I) acid, 214
Chlorides, 39—41, 74
   test for, 216
Chlorine, 27, 29, 100, 212—214
   manufacture, 88—89
   molecule, 31
   preparation, 212
   properties and reactions, 128, 143, 149, 166, 200, 229, 231, 233
   uses, 214
Chloroethane, 231
Chloroethene, 233, 240
Chloromethane, 129
Chlorophyll, 181
Chromatography
   column, 6
   paper, 7
Coal, 236—237
Cobalt(II) chloride, 49
Coke, 144, 150, 162, 179, 180
Column chromatography, 6
Combustion, 46
Compounds, 4—5
Concentration, effect on reaction rate, 127
Conservation of mass, law of, 3
Constant boiling point mixtures, 12
Constant composition, law of, 60
Contact process, 131, 209
Co-ordinate bonds, 32
Copper, 45, 90, 103, 167—168, 197
   compounds, 168—170
   test for, 170
Copper(I)
   chloride, 169, 183
   oxide, 168
Copper(II)
   carbonate, basic, 169
   chloride, 170
   hydroxide, 169
   nitrate, 170
   oxide, 60, 169, 193
   sulphate, 48, 90, 105, 170, 211
   sulphide, 170
Covalency, 30—32
   dative, 32
Covalent compounds, general properties of, 34
Cracking, 238
Crude petroleum, 237—238
Cryolite, 150
Crystal lattice, 25, 33
Crystals
   ionic, 33
   macromolecular, 34
   molecular, 33
   structures of, 33—34
Crystallisation, 9

Dalton, John, 15, 27
Daniell cell, 103
Dative covalency, 32
Deliquescence, 45
Desiccator, 10
   vacuum, 10
Detergents, 56
Diamminesilver chloride, 216
Diamond, 177—179
   lattice, 34, 178
1,2-Dibromoethane, 231
1,2-Dichloroethane, 231
Dichloromethane, 229
Diffusion
   in liquids, 24
   of gases, 19—20
Dilead(II) lead(IV) oxide, 114, 156
Dinitrogen oxide, 198
Dinitrogen tetraoxide, 132, 199
Distillation, 5
   fractional, 8
Disodium oxide, 38, 138
'Double decomposition', 76
Dreschel bottle, 11
Dry ice, 185

Efflorescence, 139
Electric currents, 87
Electrochemical
   equivalent, 92
   series, 104—105
Electrode, 87
Electrolysis, 86—94
Electrolyte, 87
Electronegativity, 38
Electronic structure, 28
Electrons, 27
Electroplating, 91
Electropositivity, 37
Electrovalency, 29
Elements, 3, 12
Emerald, 152
Empirical formula, 225
Endothermic reactions, 121
Enthalpy
   of fusion, 25
   of vaporisation, 24
Enzyme, 238
Epsom salt, 146
Equations, 17
   ionic, 71
Equilibria, 129—132
Esterification, 234
Ethane, 228
Ethanedioic acid, 182
Ethanioc acid, 234, 235—236
Ethanol, 12, 230, 233—235, 238—239
Ethene, 230—231, 239
   molecule, shape of, 36
Ethyl ethanoate, 234
Ethyne, 117, 150, 232—233
Evaporation, 23
Exothermic reactions, 121

Faraday constant, 92

Faraday's laws of electrolysis, 92—94
Fats, 239
Fermentation, 238—239
Filtration, 7
Flame test, 141, 143, 150
Flint, 187
Fluorine, 40—42
Formic acid, 181
Formulae, 15
    empirical, 225
    molecular, 225
    structural, 225
Fountain experiment, 192, 214
Fractional distillation, 8
Fractionating column, 8
Frasch process, 202

Galena, 154
Galvanising, 105
Gases, 1, 18
    drying of, 11
    ideal, 19
    real, 19
    solubility in water, 55
    velocities of, 19
Gay Lussac's law, 61
Glass, 187
Glauber's salt, 141
Glucose, 238
Glyceryl stearate, 239
Graduated flasks, 77
Graham's law of diffusion, 19
Graphite, 177—179
Groups, 37
Gunpowder, 143
Gypsum, 202

Haber process, 130—131
Haematite, 162
Half equations, 99
Halogens, 40—42, 212
Hard water, 56—57
Heat of
    combustion, 122
    fusion, 125
    neutralisation, 125
    reaction, 123
    solution, 124
    vaporisation, 125
Hofmann's voltameter, 49
Homologous series, 225
Hydration, 52, 124
Hydrides, 40
Hydriodic acid, 142, 218
Hydrobromic acid, 217
Hydrocarbons, 226
    saturated, 227
    unsaturated, 230, 232
Hydrochloric acid, 12, 212
    preparation, 215
    properties, 215—216
Hydrogen, 37, 107—112
    bonding, 52
    manufacture, 107, 180, 237
    molecule, 31
    occurrence, 107
    preparation, 50—51, 107—110
    properties and reactions, 41, 101, 110—111, 128, 130
    test for, 112
    uses, 112
Hydrogen bromide
    preparation, 216
    properties, 217
Hydrogen chloride
    preparation, 214
    properties and reactions, 166, 214—215, 231, 233
    solubility, 55
Hydrogen iodide
    preparation, 217
    properties and reactions, 129, 218
Hydrogen peroxide, 100, 113, 118—119, 158
    volume strength of, 118
Hydrogen sulphide, 102, 158, 204—206
Hydrogenation, 231, 232
Hydrogencarbonates, 185—186
Hydrolysis, 39, 153, 161
Hydroxides, 72, 74
Hypochlorous acid (chloric(I) acid), 214

Ideal gases, 19
Indicators, 79
Inert gases, 29
Inks, separation of, 7
Invar steel, 163
Iodides, test for, 218
Iodine, 10, 40—42, 129, 142, 217, 238
    lattice, 33
Ionic compounds, general properties of, 34
Ions, 29
    formation of, 29—30
Iron, 108, 116, 162—164
    cast, 163
    compounds, 164—167
        tests for, 167
    extraction, 162—163
    occurrence, 162
    pig, 163
    properties, 51, 105, 164
    pyrites, 202
    wrought, 163
Iron(II)
    carbonate, 162
    chloride, 165
    compounds, test for, 167
    diiron(III) oxide, 162, 165
    disulphide, 202
    hydroxide, 165
    oxide, 165
    sulphate, 118, 166
    sulphide, 167, 204
Iron(III)
    chloride, 166
    compounds, test for, 167
    hydroxide, 165
    oxide, 151, 162, 165
    sulphate, 167
Isomerism, structural, 229—230
Isotopes, 27, 62

Jeweller's rouge, 165

Kinetic energy, 25
Kinetic theory
    of gases, 18

of liquids, 23
of solids, 25
Kipp's apparatus, 110

Latent heat
  of fusion, 25
  of vaporisation, 24
Law
  Avogadro's, 61
  Boyle's, 20
  Charles', 20
  of conservation of mass, 3
  of constant composition, 60
  Gay Lussac's, 61
  Graham's, 19
  of multiple proportions, 61
Le Chatelier's principle, 130
Lead, 154—155
  Chamber process, 128
  compounds, 155—158
    test for, 158
Lead(II)
  carbonate, 157
    basic, 157
  chloride, 157
  ethanoate, 206
  hydroxide, 156
  iodide, 157
  nitrate, 157, 199
  oxide, 156
  sulphate, 157
  sulphide, 154, 158
Lead(IV) oxide, 114, 156, 212
Light, effect on reaction rate, 128
Limestone, 139, 148, 162, 184
Limewater, 45, 148, 185
Liquids, 1, 23
  mixtures of, 8
Litmus, 79
Lone pairs of electrons, 32

Macromolecular crystals, 34
Magnesium, 45, 51, 108, 116, 143—145, 184
  carbonate, 146
  chloride, 39, 56, 144
  compounds, 145—7
    test for, 147
  hydrogencarbonate, 56
  hydroxide, 146
  nitride, 45
  occurrence and extraction, 144
  oxide, 38, 144, 145—6
  sulphate, 11, 56, 146
  uses, 145
Magnetite, 162
Manganese(IV) oxide, 101, 113, 114, 212
Marble, 149, 183
Mass number, 27
Melting point, 12, 25
Mercury(II)
  nitrate, 198
  oxide, 2, 114
Metallic latices, 34
Metals, 12, 135—172
Methane, 31, 35, 128, 228—229
Methanoic acid, 181
Methanol, 183, 233
Methyl orange, 79

screened, 79
Mixtures, 4—5
  constant boiling point, 12
  liquid — liquid, 8
  liquid — solid, 5—7
  separation of, 5—10
  solid — solid, 9—10
Molar latent heat
  of fusion, 25
  of vaporisation, 24
Molar solutions, 80
  volume, 65
Molarity, 80
Mole, 62, 63
Molecular crystals, 33
Molecular formula, 225
Molecules, 15
  shapes of, 35—37
Multiple proportions, law of, 61

Naphthalene, 10
Natural gas, 237
Neutralisation, 72
Neutrons, 27
Nitrates, 198—199
  test for, 166
Nitric acid
  manufacture, 196
  preparation, 196
  properties and reactions, 12, 101, 155
    166, 168, 197—198
Nitrogen, 45, 130, 188—190
  cycle, 190
  dioxide, 190, 196, 199
  oxide, 128, 131, 166, 190, 193, 196, 197
  solubility, 55
Noble gases, 29
Non-electrolyte, 87
Non-metals, 12—13, 177—218
Nucleus, 28

Oil drop experiment, 14
Oleum, 209, 210
Opal, 187
Organic chemistry, 225—40
Ostwald process, 196
Oxalic acid, 182
Oxidation, 98—105
Oxides, 38—39, 72, 116—117
Oxidising agents, 100—101
  test for, 102—103
Oxonium ion, 32, 69
Oxygen, 3, 100, 112—117
  manufacture, 112
  molecule, 31
  occurrence, 112
  percentage in air, 44—45
  preparation, 113—115
  properties and reactions, 115—116
  solubility, 55
  test for, 117
  uses, 117

Paper chromatography, 7
Particle size, effect on reaction rate, 129
Partition chromatography, 7
Periodic Table, 37, 280

Periods, 37
Permanent hardness in water, 56—57
Permutit, 57
Petroleum, crude, 237—238
pH, 83
Phenolphthalein, 79
Phosphates, 200
Phosphine, 40
Phosphonic acid, 39, 40
Phosphoric acid, 39, 40, 200
Phosphorous acid (phosphonic acid), 39, 40
Phosphorus, 45, 116, 188, 216, 217
    pentachloride, 40, 200—201, 235
    pentoxide (see also phosphorus(V) oxide), 10, 200
    trichloride, 40, 200—201
    oxide, 40, 201
    trioxide, 39
Phosphorus(III) oxide, 38, 39
Phosphorus(V) oxide, 10, 38, 39, 200
Phosphoryl chloride (phosphorus trichloride oxide), 40
Photosynthesis, 181
Physical changes, 1—2
Pig-iron, 163
Pipette, 77
Plaster of Paris, 149
Platinised asbestos, 131, 209
Platinum, 91, 193, 196, 231
Polarisation
    in cells, 104
    of bonds, 52
Poly(chloroethene), 240
Poly(ethene), 239
Polymerisation, 239—240
Potassium, 111, 135, 142—143
    chlorate(V), 114, 143, 214
    chloride, 142
    compounds, 142—143
        test for, 143
    dichromate(VI), 101, 208, 235
    hydroxide, 142, 143
    iodide, 102, 118, 142
    manganate(VII), 14, 101, 212, 235
    nitrate, 114, 143, 196
    nitrite, 143
    permanganate (see also potassium manganate(VII)), 14
    triiodide, 142
Precipitation, 75
Pressure
    effect on reaction rate, 127
    standard, 23
Primary standards, 77
Producer gas, 180
Propane, 228
Propene, 230
Protons, 27
Pyrogallol, 45

Quartz, 187
Quenching, 164
Quicklime, 11, 148

Radicals, 16
Rate of change, 125—129
Real gases, 19
Red lead, 156

Redox reactions, 98
Reducing agents, 101—102
    tests for, 103
Reduction, 98—105
Relative
    atomic masses, 62
        table of, 279
    density, 63, 182
    molecular mass, 63
Respiration, 46—47
Rocksill wool, 230
Ruby, 152
Rusting, 48

Saltpetre, 141
Salts, 73—76
Sand, 187
Saponification, 239
Sapphire, 152
Saturated
    hydrocarbons, 227
    solutions, 53
Screened methyl orange, 79
Seawater, 144
Seeding, 54
Separating funnel, 8
Shapes of molecules, 35—37
Silane, 40
Silica (see also silicon(IV) oxide), 39
    gel, 10
Silicon, 3, 177
    dioxide (silicon(IV) oxide), 39
    tetrachloride, 39
Silicon(IV) oxide, 39, 148, 163, 187
Silver, 171
    chloride, 171
    compounds, 171—172
        test for, 172
    nitrate, 172, 198, 216, 217, 218
    oxide, 171
Simple cells, 103—104
Slaked lime, 148
Soap, 56, 239
Soda-lime, 228
Sodium, 11, 29, 50, 107, 111, 115, 135—138, 234
    aluminate, 151, 153
    carbonate, 57, 77, 139—140, 146
    chlorate(I), 214
    chlorate(V), 214
    chloride, 39, 52, 88—89, 139, 140—141, 214
        lattice, 33
        structure, 29
    compounds, 137—141
        test for, 141
    ethanoate, 228
    hexafluoroaluminate, 150
    hydride, 40
    hydrogencarbonate, 139, 140, 146, 157
    hydrogensulphate, 141
    hydrogensulphite, 207
    hydroxide, 91, 138, 159, 160, 214, 234, 239
    monoxide (disodium oxide), 38, 138
    nitrate, 114, 141, 196
    nitrite, 115, 188
    peroxide, 137
    plumbate(II), 156

silicate, 187
sulphate, 53, 141
sulphite, 206, 207
thiosulphate, 54
zincate, 159
Solid — solid mixtures, 9—10
Solids, 1, 25
    drying of, 10
Solubility, 53, 55
    curves, 53
Solute, 53
Solutions, 53
    saturated, 53
    standard, 77
    supersaturated, 54
Solvay process, 139
Solvent, 53
Stainless steel, 163
Standard solutions, 77
Starch, 238
Steel, 163
    invar, 163
    stainless, 163
Structural
    formula, 225
    isomerism, 229—230
Sublimation, 10
Sugar, 211
Sulphates, 74
    test for, 212
Sulphides, 74
Sulphur, 116, 198, 201—204
Sulphur dioxide, 39, 55, 131, 206—209, 210
Sulphur(VI) oxide, 39, 131, 209
Sulphuric acid, 10, 91, 150, 210—212
    as a dehydrating agent, 211, 230
    as an acid, 210—211
    as an oxidising agent, 101, 151, 155, 159, 167, 204, 206, 211—212, 217
    fuming, 209
    manufacture, 210
Sulphurous acid, 102, 207
Superphosphate, 150
Supersaturated solutions, 54

Temperature
    absolute, 21
    effect on reaction rate, 126
    standard, 23
Tempering, 164
Temporary hard water, 56
Terylene, 240
Tetraammine
    copper(II) hydroxide, 169
    zinc hydroxide, 160
Tetrachloromethane, 186
Thermal
    decomposition, 132
    dissociation, 132
Thermit process, 151
Thermochemistry, 121—125
Tin plating, 105
Topaz, 152
Transition
    metals, 158—172
    temperature, 203
Trichloromethane, 129
Turpentine, 213

Universal indicator, 83
Unsaturated hydrocarbons, 230, 232

Valency, 16
    and electronic structure, 28
Vanadium(V) oxide, 131, 209
Vaporisation, 23
Vapour, 23
    density, 63
    pressure, 23
Voltameter, 93
Volumetric analysis, 76—80

Wash bottle, 77
Washing soda, 139
Water, 48
    as a solvent, 52—53
    boiling point, 52
    detection of, 48—49
    gas, 180
    glass, 187
    hardness, 55—57
    molecule, shape of, 36
    of crystallisation, 57—58
    reaction with metals, 50—51, 107—108
    solubility of gases in, 55
    solubility of solids in, 53—55
    synthesis of, 50
    volumetric composition, 49
Weighing bottle, 77
Wrought iron, 163

Zinc, 103, 105, 109, 159—160
    blende, 159
    carbonate, 160
    chloride, 161
        hydroxide, 161
    compounds, 160—161
        test for, 161
    hydroxide, 160
    nitrate, 161
    oxide, 160
    sulphate, 161
    sulphide, 161